OXFORD REVISION GUIDES

AS & A Level

GEOGRAPHY

Garrett Nagle
Kris Spencer

OXFORD
UNIVERSITY PRESS
Great Clarendon Street, Oxford OX2 6DP

Oxford University Press is a department of the University of Oxford. It furthers the University's objective of excellence in research, scholarship, and education by publishing worldwide in

Oxford New York

Athens Auckland Bangkok Bogotá Bombay Buenos Aires Cape Town Chennai Dar es Salaam Delhi Florence Hong Kong Istanbul Karachi Kolkata Kuala Lumpur Madrid Melbourne Mexico City Mumbai Nairobi Paris Saõ Paulo Shanghai Singapore Taipei Tokyo Toronto Warsaw

with associated companies in Berlin Ibadan

Oxford is a registered trade mark of Oxford University Press in the UK and in certain other countries

To Angela, Rosie, Patrick, and Bethany

ISBN 0 19 913432 4 Bookshop Edition
First published 1997
Second edition 1998
Reprinted 1999 (twice), 2000
Third Edition 2001 (revisions by Garrett Nagle)

British Library Cataloguing in Publication Data

Data available

Typesetting, artwork and design by Hardlines, Charlbury, Oxon

Printed in Great Britain

Contents

It is important to revise topics more than once. Use this checklist to structure your revision. Further revision tips can be found in the final chapter.

Specification content summary tables

PLATE TECTONICS

	Edexcel A	Edexcel B	AQA A	AQA B	OCR A	OCR B	WJEC	SQA	CCEA	IB
Plate tectonics	AS	A2	AS	A2	AS/A2	A2	AS		A2	✓
Volcanoes	AS	A2	AS	A2	AS/A2	A2	AS		A2	✓
Earthquakes	AS	A2	AS	A2	AS/A2	A2	AS		A2	✓
associated features e.g. Tsunami	AS	A2	AS	A2	AS/A2	A2	AS		A2	✓
human impact/preparation	AS	A2	AS	A2	AS/A2	A2	AS		A2	✓

WEATHERING, ROCKS, AND RELIEF

	Edexcel A	Edexcel B	AQA A	AQA B	OCR A	OCR B	WJEC	SQA	CCEA	IB
Weathering processes	AS		A2		AS		A2	C		✓
Granite/limestone	AS				AS		A2	C		✓
Human impact	AS	A2				A2		C		✓
Mass movements	AS	A2		AS	AS	A2		C		✓
Slopes	AS	A2		AS	AS	A2		C		✓

RIVERS AND HYDROLOGY

	Edexcel A	Edexcel B	AQA A	AQA B	OCR A	OCR B	WJEC	SQA	CCEA	IB
Hydrological cycle	AS	AS	AS	AS	AS	AS	AS	C/A	AS	✓
Basins, regimes	AS	AS	AS	AS	AS	AS	AS	C/A	AS	✓
Hydrographs	AS	AS	AS	AS	AS	AS	AS	C/A	AS	✓
Flooding	AS	AS	AS	AS	AS	AS	AS	C/A	AS	✓
Channels, Load, Velocity, Hjulström	AS	AS	AS	AS	A2	AS	AS	C/A	AS	✓
Processes	AS	AS	AS	AS	A2	AS	AS	C/A	AS	✓
Erosion	AS	AS	AS	AS	A2	AS	AS	C/A	AS	✓
Deposition	AS	AS	AS	AS	A2	AS	AS	C/A	AS	✓
Management	AS	AS	AS	AS	AS	AS	AS	C/A	AS	✓

COASTS

	Edexcel A	Edexcel B	AQA A	AQA B	OCR A	OCR B	WJEC	SQA	CCEA	IB
Coastal environments	AS	AS	A2	AS	A2	AS	A2		A2	
Processes	AS	AS	A2	AS	A2	AS	A2		A2	
Landforms	AS	AS	A2	AS	A2	AS	A2		A2	
Sea level changes	AS	AS	A2	AS	A2	AS	A2		A2	
Human impact	AS	AS	A2	AS	A2	AS	A2		A2	

GLACIATION AND PERIGLACIATION

	Edexcel A	Edexcel B	AQA A	AQA B	OCR A	OCR B	WJEC	SQA	CCEA	IB
Periglaciation	A2		A2	AS	A2	A2	A2	C	A2	
Glaciation	A2		A2	AS	A2	A2	A2	C	A2	

WEATHER AND CLIMATE

	Edexcel A	Edexcel B	AQA A	AQA B	OCR A	OCR B	WJEC	SQA	CCEA	IB
Weather	A2	A2	AS	AS/A2	AS/A2	AS/A2	A2	C	A2	✓
Climate	A2	A2	AS	AS	AS	AS/A2	A2	C	A2	✓

ECOSYSTEMS AND SOILS

	Edexcel A	Edexcel B	AQA A	AQA B	OCR A	OCR B	WJEC	SQA	CCEA	IB
Ecosystem	A2		AS/A2		AS		AS	C/A	AS	
Succession	AS/A2	AS	AS/A2	AS/A2	AS/A2		AS	C/A	AS	✓
Biomes	A2	A2	AS/A2		A2		AS	C/A	AS	✓
Nutrient cycles	A2	AS	AS		AS/A2		AS	C/A	AS	✓
Soils	A2		AS/A2	AS/A2	AS/A2		AS	C/A	AS	✓

POPULATION

	Edexcel A	Edexcel B	AQA A	AQA B	OCR A	OCR B	WJEC	SQA	CCEA	IB
Population	AS	A2	AS/A2	AS	AS	AS	AS	C	AS	
Distribution	AS	A2	AS		AS	AS	AS	C	AS	✓
Growth	AS	A2	AS/A2	AS	AS	AS	AS	C	AS	✓
Migration	AS	AS/A2	AS	AS	AS	AS	AS	C	AS/A2	✓
Resources	AS	A2	A2	AS		AS	A2	C	AS	✓

SETTLEMENT

	Edexcel A	Edexcel B	AQA A	AQA B	OCR A	OCR B	WJEC	SQA	CCEA	IB
Urbanisation	AS/A2	AS	AS/A2	AS	AS	AS/A2	AS	C/A	AS	✓
Urban settlement	AS/A2	AS	AS/A2	AS	AS	AS/A2	AS	C/A	AS	✓
Rural settlements	AS	AS	AS		AS		AS	C	AS	✓

AGRICULTURE

	Edexcel A	Edexcel B	AQA A	AQA B	OCR A	OCR B	WJEC	SQA	CCEA	IB
Agriculture	A2		AS/A2		A2	A2	A2	C	A2	✓
Agriculture in LEDCs	A2		A2		A2	A2	A2	C/A	A2	✓

INDUSTRY AND SERVICES

	Edexcel A	Edexcel B	AQA A	AQA B	OCR A	OCR B	WJEC	SQA	CCEA	IB
Economic activity	A2	A2	AS		A2	AS	A2	C		
Models	A2		AS		A2	AS	A2	C/A	A2	✓
Manufacturing	A2	A2	AS		A2	AS	A2	C/A	A2	✓
NICs	A2	A2	AS	A2	A2	AS	A2	C/A	A2	✓
Services	A2		AS		A2	A2	A2		A2	
Retailing			AS			A2	A2			

REGIONAL INEQUALITIES

	Edexcel A	Edexcel B	AQA A	AQA B	OCR A	OCR B	WJEC	SQA	CCEA	IB
Regional	A2	A2			A2		A2	A	A2	
Core-periphery	A2	A2			A2		A2	A	A2	✓

DEVELOPMENT

	Edexcel A	Edexcel B	AQA A	AQA B	OCR A	OCR B	WJEC	SQA	CCEA	IB
Development	A2	A2					A2	A	A2	✓
Health	A2	A2	A2				A2	A	A2	✓
Trade	A2	A2		AS	A2		A2	A	A2	✓

TOURISM

	Edexcel A	Edexcel B	AQA A	AQA B	OCR A	OCR B	WJEC	SQA	CCEA	IB
Tourism	A2	A2	A2		A2	A2	A2	A	A2	
Recreation		A2	A2		A2	A2		A	A2	
Leisure	A2	A2	A2		A2	A2		A	A2	

Plate tectonics

Plate tectonics is a set of theories which describes and explains the distribution of earthquakes, volcanoes, fold mountains, and continental drift. It states that the earth's core consists of semi-molten magma (superheated semi-liquid rock), and that the earth's surface or crust moves around on the magma. The cause of the movement is radioactive decay in the core. This creates huge **convection currents** in the magma, which rise towards the earth's surface, drag continents apart, and cause them to collide.

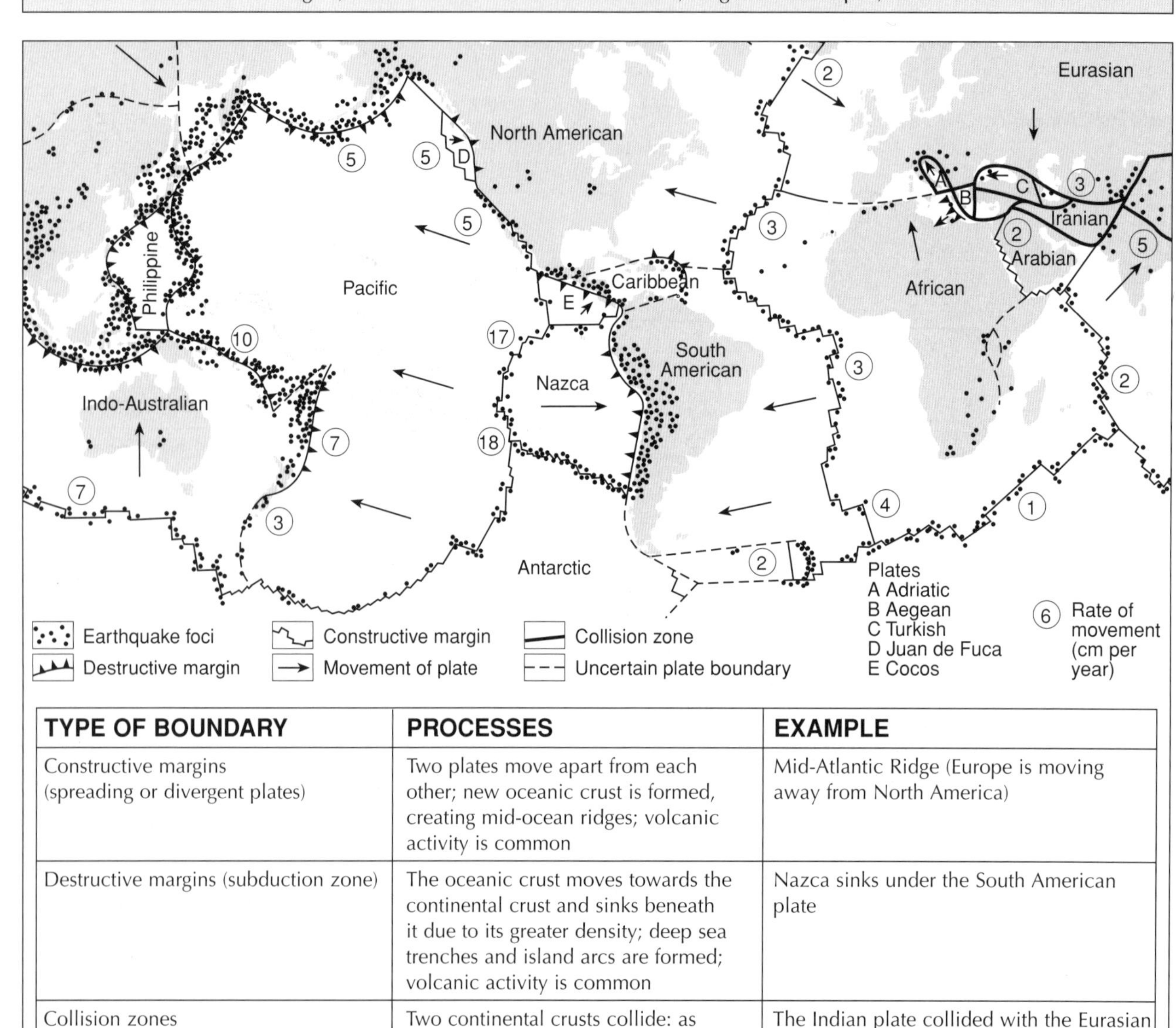

TYPE OF BOUNDARY	PROCESSES	EXAMPLE
Constructive margins (spreading or divergent plates)	Two plates move apart from each other; new oceanic crust is formed, creating mid-ocean ridges; volcanic activity is common	Mid-Atlantic Ridge (Europe is moving away from North America)
Destructive margins (subduction zone)	The oceanic crust moves towards the continental crust and sinks beneath it due to its greater density; deep sea trenches and island arcs are formed; volcanic activity is common	Nazca sinks under the South American plate
Collision zones	Two continental crusts collide: as neither can sink they are folded up into fold mountains	The Indian plate collided with the Eurasian plate to form the Himalayas
Conservative margins (passive margins or transform plates)	Two plates move sideways past each other but land is neither destroyed nor created	San Andreas fault in California

THE EVIDENCE FOR PLATE TECTONICS

The evidence to support the theories of plate tectonics includes:

- the 'fit' of the continents (North America, South America, and Africa as shown in the diagram)
- glacial deposits in Brazil match those in West Africa
- the geological sequence in India matches that of Antarctica
- fossil remains of an early reptile, mesosaurus, are found only in Brazil and the south west of Africa
- the reversal of magnetic particles is similar in rocks either side of the mid-ocean ridges.
- the age of rocks increases with distance either side of a mid-ocean ridge.

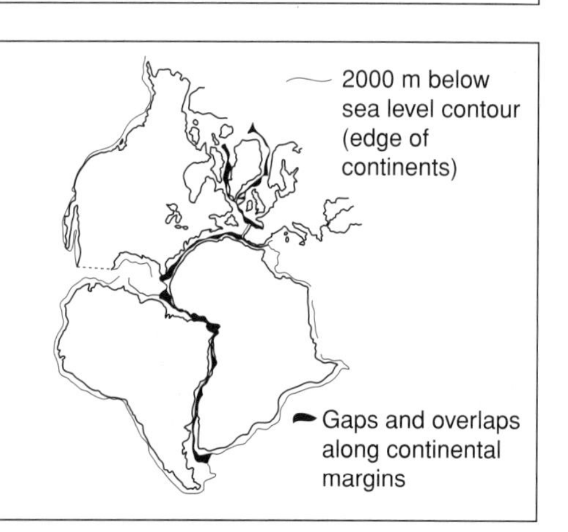

Structure of the earth

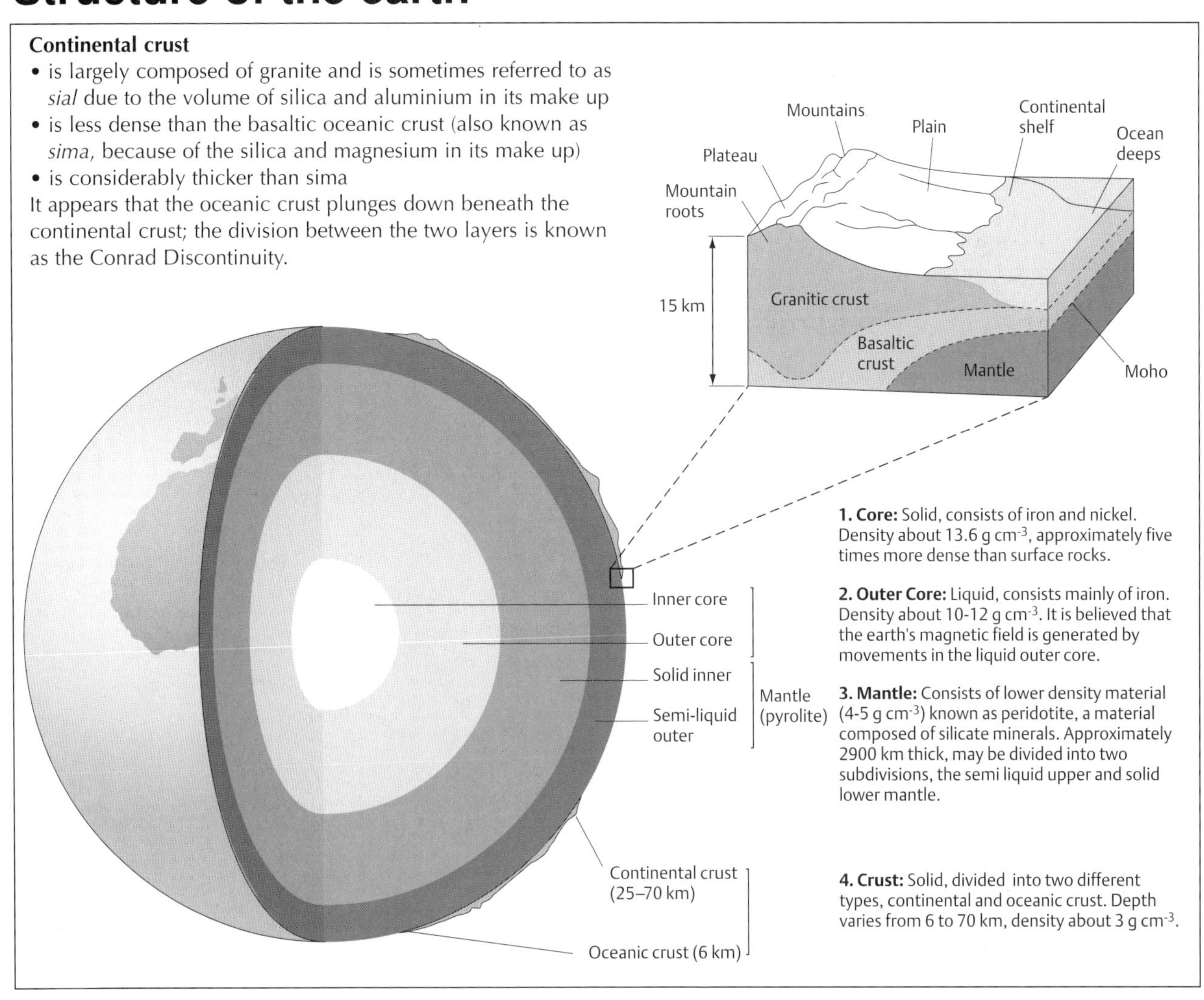

Continental crust

- is largely composed of granite and is sometimes referred to as *sial* due to the volume of silica and aluminium in its make up
- is less dense than the basaltic oceanic crust (also known as *sima,* because of the silica and magnesium in its make up)
- is considerably thicker than sima

It appears that the oceanic crust plunges down beneath the continental crust; the division between the two layers is known as the Conrad Discontinuity.

1. Core: Solid, consists of iron and nickel. Density about 13.6 g cm^{-3}, approximately five times more dense than surface rocks.

2. Outer Core: Liquid, consists mainly of iron. Density about 10-12 g cm^{-3}. It is believed that the earth's magnetic field is generated by movements in the liquid outer core.

3. Mantle: Consists of lower density material (4-5 g cm^{-3}) known as peridotite, a material composed of silicate minerals. Approximately 2900 km thick, may be divided into two subdivisions, the semi liquid upper and solid lower mantle.

4. Crust: Solid, divided into two different types, continental and oceanic crust. Depth varies from 6 to 70 km, density about 3 g cm^{-3}.

EVIDENCE RELATING TO THE INTERNAL STRUCTURE OF THE EARTH

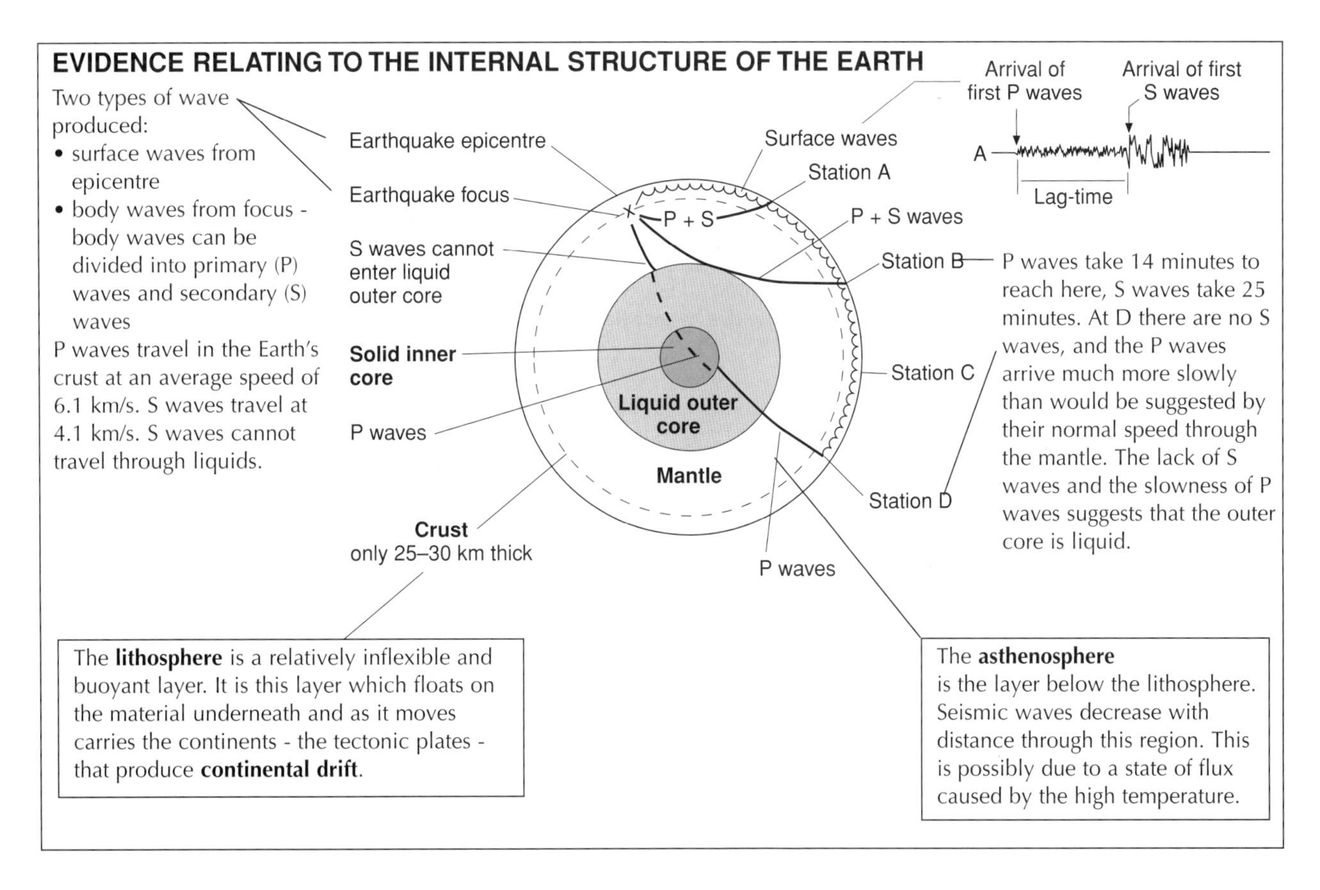

Two types of wave produced:

- surface waves from epicentre
- body waves from focus - body waves can be divided into primary (P) waves and secondary (S) waves

P waves travel in the Earth's crust at an average speed of 6.1 km/s. S waves travel at 4.1 km/s. S waves cannot travel through liquids.

P waves take 14 minutes to reach here, S waves take 25 minutes. At D there are no S waves, and the P waves arrive much more slowly than would be suggested by their normal speed through the mantle. The lack of S waves and the slowness of P waves suggests that the outer core is liquid.

The **lithosphere** is a relatively inflexible and buoyant layer. It is this layer which floats on the material underneath and as it moves carries the continents - the tectonic plates - that produce **continental drift**.

The **asthenosphere** is the layer below the lithosphere. Seismic waves decrease with distance through this region. This is possibly due to a state of flux caused by the high temperature.

Tectonic processes

LANDFORMS AND PLATE TECTONICS

The lithosphere is divided into a number of large and small rigid plates. There are three types of boundary:

- divergent - where plates are moving apart at ocean ridges or continental rifts
- convergent - where plates are moving together and one plate is forced beneath another forming ocean trenches
- transform or transcurrent - where plates are moving past each other and are neither constructive nor destructive

Diverging plates spread apart, splitting the crust. This is followed by the formation of new crust. They are therefore **constuctive**. Converging plates involve major mountain building and subduction of the crust. They are known as **destructive**.

WHAT HAPPENS AT CONSTRUCTIVE PLATE MARGINS?

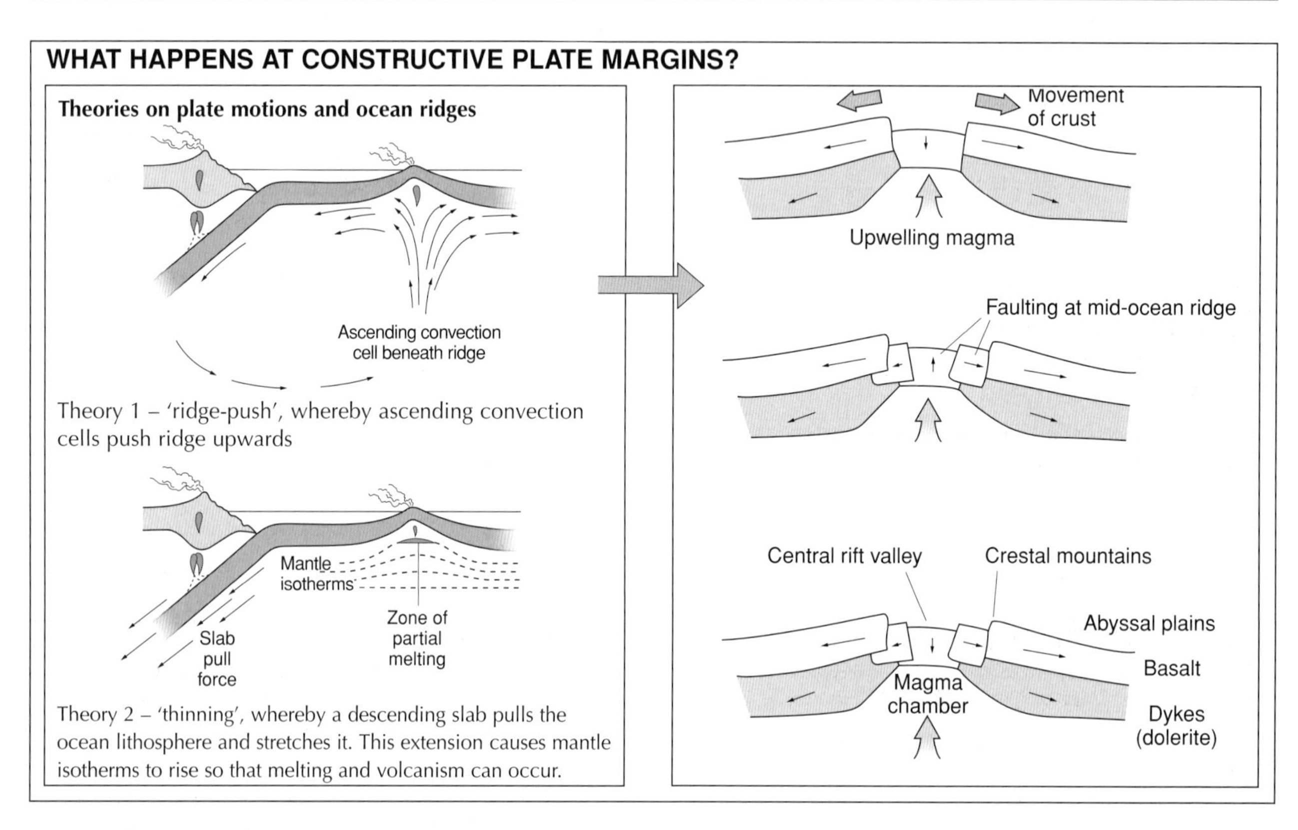

Theory 1 – 'ridge-push', whereby ascending convection cells push ridge upwards

Theory 2 – 'thinning', whereby a descending slab pulls the ocean lithosphere and stretches it. This extension causes mantle isotherms to rise so that melting and volcanism can occur.

WHAT HAPPENS AT DESTRUCTIVE PLATE MARGINS?

Young fold mountains – buckling and faulting, metamorphism

Deep ocean trench

Andesitic volcanoes

Sea-level

Subduction: DENSE NAZCA PLATE

Batholiths SOUTH AMERICAN PLATE

700 km

Asthenosphere

Plutons: rising bubbles of magma

Molten

Ⓐ Accretionary wedges – sediments scraped off descending plate incorporated into new continental crust → sedimentary rocks

✖ Benioff zone – frictional contact between descending plate and overlying material

→ Earthquake focus

Fold mountains

Fold mountains are formed by extreme pressure which has folded and uplifted sedimentary rocks, often changing them metamorphically.

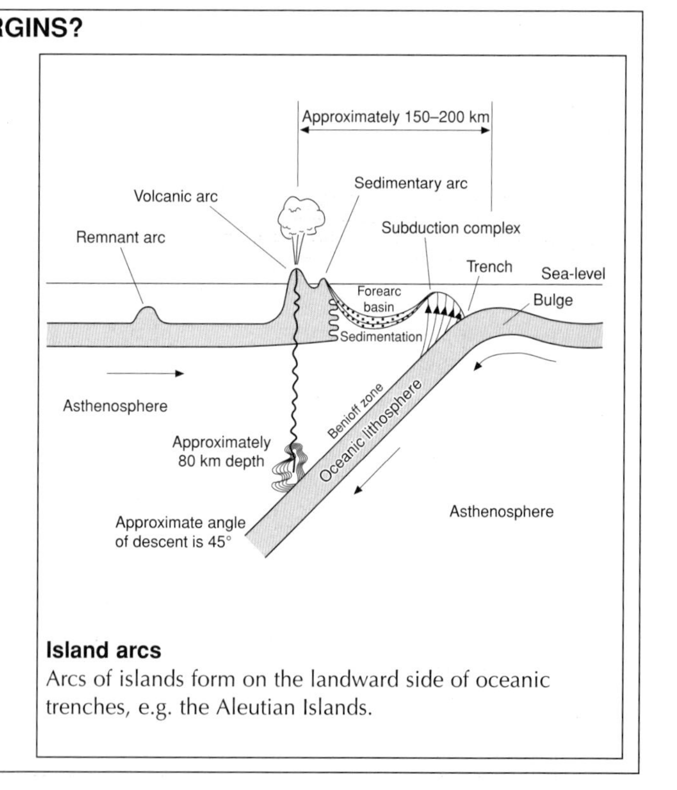

Island arcs

Arcs of islands form on the landward side of oceanic trenches, e.g. the Aleutian Islands.

Earthquakes

CAUSES

Earthquakes occur when normal movements of the crust are concentrated into a single shock or a series of sudden shocks. Aftershocks occur later as stresses are redistributed. The sequence is as follows:

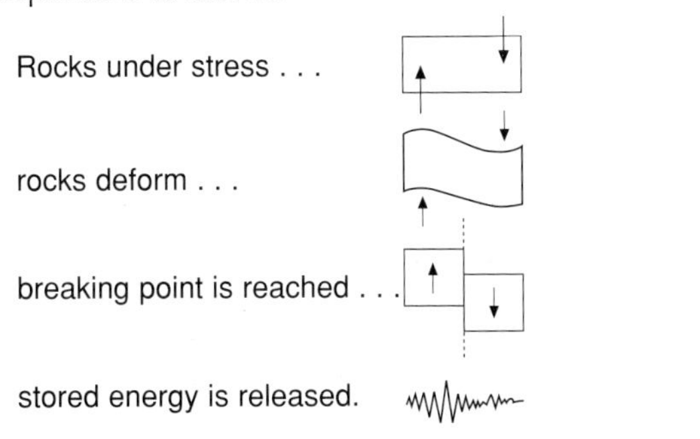

EARTHQUAKE DAMAGE

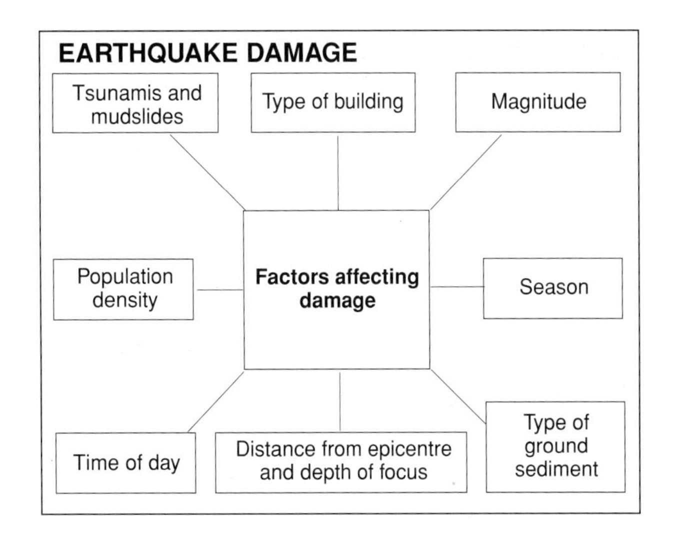

SHOCKWAVES

Waves associated with the focus (sub-surface)

P waves: fast/compression
S waves: slower/distortion } travel through the interior

Waves associated with the epicentre (surface)

Love and Rayleigh waves which travel on the surface and cause the most damage.

PREDICTION

- Crustal movement.
- Historical evidence.
- Seismic activity.
- Minor quakes before 'The Big One'.
- Change in properties of ocean crust.
- Gas omissions from ground.
- Changes in electrical conductivity.
- Unusual animal behaviour.

CASE STUDY: THE KOBE EARTHQUAKE

Details: 17th January 1995, killed over 5000 people, injured over 30 000, and made almost 750 000 homeless.

Causes: Pacific plate is being subducted beneath the Philippine plate. Kobe is situated in a geographically complex area near the northern tip of the Philippine plate.

Secondary factors: Rain and strong winds increased landslide risk; damp, unhygienic conditions encouraged disease; fires, broken glass, broken water pipes, and lack of insurance meant that many lost their livelihood.

HUMAN IMPACT

- Mining - gold-mining in the Witwatersand area of South Africa has been blamed for frequent seismic activity because of changed rock stress.
- Reservoirs - previously an area free from tectonic tremors, the states of Nevada and Arizona in the USA experienced over 100 tremors in 1937 following the construction of the Hoover Dam and the creation of Lake Mead due to seepage.

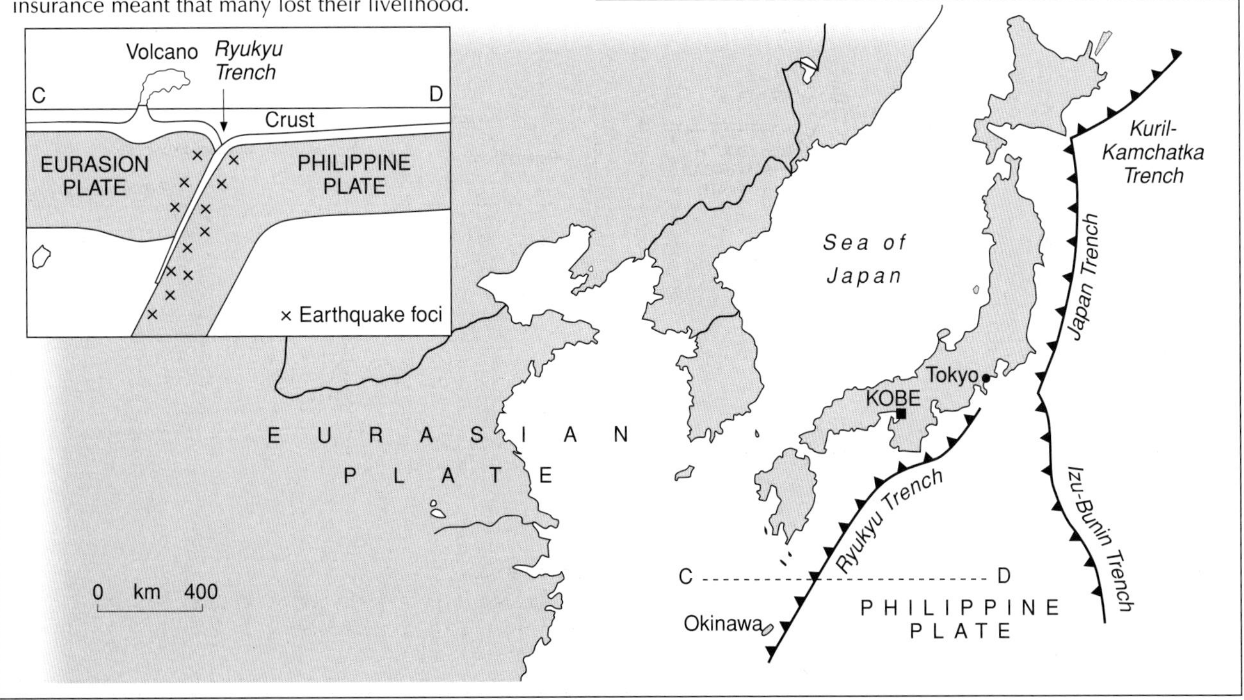

Volcanoes

A volcano is an opening in the earth's crust through which magma, molten rock, and ash are erupted on to the land. Volcanoes tend to be conical in shape, although there are a variety of forms depending upon:

- the nature of the material erupted
- the type of eruption
- the amount of change since the eruption.

Most volcanoes are located at the edges of plate boundaries, although some are found in the interior of plates.

Type of flow	Nature of lava
Aa flow is a few metres thick; consists of two distinct zones - an upper rubbly part, and a lower massive part of solid lava which cools slowly.	Aa surfaces are a loose jumble of irregularly shaped cindery blocks with razory sides.
Pahoehoe flow is the least viscous of all lavas; rates of advance can be slow; cool surface, flow occurs underneath.	Pahoehoe surfaces can be smooth and glossy but may also have cavities; surface may also be crumpled with channels.

TYPES OF VOLCANIC ERUPTIONS

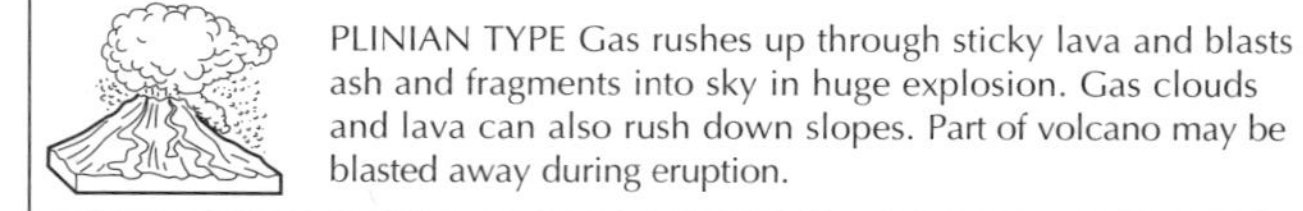

HAWAIIAN TYPE Runny basaltic lava which travels down sides in lava flows. Gases escape easily.

VULCANIAN TYPE Violent gas explosions blast out plugs of sticky or cooled lava. Fragments build up cone of ash and pumice.

PLINIAN TYPE Gas rushes up through sticky lava and blasts ash and fragments into sky in huge explosion. Gas clouds and lava can also rush down slopes. Part of volcano may be blasted away during eruption.

THE PACIFIC RING OF FIRE

Three-quarters of the earth's 550 historically active volcanoes lie along the Pacific Ring of Fire. This includes most of the world's recent eruptions, including Mount Pinatubo in the Philippines, which erupted in 1991. Without volcanic activity the Philippines would not exist: they comprise the remains of previous eruptions.

WORLD'S EXPLOSIVE VOLCANOES

Place	Date
Tambura, Indonesia	1815
Krakatoa, Indonesia	1883
Katmai, US	1912
Mt St Helens, US	1980
Mt Pinatubo, Philippines	1991

PREDICTING VOLCANOES

Scientists are increasingly successful in predicting volcanoes. Since 1980 they have correctly predicted 19 of Mt St Helen's 22 eruptions and Alaska's Redoubt volcano in 1989. The main ways of prediction include:

- **seisometers** to record swarms of tiny earthquakes that occur as the magma rises
- **chemical sensors** to measure increased sulphur levels
- **lasers** to detect the physical swelling of the volcano
- **ultra sound** to monitor low frequency waves in the magma, resulting from the surge of gas and molten rock, as happened at Pinatubo, El Chichon, and Mt St Helens

CASE STUDY: MOUNT PINATUBO

Details: 9 June 1991; eruption after 600 years; between 12 and 15 June ash and rock was scattered over a radius of 100 km; killed 350 people and made 200 000 people homeless, largely due to mudslides.

Effects: Mudslides covered 50 000 ha of cropland and destroyed 200 000 homes; 600 000 people lost their jobs.

Causes:

a) Earthquake 16 July 1990 (7.7 on the Richter Scale; 1 600 dead)

b) Basalt from the upper mantle squeezed into the magma chamber of the dormant volcano

c) Basalt reactivated viscous lava and created gas-charged magma (andesite)

d) This rose towards the surface causing volcano to bulge

e) Pressure blasted away the dome spewing 20 million tonnes of material into the atmosphere

5 Clouds of gas, steam, and dust rise high into the atmosphere

Crater

300–400°C

Ash and lava forms a cone

Side vent

4 Volcanic ash, a mixture of gas, ash, and molten rock, flows down mountain at 100 km per hour or more

3 Build-up of pressure over hundreds of years finally released in volcanic eruption

Magma chamber 650–1200°C

Eurasian plate

Philippine plate

2 Rocks of Philippine plate melt in high temperatures below Earth's crust, creating liquid magma which is forced up through cracks

1 Thin Philippine plate driven beneath thicker Eurasian plate by continental drift

Folds and faults

A fault is a fracture in a rock accompanied by a movement along one side or both sides. The total movement is called a **shift**. Displacement can be vertical (a **throw**) or horizontal (a **heave**). Faulting movement is usual during an earthquake. The degree of **folding** depends on the relative force on the rock from each direction. Folding can create fold mountains, escarpments (such as the North and South Downs), and gentle undulations in the land.

MAIN TYPES OF FAULT

Normal fault

This fault is the result of tension. Rock strata are pulled apart and one side is thrown down. Movement is down the dip of

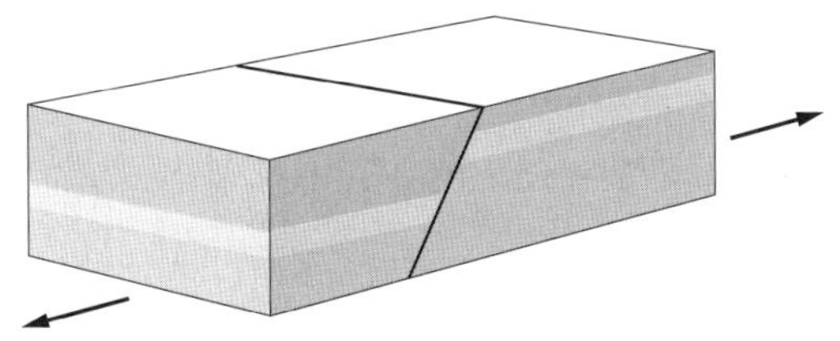

the fault plain. Land area is increased at the surface and a fault scarp is produced.

Reverse or thrust fault

This is the result of compression. Beds on one side of the fault plain are thrust over the other, i.e. overthrusting up the

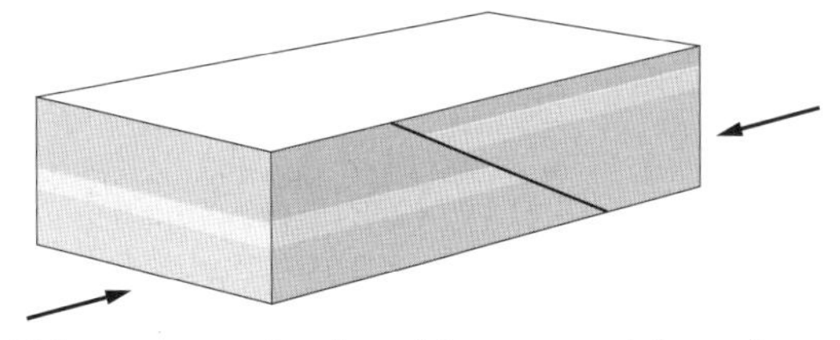

dip. This causes overlapping of the strata and the surface area is decreased. An overhanging fault scarp is formed which is usually reduced by erosion, e.g the Mid-Craven fault in the Pennines

Tear or wrench fault

This is formed where the shift is horizontal although the fracture is vertical.

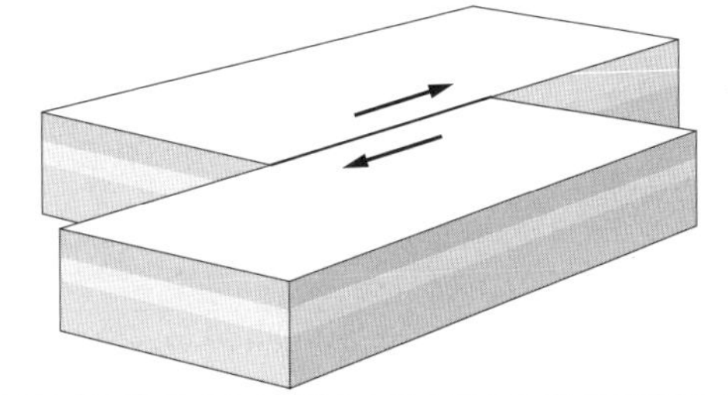

This is always a product of an earthquake, usually by a plate boundary, e.g. the San Andreas Fault.

LANDFORMS PRODUCED BY FAULTS

- **Horst** – upward-faulted blocks produced by a sequence of faults. They stand out above relatively low land on either side, bounded by fault scarps on either side. Extensive horst produce plateau areas or **block mountains**. They are bounded by **fault scarps**. If further earth movements have tilted the blocks then they are referred to as tilted **fault blocks**, e.g. the North Pennines in the UK.
- **Rift valley** (graben) - the rift valley is a faulted trough let down by parallel faults.

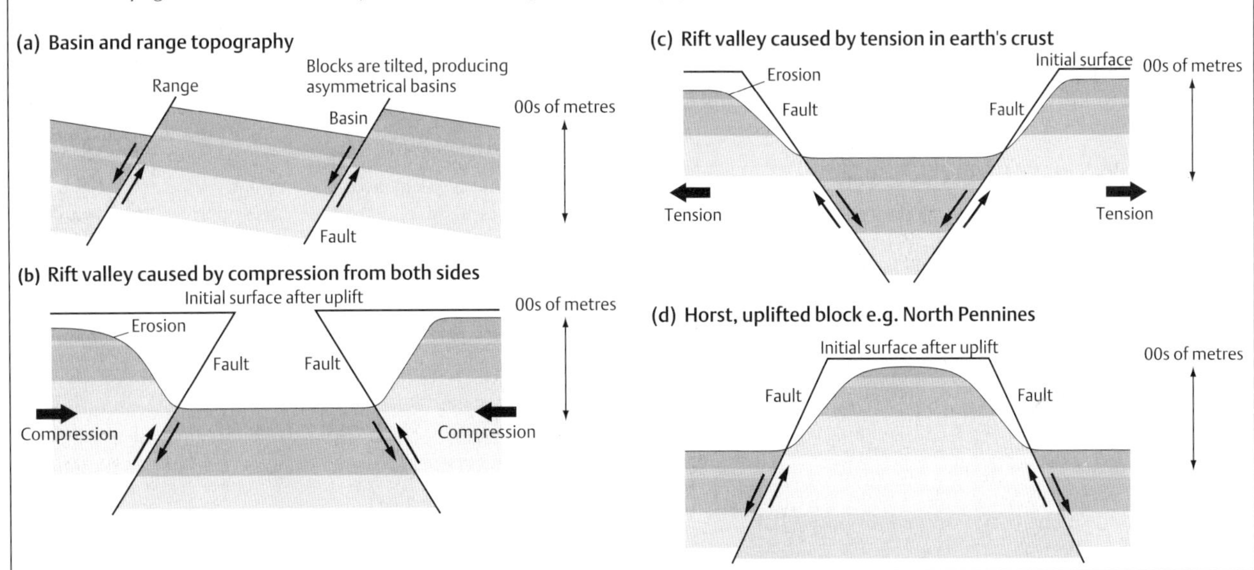

MAIN TYPES OF FOLDING

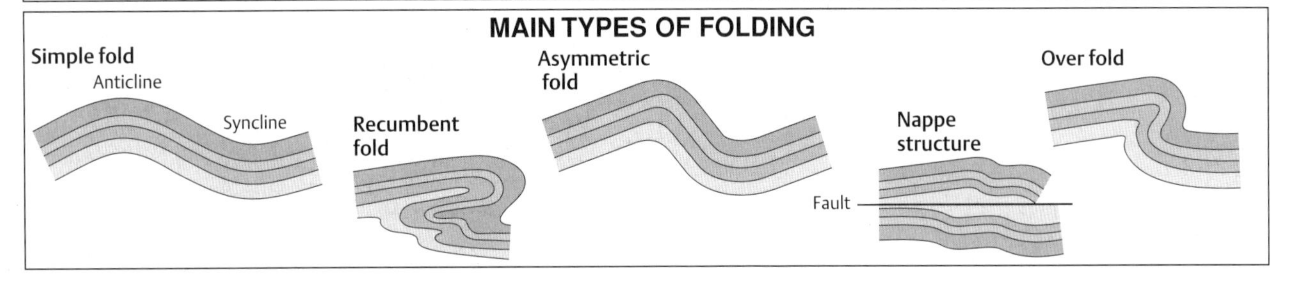

Weathering, rocks, and relief

Weathering is the decomposition and disintegration of rocks in situ. Decomposition refers to the chemical process and creates altered rock substances whereas disintegration or mechanical weathering produces smaller, angular fragments of the same rock. Weathering is important for landscape evolution as it breaks down rock and facilitates erosion and transport.

MECHANICAL (PHYSICAL) WEATHERING

There are four main types of mechanical weathering: freeze-thaw (ice crystal growth), salt crystal growth, disintegration, and pressure release.

Freeze-thaw occurs when water in joints and cracks freezes at 0°C and expands by 10% and exerts pressure up to 2100 kg/cm^2. Rocks can only withstand a maximum pressure of about 500 kg/cm^2. It is most effective in environments where moisture is plentiful and there are frequent fluctuations above and below freezing point, e.g. periglacial and alpine regions.

Salt crystal growth occurs in two main ways: first, in areas where temperatures fluctuate around 26-28°C, sodium sulphate (Na_2SO_4) and sodium carbonate (Na_2CO_3) expand by 300%; second, when water evaporates, salt crystals may be left behind to attack the structure. Both mechanisms are frequent in hot desert regions.

Disintegration is found in hot desert areas where there is a large diurnal temperature range. Rocks heat up and expand by day and cool and contract by night. As rock is a poor conductor of heat, stresses occur only in the outer layers and cause peeling or *exfoliation* to occur. Griggs (1936) showed that moisture is essential for this to happen.

Pressure release is the process whereby overlying rocks are removed by erosion thereby causing underlying ones to expand and fracture parallel to the surface. The removal of a great weight, such as a glacier, has the same effect.

CHEMICAL WEATHERING

There are four main types of chemical weathering: carbonation-solution, hydrolysis, hydration, and oxidation.

Carbonation-solution occurs on rocks containing calcium carbonate, e.g. chalk and limestone. Rainfall and dissolved carbon dioxide forms a weak carbonic acid. (Organic acids also acidify water.) Calcium carbonate reacts with an acid water and forms calcium bicarbonate, or calcium hydrogen carbonate, which is soluble and removed by percolating water.

Hydrolysis occurs on rocks containing orthoclase feldspar, e.g. granite. Orthoclase reacts with acid water and forms kaolinite (or kaolin or china clay), silicic acid, and potassium hydroxyl. The acid and hydroxyl are removed in the solution leaving china clay behind as the end product. Other minerals in the granite, such as quartz and mica, remain in the kaolin.

Hydration is the process whereby certain minerals absorb water, expand, and change, e.g. gypsum becomes anhydrate.

Oxidation occurs when iron compounds react with oxygen to produce a reddish brown coating.

LIMESTONE WEATHERING

Factors controlling the amount and rate of limestone solution

- The amount of CO_2, which is controlled by:
  - the amount of atmospheric CO_2
  - the amount of CO_2 in the soil and groundwater
  - the amount of CO_2 in caves and caverns
  - temperature - CO_2 is more soluble at low temperatures.
- The amount of water in contact with the limestone.
- Water temperature.
- The turbulence of the water.
- The presence of organic acids.
- The presence of lead, iron sulphides, sodium, or potassium in the water.

Rates of limestone solution

Rates of solution vary. In the Burren, western Ireland, the average depth of solution is 8 cm on bare ground, 10 cm under soil, and 22 cm underground. This has taken place over the last 10 000 years, representing an average of 15 cm in 10 000 years or 0.0152 mm per year.

Accelerated solution

Accelerated solution occurs under certain conditions:

- Impermeable rocks adjoin limestone - waters from non-karstic areas have aggressive waters and will cause above average rates of solution.
- Alluvial corrosion - intense solution takes place by water which passes through alluvium and morainic sands and gravels.
- Corrosion by mixture - this occurs when waters of different hardness mix.
- At the margins of snow and ice fields - snow meltwater is able to dissolve more limestone than rain-water.
- Limestone denudation increases as annual rainfall and run-off increase.
- Limestone weathers more quickly under soil cover than on bare surfaces.

Human activity also impacts on the nature and rate of limestone denudation:

- The burning of fossil fuels and deforestation has increased atmospheric levels of carbon dioxide and thus the weathering of limestone might accelerate.
- Acid rain is increasing levels of acidity (sulphur dioxide and nitrogen oxides) in rain-water.
- Agriculture and forestry are affecting soil acidity and carbon dioxide levels.

Controls on weathering

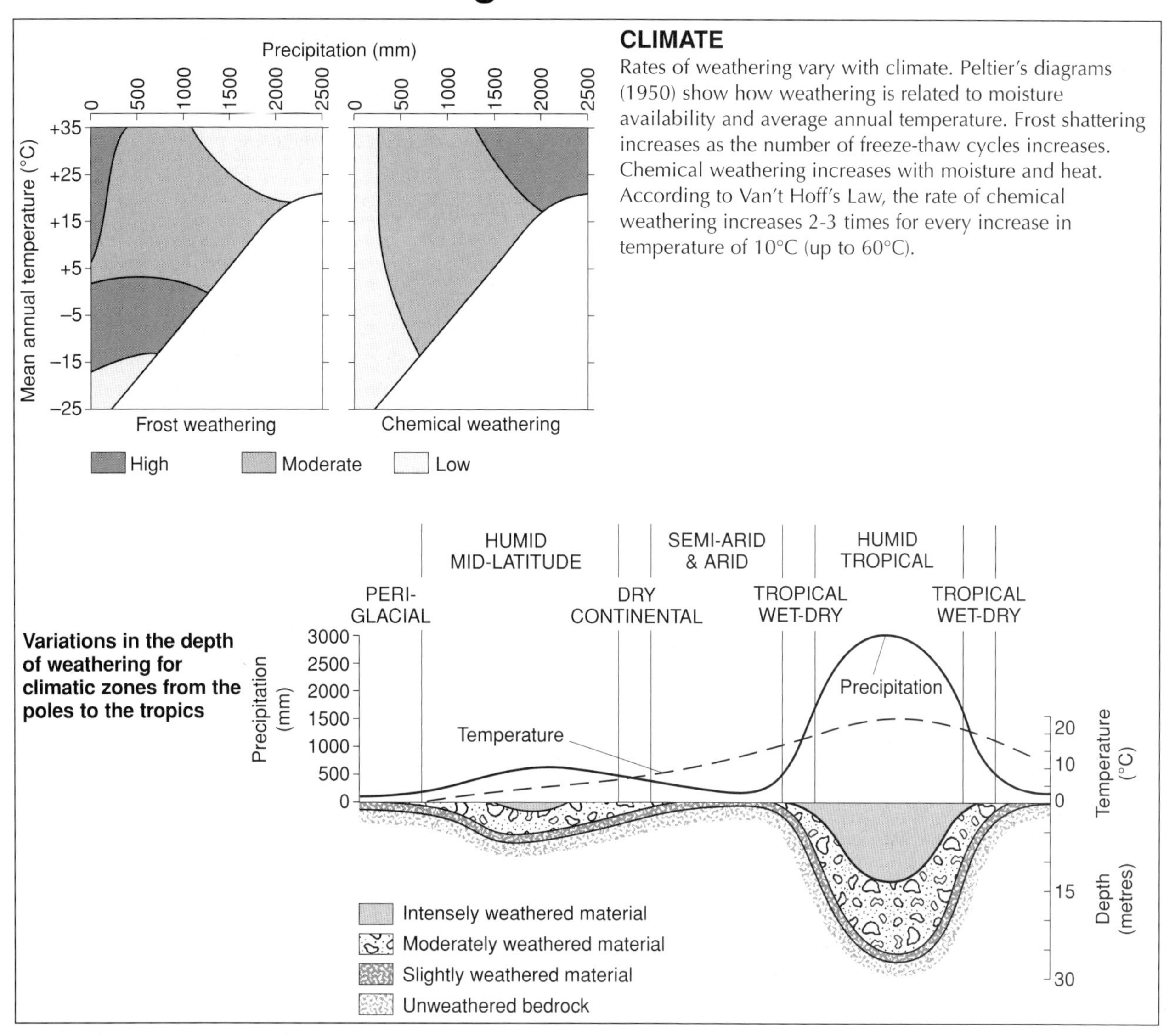

CLIMATE

Rates of weathering vary with climate. Peltier's diagrams (1950) show how weathering is related to moisture availability and average annual temperature. Frost shattering increases as the number of freeze-thaw cycles increases. Chemical weathering increases with moisture and heat. According to Van't Hoff's Law, the rate of chemical weathering increases 2-3 times for every increase in temperature of 10°C (up to 60°C).

GEOLOGY

Rock type influences the rate and type of weathering in many ways due to:

- chemical composition
- the nature of cements in sedimentary rock
- joints and bedding planes.

For example, limestone consists of calcium carbonate and is therefore susceptible to carbonation-solution, whereas granite with orthoclase feldspar is prone to hydrolysis. In sedimentary rocks, the nature of the cement is crucial: iron-oxide based cements are prone to oxidation whereas quartz cements are very resistant.

The importance of joints and bedding planes: the formation of tors

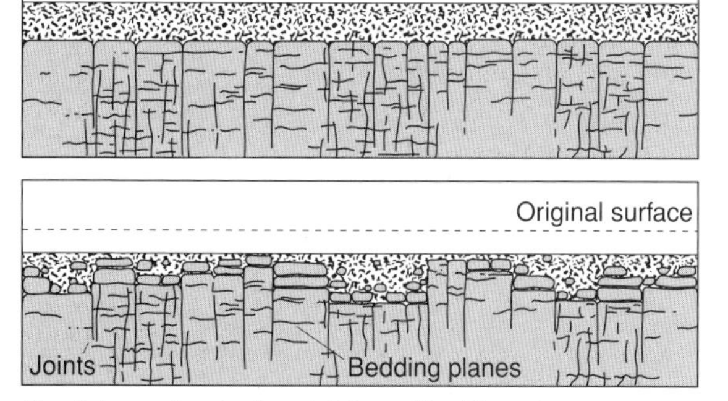

Breakdown of rock along joints and bedding planes

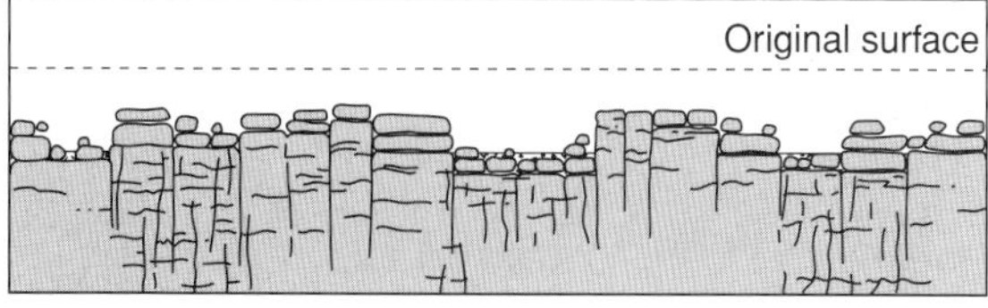

Removal of weathered material to expose tors, e.g. Hay Tor, Yes Tor on Dartmoor

Slopes (1)

Slopes can be defined as any part of the solid land surface. They can be **sub-aerial** (exposed) or **sub-marine** (underwater), **aggradational** (depositional), **degrational** (eroded), or **transportational**, or any mixture of these. Geographers study the **hillslope**, which is the area between the **watershed** (or drainage basin divide) and the **base**, that may or may not contain a river or stream. Slope **form** refers to the shape of the slope in cross-section; slope **processes** the activities acting on the slopes; and slope **evolution** the development of slopes over time.

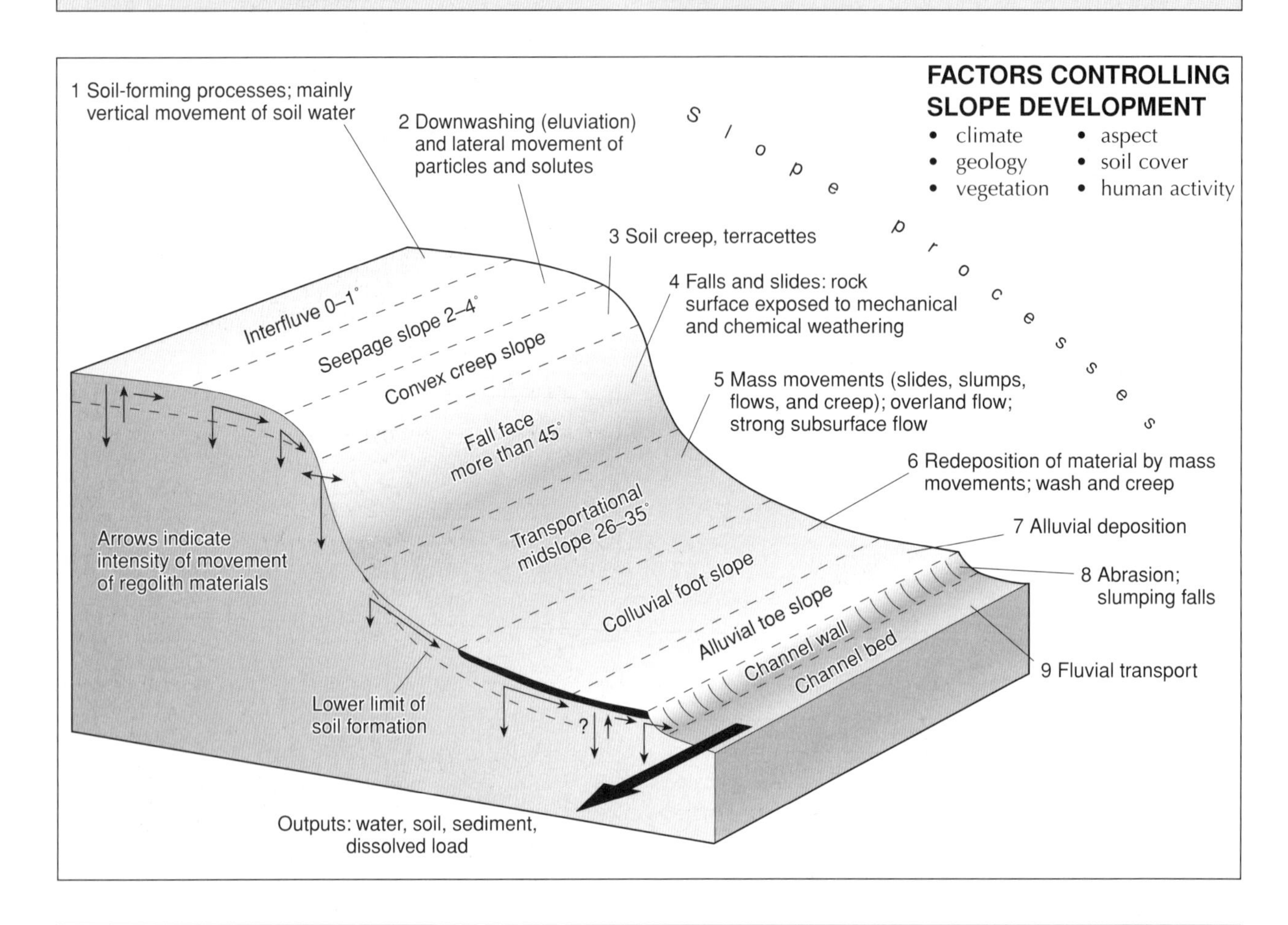

FACTORS CONTROLLING SLOPE DEVELOPMENT

- climate
- geology
- vegetation
- aspect
- soil cover
- human activity

SLOPE PROCESSES: soil creep

Individual soil particles are pushed or heaved to the surface by wetting, heating, or freezing of water. They move at right angles to the surface (2) as it is the zone of least resistance. They fall (5) under the influence of gravity once the particles have dried, cooled, or the water has thawed. Net movement is downslope (6). Rates are slow - 1 mm/year in the UK and up to 5 mm/year in the tropical rainforest. They form **terracettes**, e.g. those at Maiden Castle, Dorset, and The Manger, Vale of the White Horse.

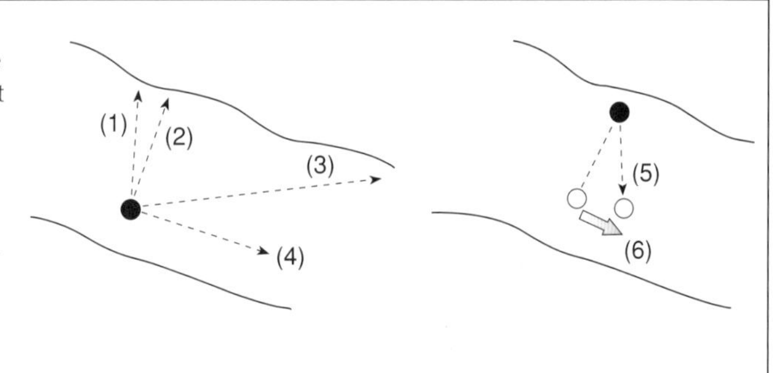

A
Raindrop
Splash
Ground surface
Compaction

B
Raindrop
(a)
Splash
(b)
Pushing component
Compaction component

SLOPE PROCESSES: rain-splash erosion

On flat surfaces (A) raindrops compact the soil and dislodge particles equally in all directions. On steep slopes (B) the downward component (b) is more effective than the upward motion (a) due to gravity. Hence erosion downslope increases with slope angle.

Slopes (2)

GEOLOGY

Rock type affects slopes through its strength, dip, and orientation of joints and bedding planes.

Dip of rock

(i) Steeply dipping, e.g. Hogsback near Guildford

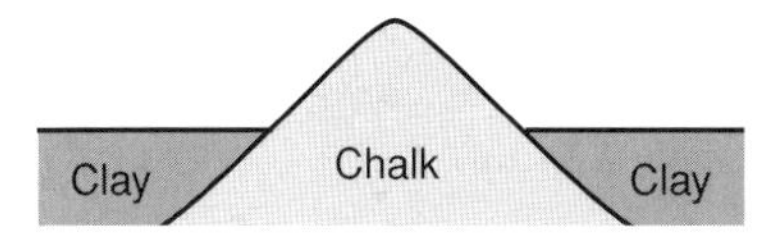

(ii) Gently dipping, e.g. North Downs

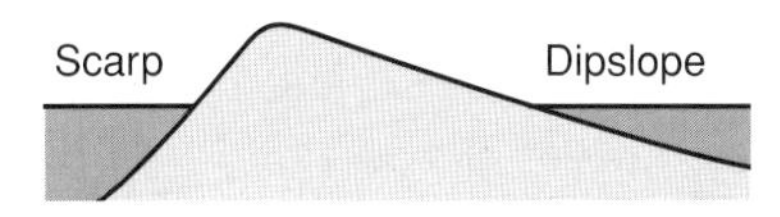

(iii) Horizontal strata, e.g. Salisbury Plain

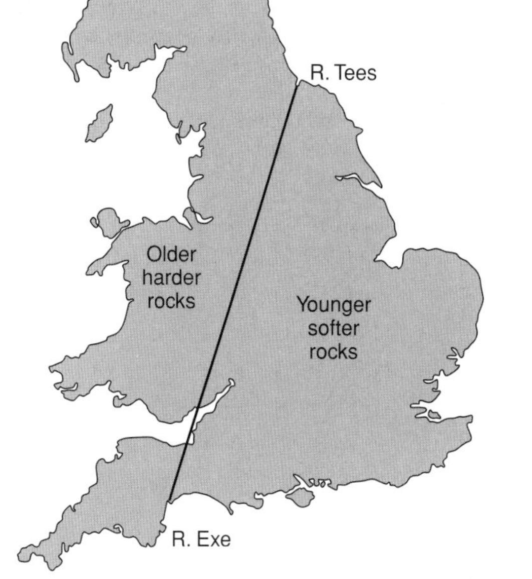

The **Tees-Exe line** is an imaginary line running from the River Tees to the River Exe. It divides Britain into hard and soft rock. To the north and west are old, hard, resistant rocks, e.g. granite, basalt, and carboniferous limestone, forming upland rugged areas. To the south and east are younger softer rocks, such as chalk and clay, forming more subdued low-lying landscapes.

Climate

L.C. Peltier's classification of morphogenetic regions (1950)

	Annual temp. (°C)	Annual rainfall (mm)	Processes
Glacial	–20 to –5	0 - 1100	Ice erosion Nivation Wind action
Periglacial	–15 to 0	125 - 1300	Mass movement Wind action Weak water action
Boreal	–10 to 5	250 - 1500	Moderate frost action Slight wind action Moderate water action
Maritime	5 to 20	1250 - 1500	Mass movement Running water
Selva (rainforest)	15 to 30	1400 - 2250	Mass movement Slight slope wash
Moderate (temperate)	5 to 30	800 - 1500	Strong water action Mass movement Slight frost action
Savanna	10 to 30	600 - 1250	Running water Moderate wind action
Semi-arid	5 to 30	250 - 600	Strong wind action Running water
Arid	15 to 30	0 - 350	Strong wind action Slight water action

Slopes vary with climate. In general, humid slopes are rounder, due to chemical weathering, whereas arid slopes are jagged or straight owing to mechanical weathering and overland run-off. **Climatic geomorphology** is a branch of geography which studies how different processes operate in different climatic zones and thereby produce different slope forms or shapes.

Aspect

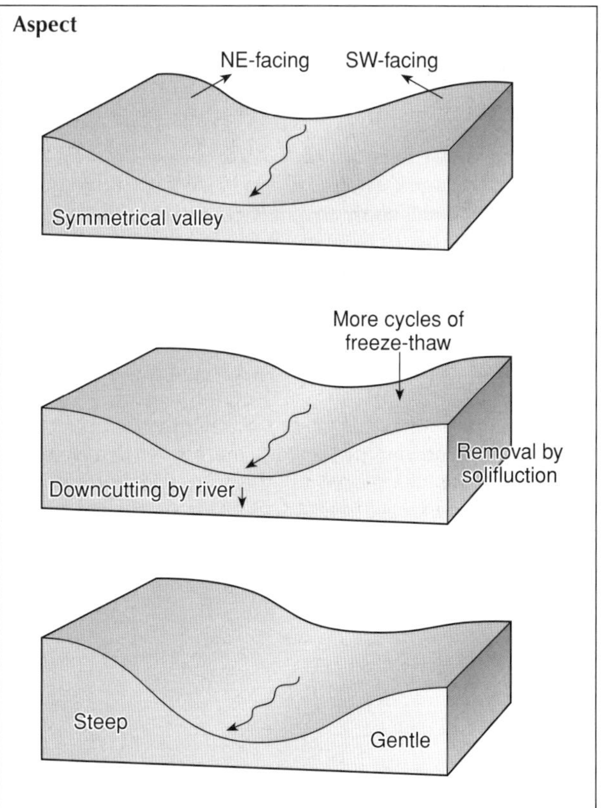

The NE-facing slope remains in the shade. Under periglacial conditions, temperatures rarely rise above freezing. By contrast, the SW-facing slope is subjected to many cycles of freeze-thaw. Solifluction and overland run-off lower the level of the slope, and streams remove the debris from the valley. The result is an asymmetric slope, e.g. the River Exe in Devon, and Clatford Bottom in Wiltshire.

Slope theories

SLOPE EVOLUTION

Slope evolution refers to the change in slope form (shape) over time.

Slopes can be divided into those that are (i) time independent, in which the slope retains a constant angle, although altitude may be lowered, and (ii) time dependent, in which slope angle and altitude decline progressively over time.

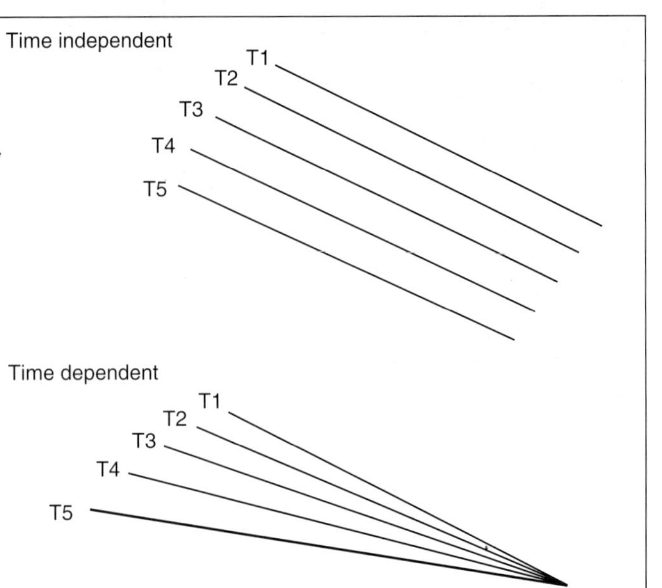

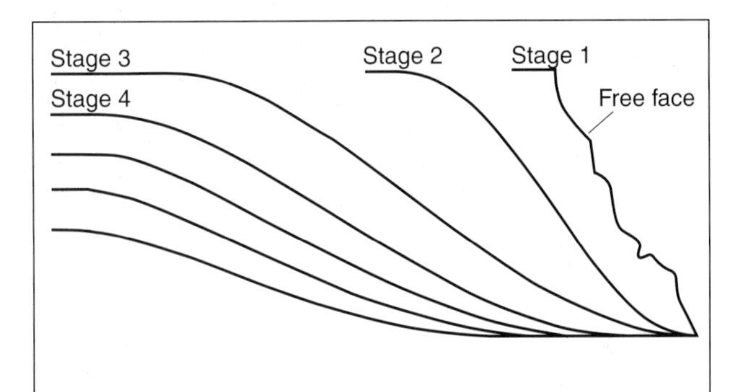

SLOPE DECLINE: W. M. DAVIES

The main processes involved are soil creep, solution, overland run-off, weathering, and fluvial transport at the base. Slopes decline progressively over time. The free face is changed by falls and slumps and develops a regolith. Weathered material is transported by overland run-off and surface wash, eventually producing a convex-concave profile.

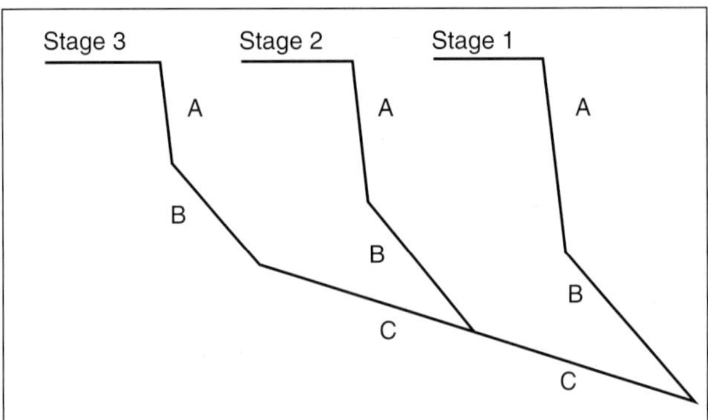

SLOPE REPLACEMENT: WALTHER PENCK

Slope A is replaced by slope B, which is in turn replaced by slope C. Replacement is by lower angle slopes which extend upwards at a constant angle. The segments become increasingly longer as the slope develops. Some free faces may be completely removed. It is common in tectonically active and coastal areas. The size of sediment decreases from A to B to C.

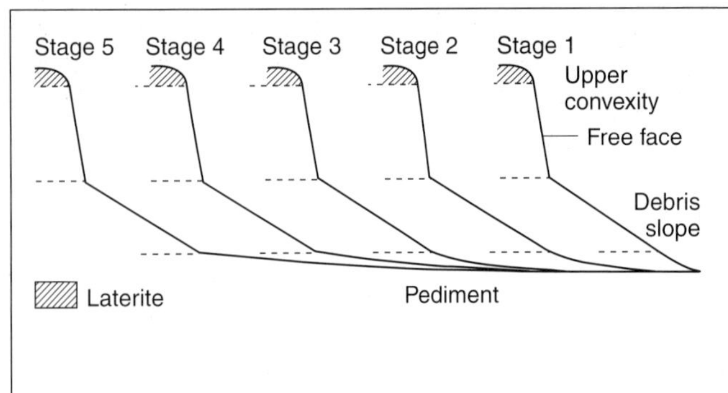

SLOPE RETREAT: L.C. KING

The key elements are a very hard lateritic cap rock controlling the rate of slope evolution. Mechanical weathering and sheet wash are the dominating processes in a semi-arid climate. All elements, except the pediment, retain a constant angle. Pediments vary from 1-2 degrees on fine material to 3-5 degrees on gravel and stony material. It is common in dry areas such as Monument Valley, Arizona.

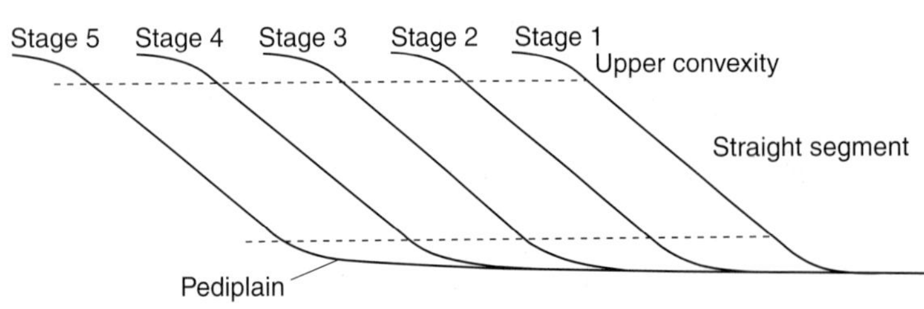

Limestone

Limestone consists mainly of CaCO (calcium carbonate). It is formed from the remains of organic matter, notably plants and shells. Limestone scenery is unique, because of its *permeability* and *solubility* in rain and ground-water.

DISTRIBUTION OF LIMESTONE IN THE BRITISH ISLES

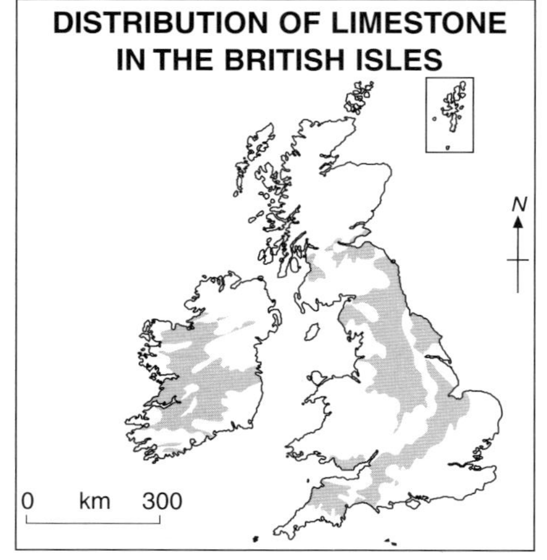

LIMESTONE SCENERY

Variations exist between the different types of limestone because of their
- hardness
- chemical composition
- jointing
- bedding planes.

In the UK there are three main types of contrasting limestone scenery:
- **Carboniferous limestone** (220-280 million years old), such as the Mendips and the Pennines. It has a distinctive bedding plane and joint pattern known as massively jointed. These act as weaknesses allowing water to percolate into the rock and dissolve it, a process known as **carbonation-solution**
- **Jurassic limestone** (120-150 million years old) such as the Cotswolds
- **Cretaceous limestone** or chalk (70-100 million years old), such as the North and South Downs.

Limestone pavement large areas of bare exposed rock e.g. the Burren, County Clare and at Malham, Yorkshire.

Clints elevated blocks between grikes.

Karren or **lapies** small-scale solution grooves, only a few centimetres deep, caused by run-off and solution on limestone.

Grikes deep grooves formed by acid rainwater running over surface limestone.

Scar

Gorge

Plateau

Fault

Impermeable rock

Impermeable rock

Dry valleys - a river valley without a river, a common feature on chalk and limestone, e.g. Cheddar Gorge. It is likely that surface streams running over the impermeable limestone formed the Gorge during cold periglacial periods. Intense weathering, erosion, and mass movements took place during this period.
- Freeze-thaw was rapid, helped by numerous exposed joints and bedding planes.
- Carbonation was increased owing to the increased solubility of CO_2 with low temperatures.
- Snowmelt caused river discharges to rise to a level more than 50 times greater than today.
- Mass movements and overland run-off all helped remove the denuded limestone.

When conditions warmed at the end of the periglacial period surface water sank into the limestone. Cheddar Gorge was left as a dry valley.

Swallow holes (or **sinks**)
- depressions in the landscape caused by limestone solution (smaller than dolines).

They can also be formed by:
- the enlargement of a grike system
- carbonation
- fluvial activity
- the collapse of a cave, such as at Gaping Ghyll near Malham.

Often a river disappears down a hole, hence the term sink.

Stalactites and **stalagmites** are both deposits of calcium carbonate. The former hang down from a cave ceiling, the latter are formed at the base of a cave. Rates of deposition are slow, about 1mm/100yrs. The speed with which water drips from a cave ceiling appears to have some influence on whether stalactites (slow drip) or stalagmites (fast drip) are formed.

Resurgent streams arise when the limestone is underlain by an impermeable rock, such as clay, such as the River Axe at Wookey Hole.

Caves and **tunnels** - underground features formed by carbonation-solution and erosion by rivers. Along with stalactites and stalagmites these features are characteristic of a **karst** landscape, formed from dry carboniferous limestone.

Chalk and clay landscapes

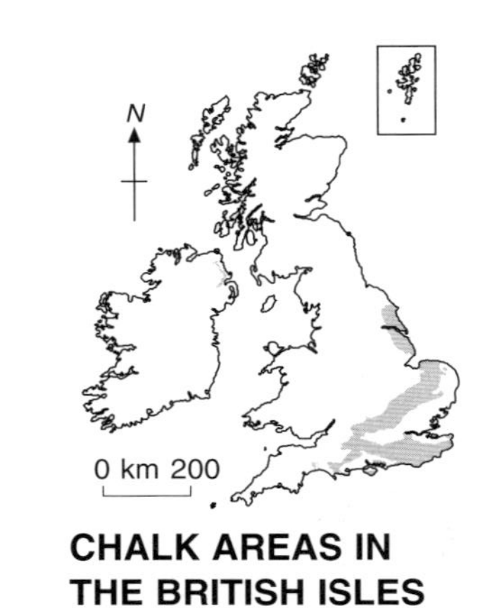

CHALK AREAS IN THE BRITISH ISLES

Dry valleys, such as Scratchy Bottom near Lulworth Cove and the Devil's Dyke near Brighton, are common, especially on dip slopes. In some cases, such as in the Vale of the White Horse, Uffington, periglacial avalanches have been suggested as a mechanism for their formation.

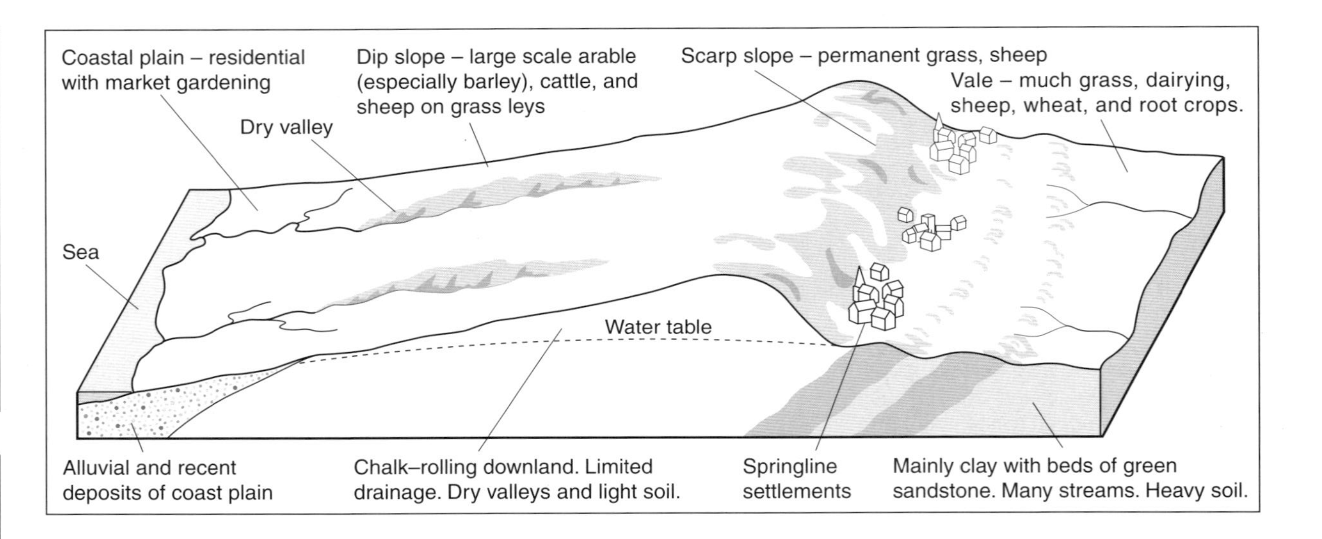

CHALK

Unlike carboniferous limestone, which is hard, grey, angular, and jointed, chalk is soft, white, rounded, and porous. Chalk in southern England is best illustrated by the escarpments of the North and South Downs. Other main types of chalk slopes are the Hogsback of Dorset and Guildford, and the flat Salisbury Plain. However, there are other famous chalk landforms such as the Folkestone Warren, the Seven Sisters, and the Needles.

Escarpments are not unique to chalk, but they are generally most easily identified on chalk. They have a steep **scarp** slope and a gentle **dip** slope. The steepness of the scarp slope depends upon weathering, erosion, and mass movement on the slope and removal by a river at the base. Sometimes escarpments are called **cuestas**.

CLAY

Clay is a fine grained, soft rock that is easily eroded. It is the end result of chemical weathering and river erosion. Clay is very porous but it is impermeable. It is impermeable because when it is wet the individual particles expand (hydrate) and pack very tightly. This seals off the surface and makes it impermeable. Because of its softness, clay forms undulating lowlands with lots of surface drainage, such as rivers, marshes, and moors. When drained, clay provides fertile soils, such as those in East Anglia. Settlements are usually found on the higher ground to avoid the risk of flooding. The villages on the edge of Otmoor near Oxford are a good example.

SLOPES

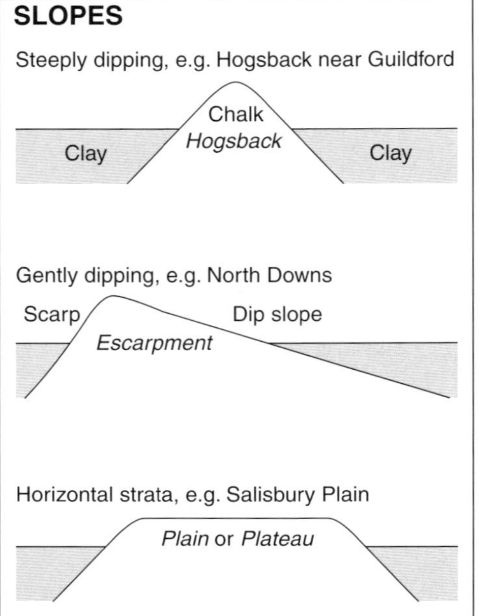

HOLLOWS IN CHALK

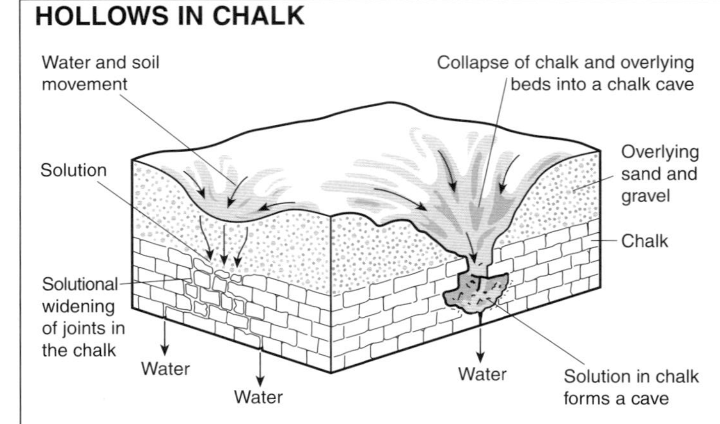

A feature very common in Dorset's heathlands are hollows, generally 10-20 m wide and 3-4 m deep. However, one, Culpepper's Dish, is over 80 m wide and 20 m deep. Two explanations are:

- solution at the boundary of the chalk and the overlaying material, and
- the collapse of a small cave system in the chalk and the sinking of the overlaying beds.

Granite

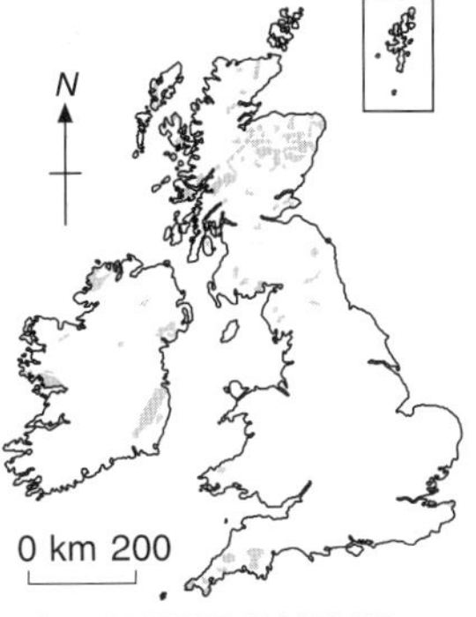

GRANITE IN THE BRITISH ISLES

Granite is an igneous, crystalline rock. It has great physical strength and is very resistant to erosion. There are many types of granite but all share certain characteristics. They contain quartz, mica, and felspar. These are resistant minerals. The main processes of weathering that occur on granite are freeze-thaw and hydrolysis.

Characteristic granite landscapes include exposed large-scale **batholiths**, which form mountains. An example is the Wicklow Mountains. **Tors** are isolated masses of bare rock. They can be up to 20 m high, such as Hay Tor and Yes Tor. Some of the boulders of the mass are attached to part of the bedrock. Others merely rest on the top.

Bodmin Moor is one exposure of a **granitic dome** in the South West peninsula. Denudation has slowly removed rocks to expose the granite masses. Due to its greater resistance, the granite has remained as upland areas.

Due to granite's resistance, weathering results in a thin, gritty soil cover. Such soils are generally infertile, so rough grazing is the dominant land use. Granite is an impermeable rock and many marshy hollows at the heads of the valleys indicate the limited downward movement of water.

THE FORMATION OF TORS

Tors are a good example of **equifinality**. This means that different processes can produce the same end result. Thus it is highly debatable whether tors are formed by chemical weathering or mechanical weathering, or by a combination of the two. What is clear, is that the joints and bedding planes, and the great strength and resistance of the rocks have determined the distribution of tors on the landscape.

After Linton

Linton (1955) argued that the well-developed jointing system (of irregular spacing) was chemically weathered. This occurred under humid conditions during warm, wet periods in the Tertiary era. Decomposition was most rapid along joint planes. Where the distance between the joint planes was largest, masses of granite remained relatively unweathered and formed, essentially, embryonic tors. Subsequent denudation, perhaps under periglacial conditions, removed the residue of weathering, leaving the unweathered blocks as tors.

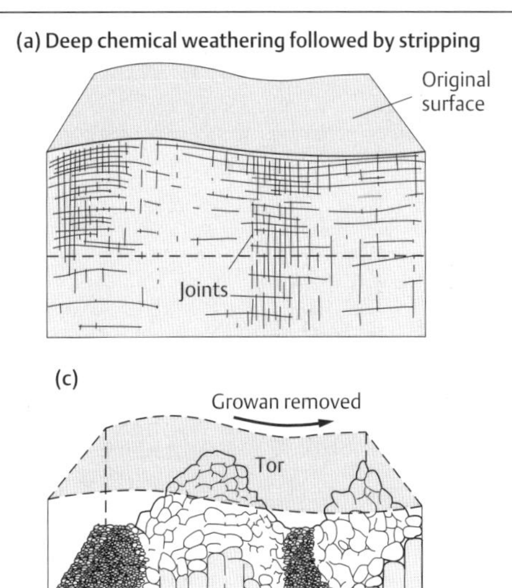

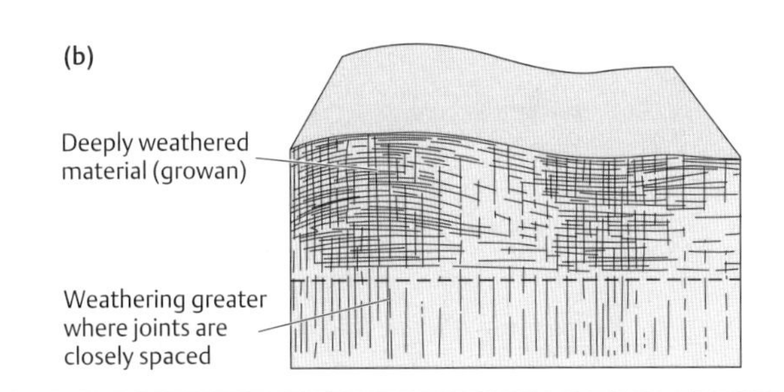

After Palmer and Nielson

An alternative theory by Palmer and Neilson (1962) also relates tor formation to the varied spacing of joints within the granite. They believe that frost action under periglacial conditions was the dominant process. This led to the removal of the more closely jointed portions of the rock. The evidence tends to support their idea, as the amount of kaolin in the joints is limited; so too is the amount of rounding that has occurred. Both of these features are expected to be dominant if chemical weathering were the main process in operation. Palmer and Neilson suggest that intense frost shattering followed by solifluction, removed the finer material and left the tors standing.

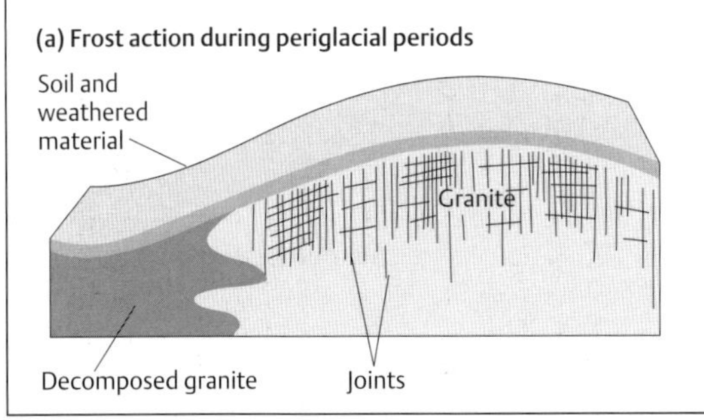

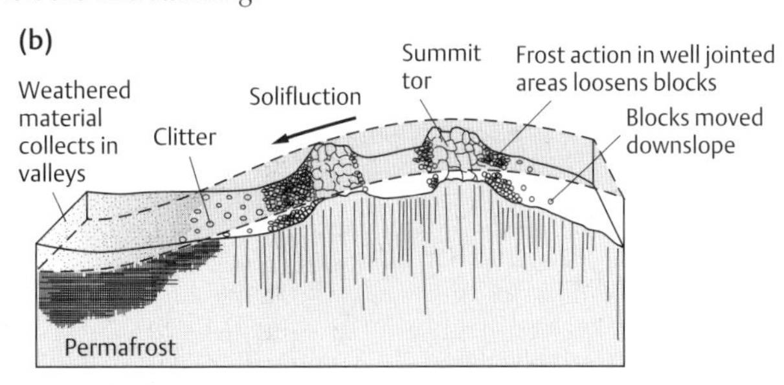

Mass movements (1)

Mass movements are any large-scale movement of the earth's surface that are not accompanied by a moving agent such as a river, glacier, or ocean wave. They include very small movements, such as soil creep, and fast movements, such as avalanches. They vary from dry movements, such as rock falls, to very fluid movements, such as mudflows.

A CLASSIFICATION OF MASS MOVEMENTS

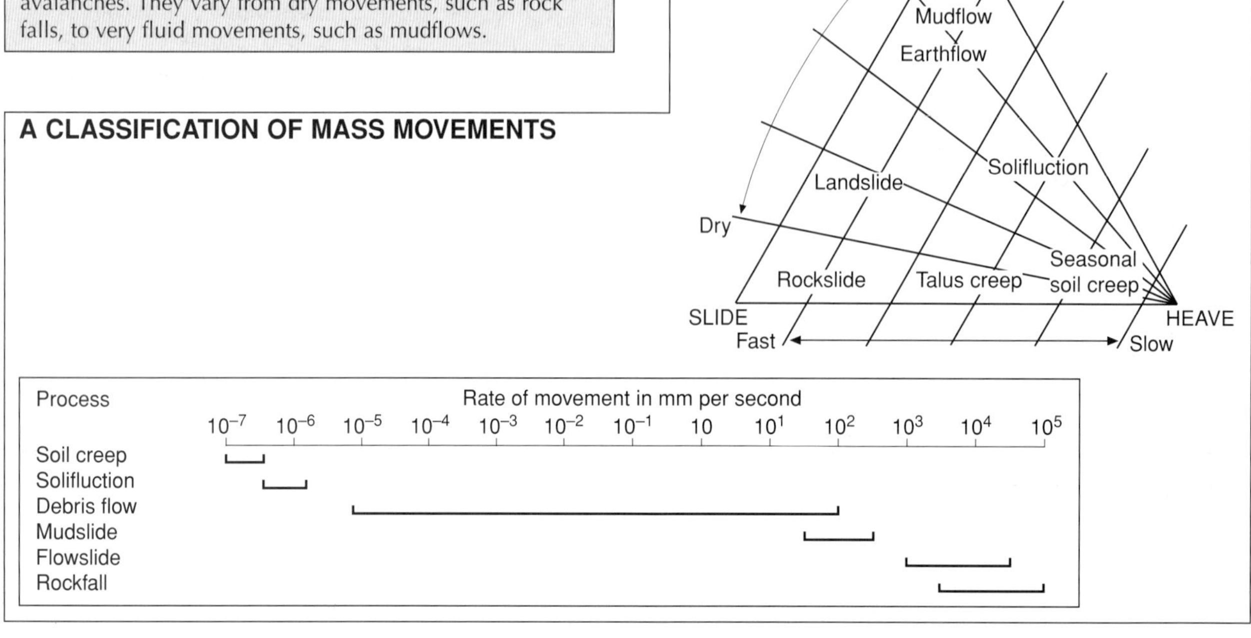

FLOW

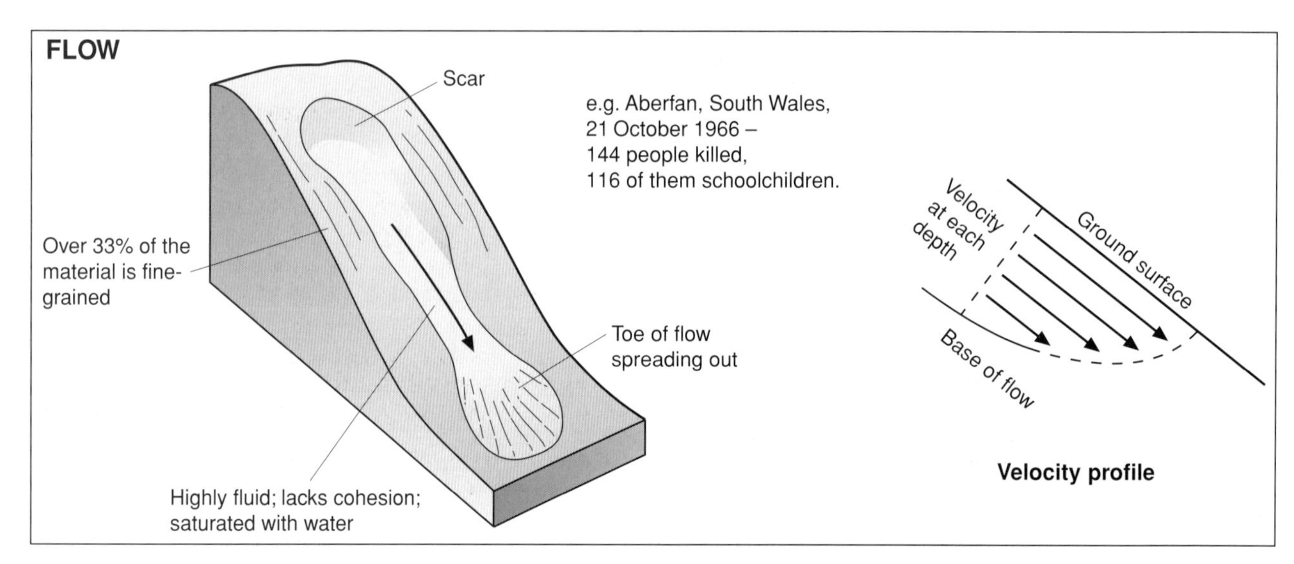

SLIDE

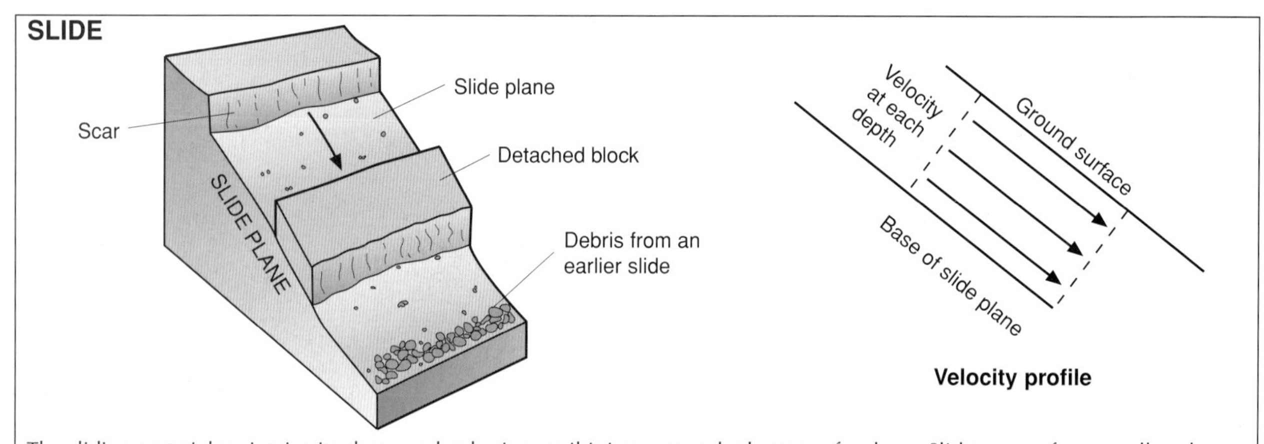

The sliding material maintains its shape and cohesion until it impacts at the bottom of a slope. Slides range from small-scale events close to roads to large-scale movements killing thousands of people, e.g. the Vaiont Dam disaster in Italy where more than 2000 people died on 9 October 1963.

Mass movements (2)

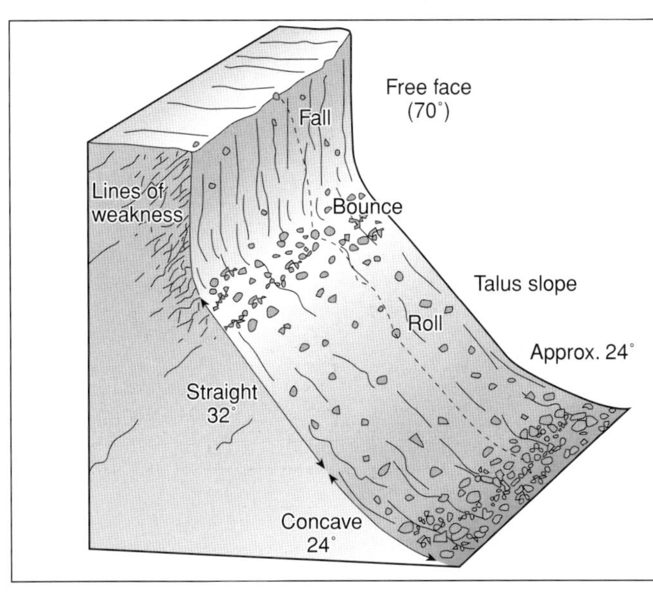

FALLS

Falls occur on steep slopes (> 70°). The initial cause of the fall may be weathering, e.g. freeze-thaw or disintegration, or erosion prising open lines of weakness. Once the rocks are detached they fall under the influence of gravity. If the fall is short it produces a relatively straight scree; if it is long, it forms a concave scree. A good example of falls and scree is Wastwater in the Lake District.

SLUMPS

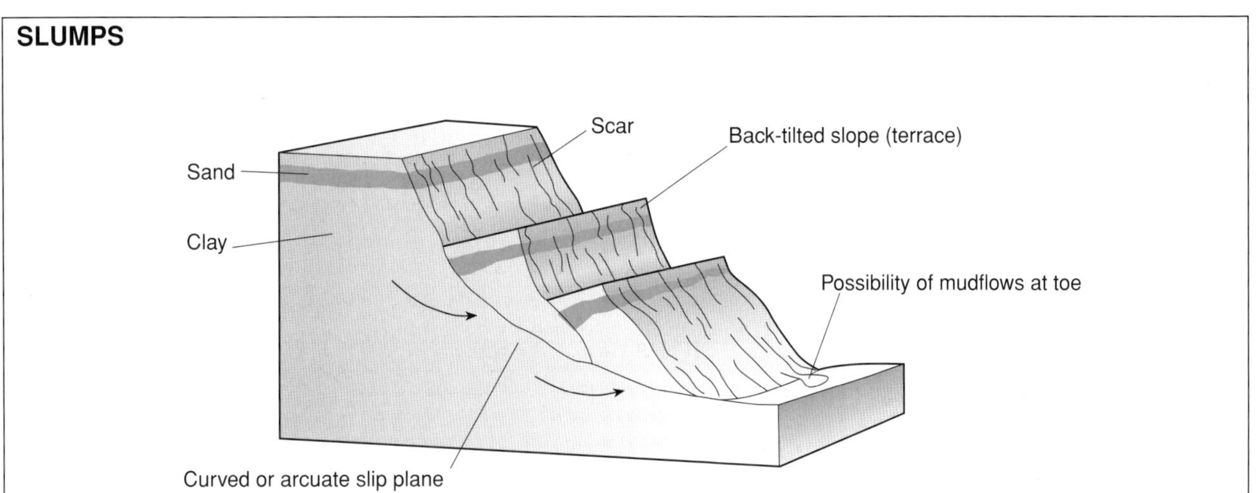

Slumps occur on weaker rocks, especially clay, and have a rotational movement along a curved slip plane. Clay absorbs water, becomes saturated, and exceeds its liquid limit. It then flows along a slip plane. Frequently the base of a cliff has been undercut and weakened by erosion, thereby reducing its strength, e.g. Folkestone Warren. Human activity can also intensify the condition by causing increased pressure on the rocks, e.g. the Holbeck Hall Hotel, Scarborough.

AVALANCHES

Avalanches are rapid movements of snow, ice, rock, or earth down a slope. They are common in mountainous areas: newly-fallen snow may fall off older snow, especially in winter (a **dry avalanche**), while in spring partially-melted snow moves (a **wet avalanche**), often triggered by skiing. Avalanches frequently occur on steep slopes over 22°, especially on north-facing slopes where the lack of sun inhibits the stabilisation of snow. **Debris avalanches** are a rapid mass movement of sediments, often associated with saturated ground conditions.

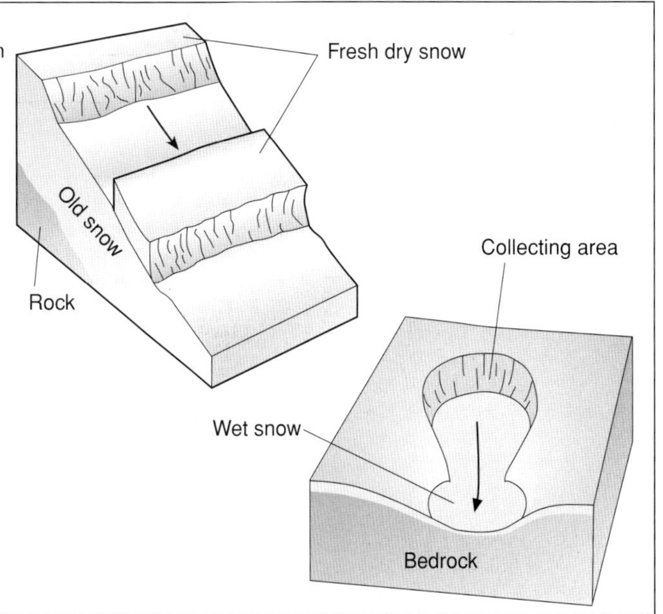

Slope failure

SHEAR STRENGTH AND SHEAR RESISTANCE

Slope failure is caused by two factors:

1 a reduction in the internal resistance, or **shear strength**, of the slope, or

2 an increase in **shear stress**, that is the forces attempting to pull a mass downslope.

Both can occur at the same time.

Slope instability caused by road construction

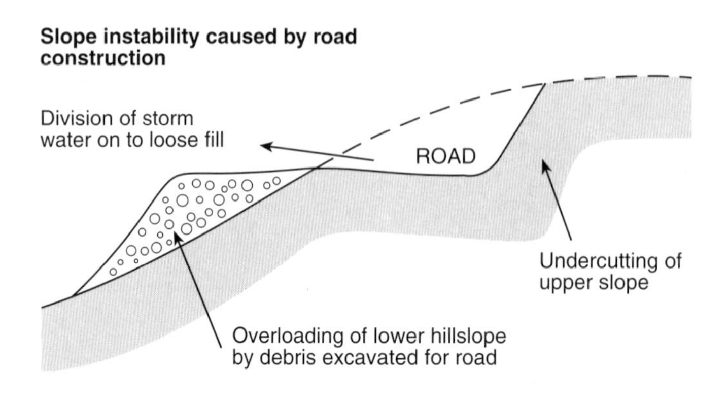

FACTORS INCREASING STRESS AND DECREASING RESISTANCE

Factors	Example
Factors that contribute to increased shear stress	
Removal of lateral support through undercutting or slope steepening	Erosion by rivers and glaciers, wave action, faulting, previous rock falls or slides
Removal of underlying support	Undercutting by rivers and waves, sub-surface solution, loss of strength by extrusion of underlying sediments
Loading of slope	Weight of water, vegetation, accumulation of debris
Lateral pressure	Water in cracks, freezing in cracks, swelling (especially through hydration of clays), pressure release
Transient stresses	Earthquakes, movement of trees in wind
Factors that contribute to reduced shear strength	
Weathering effects	Disintegration of granular rocks, hydration of clay minerals, dissolution of cementing minerals in rock or soil
Changes in pore-water pressure	Saturation, softening of material
Changes of structure	Creation of fissures in shales and clays, remoulding of sand and sensitive clays
Organic effects	Burrowing of animals, decay of tree roots

THE ROLE OF WATER

Water can weaken a slope by increasing shear stress and decreasing shear resistance. The weight of a potentially mobile mass is increased by:

- an increase in the volume of water
- heavy or prolonged rain
- rising water tables
- saturated surface layers increase.

Moreover, water reduces the cohesion of particles by saturation. Water pressure in saturated soils (pore water pressure) decreases the frictional strength of the solid material. This weakens the slope. Over time the safety factor for a particular slope will change. These changes may be gradual. For example, percolation carrying away finer material. By contrast, some changes are rapid. In Britain, for example, shear stresses increase in winter.

WHAT KEEPS SLOPES IN PLACE?

There are a number of ways that downslope movement can be opposed:

- *Friction* will vary with the weight of the particle and slope angle. Friction can be overcome on very gentle slope angles if water is present. For example solifluction can occur on slopes as gentle as 3°.
- *Cohesive forces* act to bind the particles on the slope. Clay may have high cohesion, but this may be reduced if the water content becomes so high that the clay liquefies, when it loses its cohesive strength.
- *Vegetation* binds the soil and thereby stabilises slopes. However, vegetation may allow soil moisture to build up and make landslides more likely.

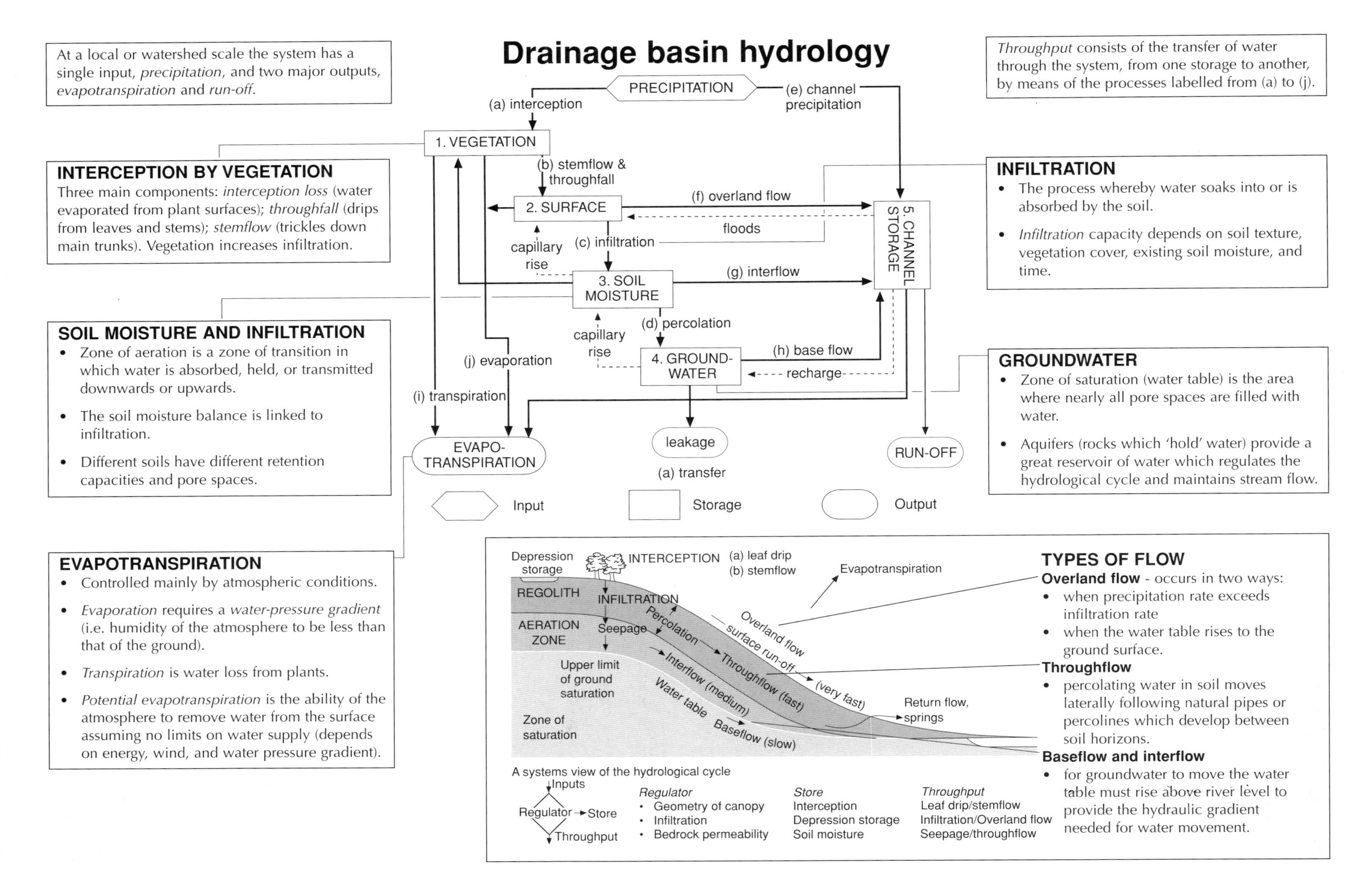
Drainage basin hydrology
At a local or watershed scale the system has a single input, *precipitation*, and two major outputs, *evapotranspiration* and *run-off*.
Throughput consists of the transfer of water through the system, from one storage to another, by means of the processes labelled from (a) to (j).
PRECIPITATION
(a) interception
(e) channel precipitation
1. VEGETATION
(b) stemflow & throughfall
2. SURFACE
(f) overland flow
floods
capillary rise
(c) infiltration
3. SOIL MOISTURE
(g) interflow
(d) percolation
capillary rise
4. GROUND-WATER
(h) base flow
recharge
5. CHANNEL STORAGE
(j) evaporation
(i) transpiration
EVAPO-TRANSPIRATION
leakage
(a) transfer
RUN-OFF
Input
Storage
Output
INTERCEPTION BY VEGETATION
Three main components: *interception loss* (water evaporated from plant surfaces); *throughfall* (drips from leaves and stems); *stemflow* (trickles down main trunks). Vegetation increases infiltration.
SOIL MOISTURE AND INFILTRATION
• Zone of aeration is a zone of transition in which water is absorbed, held, or transmitted downwards or upwards.
• The soil moisture balance is linked to infiltration.
• Different soils have different retention capacities and pore spaces.
EVAPOTRANSPIRATION
• Controlled mainly by atmospheric conditions.
• *Evaporation* requires a *water-pressure gradient* (i.e. humidity of the atmosphere to be less than that of the ground).
• *Transpiration* is water loss from plants.
• *Potential evapotranspiration* is the ability of the atmosphere to remove water from the surface assuming no limits on water supply (depends on energy, wind, and water pressure gradient).
INFILTRATION
• The process whereby water soaks into or is absorbed by the soil.
• *Infiltration* capacity depends on soil texture, vegetation cover, existing soil moisture, and time.
GROUNDWATER
• Zone of saturation (water table) is the area where nearly all pore spaces are filled with water.
• Aquifers (rocks which 'hold' water) provide a great reservoir of water which regulates the hydrological cycle and maintains stream flow.
Depression storage
INTERCEPTION
(a) leaf drip
(b) stemflow
Evapotranspiration
REGOLITH
INFILTRATION
AERATION ZONE
Seepage
Percolation
Overland flow
surface run-off
Throughflow (fast)
(very fast)
Interflow (medium)
Upper limit of ground saturation
Water table
Baseflow (slow)
Zone of saturation
Return flow, springs
A systems view of the hydrological cycle
Inputs
Regulator
Store
Throughput
Regulator
• Geometry of canopy
• Infiltration
• Bedrock permeability
Store
Interception
Depression storage
Soil moisture
Throughput
Leaf drip/stemflow
Infiltration/Overland flow
Seepage/throughflow
TYPES OF FLOW
Overland flow - occurs in two ways:
• when precipitation rate exceeds infiltration rate
• when the water table rises to the ground surface.
Throughflow
• percolating water in soil moves laterally following natural pipes or percolines which develop between soil horizons.
Baseflow and interflow
• for groundwater to move the water table must rise above river level to provide the hydraulic gradient needed for water movement.

River regimes

The pattern of seasonal variation in the flow of a river is known as the *regime*. The regime is related to a number of factors, notably the seasonality of *rainfall, temperature*, and *evapotranspiration*. An equatorial river will have a regular regime, but rivers in climates with marked seasonal contrast may have one or more peaks. This can be related to climatic zones.

SIMPLE REGIMES

Simple regimes are where a simple distinction can be made between one period of high water levels and run-off and one period of low water levels and run-off.

Glacier melt
European mountain rivers have a high-water period (July-August) when glaciers feeding them melt most rapidly.

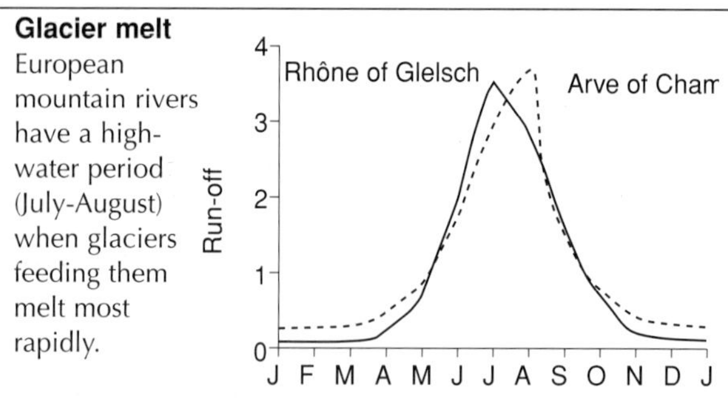

Snowmelt
Melting of snow cover either in mountainous areas during early summer or over the Great Plains of North America in early/late spring.

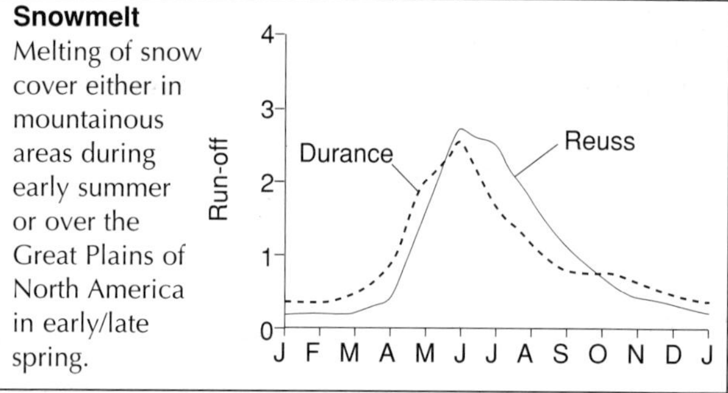

Tropical seasonal rainfall (monsoonal)
In tropical areas, evapotranspiration tends to be stable (high) but summer rains cause a peak.

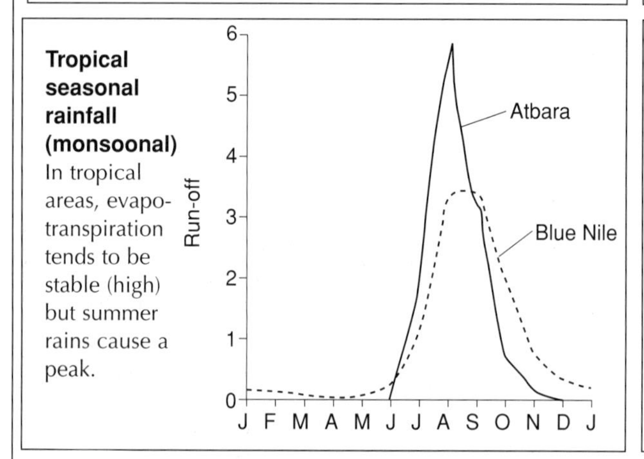

Oceanic rainfall/evapotranspiration
In many oceanic areas of Europe, rainfall is evenly distributed but high evapotranspiration in summer leads to low run-off.

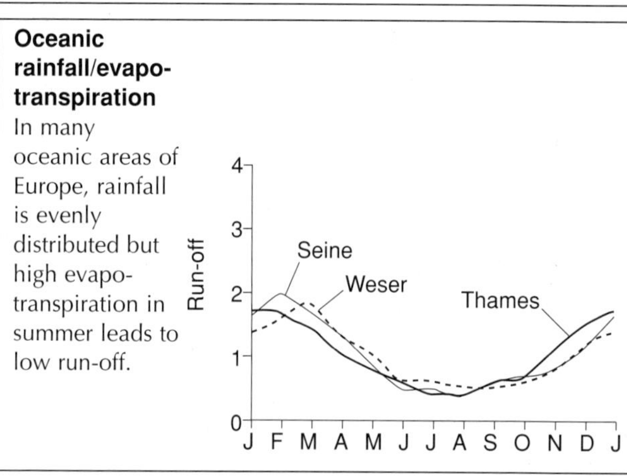

Equatorial
Relatively constant flow throughout the year due to year-round rainfall. Slight maxium after the equinoxes in March and September.

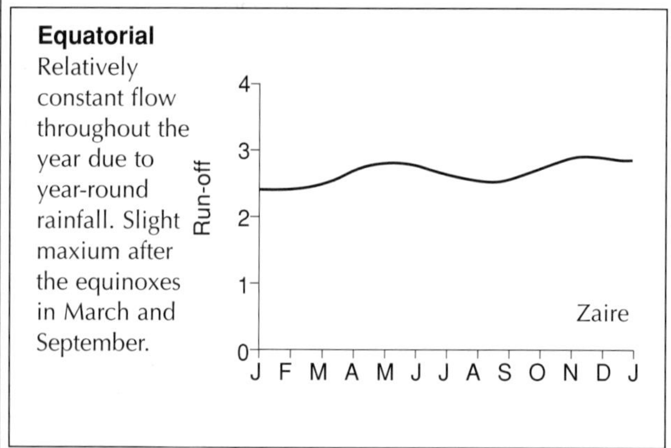

Mediterranean stream
Flow is higher in winter due to higher rainfall, low temperatures, and less evaporation. Drought is common in summer causing low flow.

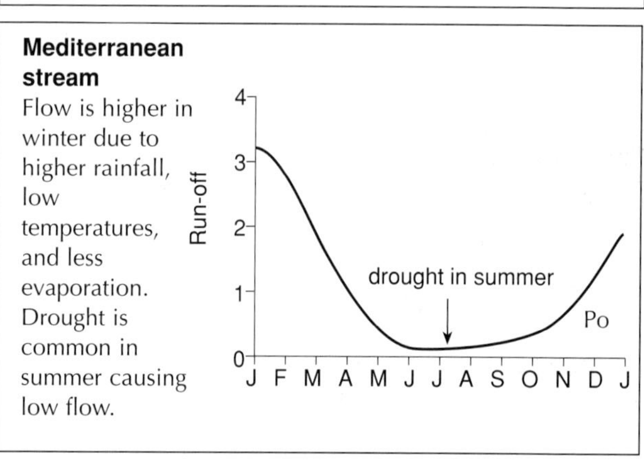

COMPLEX REGIMES

Some rivers are characterised by at least four hydrological phases (two low, two high) which give a more complex regime.

Other rivers, like the Rhine, flow through several distinctive relief regions and receive water from large tributaries which themselves flow over varied terrain. Rivers in this group normally have a single regime in their upper reaches.

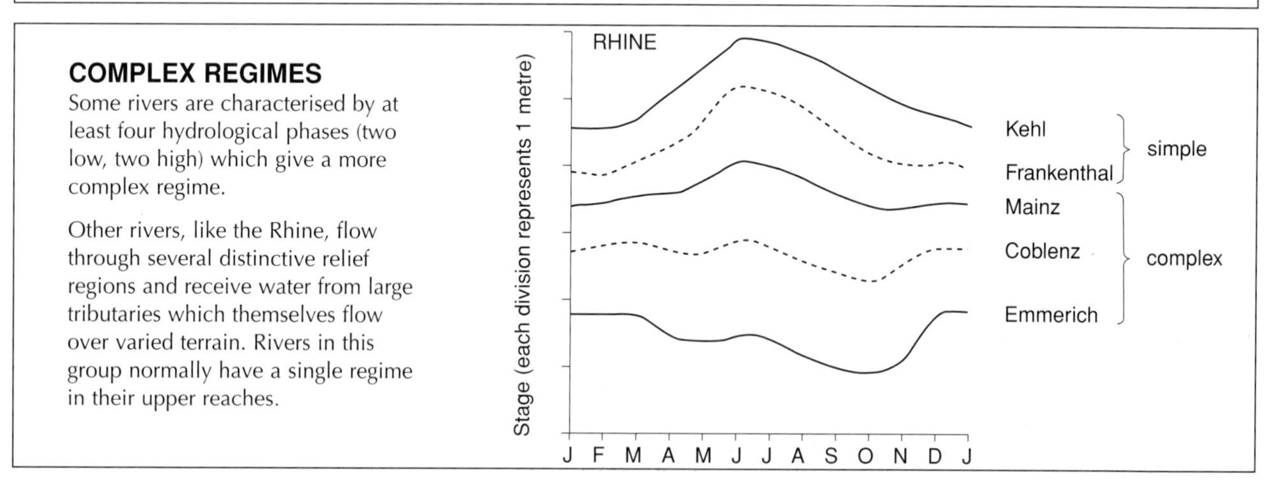

Storm hydrographs

DEFINITIONS

- A *storm hydrograph* measures the speed at which rainfall falling on a drainage basin reaches the river channel. It is a graph on which river discharge during a storm or *run-off* event is plotted against time.
- *Discharge* (Q) is the volume of flow passing through a cross-section of a river during a given period of time (usually measured in cumecs - m^3/s).

READING A STORM HYDROGRAPH

Discharge peak

- higher in larger basins
- 'steep catchments' will have lower infiltration rates so high peaks
- 'flat catchments' will have high infiltration so more throughflow and lower peaks

Hydrograph size (area under the graph)

- the higher the rainfall, the greater the discharge
- the larger the basin size, the greater the discharge

Recession limb

- indicates the amount of groundwater depletion caused by throughflow
- influenced by geological composition and behaviour of local aquifers

Run-off: discharge in cumecs (m^3/sec)
50
40
30
20
10
Peak flow or discharge
River in flood
Rising limb
Falling limb or recession
rainfall peak
Bankfull discharge
Lag time
approach segment
Run-off or storm
Throughflow
Rainfall
Baseflow
time of rise
Rainfall in mm
50
40
30
20
10
0
1200 (day 1)
0000 (day 2)
1200 (day 1)
0000 (day 3)
Time (hours)

Lag time

- time interval between peak rainfall and peak discharge
- influenced by basin shape, steepness, stream order
- relationship between run-off and throughflow is main determinant

Run-off

- reveals the relationship between overland flow and throughflow
- where infiltration is low and rainfall strong overland flow will dominate

Baseflow

- the seepage of groundwater into the channel
- a slow movement which is the main, long-term supplier of the river's discharge

INFLUENCES

Climate

Q (discharge)
Semi-arid
Run-off
Baseflow
Time (t)

- short but high intensity rain
- low infiltration
- overland flow is dominant

Q (discharge)
Humid temperate
Time (t)

- deep soils and abundant vegetation
- high infiltration
- throughflow is dominant

Basin shape

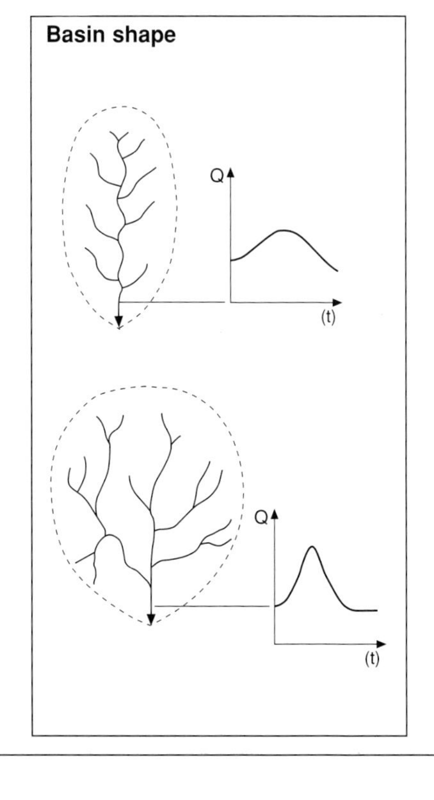

Land use

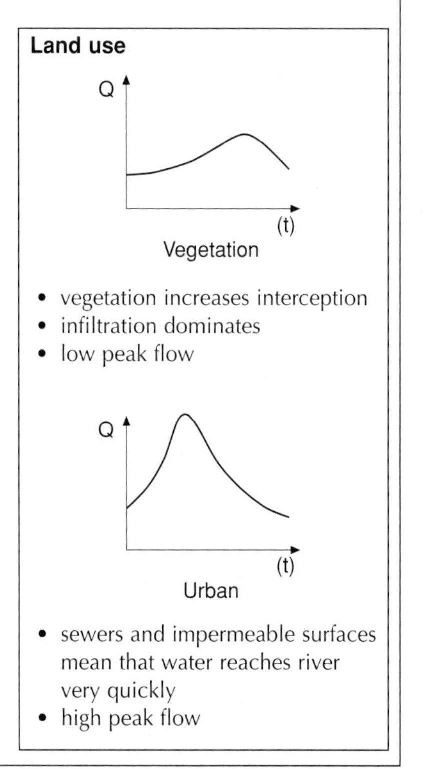

- vegetation increases interception
- infiltration dominates
- low peak flow

- sewers and impermeable surfaces mean that water reaches river very quickly
- high peak flow

Urban hydrology

Storm-water sewers
- reduce the distance storm-water must travel before reaching a channel
- increase the velocity of flow because sewers are smoother than natural channels
- reduce storage because sewers are designed to drain quickly

THE EFFECTS OF URBANISATION ON HYDROLOGICAL PROCESSES

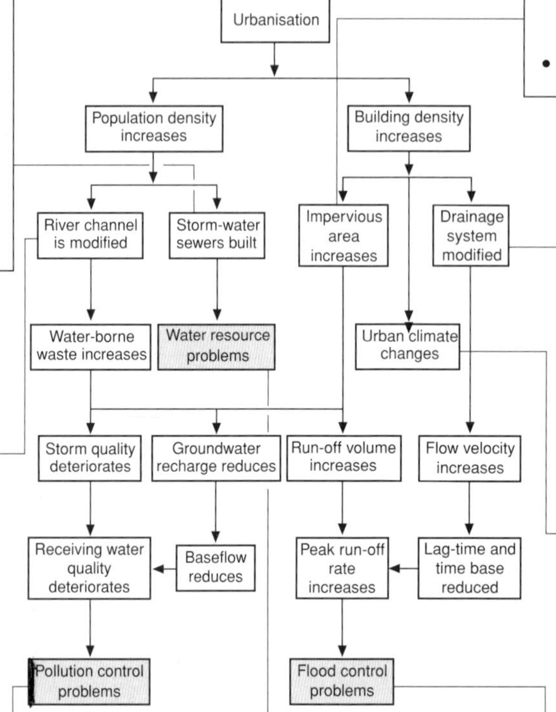

Replacement of vegetated soils with impermeable surfaces
- reduces storage and so increases run-off
- increases velocity of overland flow
- decreases evapotranspiration because urban surfaces are usually dry
- reduces infiltration and percolation

Building activity
- clears vegetation which exposes soil and increases overland flow
- disturbs and dumps the soil, increasing erodibility
- eventually protects the soil with armour of concrete or tarmac

Encroachment on the river channel
- embankments, reclamation, and river-side roads
- usually reduces channel width leading to higher floods
- bridges can restrict free discharge of floods and increase levels upstream

Rainfall climatology of urban areas
- greater aerodynamic roughness and urban heat island
- more rainfall, especially in summer
- heavier and more frequent thunderstorms

Pollution control problems
- storm-water that washes off roads and roofs can contain heavy metals, volatile solids, and organic chemicals
- annual run-off from 1 km of the M1 included 1.5 tons of suspended sediment, 4 kg of lead, 126 kg of oil, and 18 kg of aromatic hydrocarbons

Water resource problems
- groundwater recharge may be reduced because sewers bypass the mechanisms of percolation and seepage
- groundwater abstraction through wells may also reduce the store locally
- irrigation can draw on water resources leading not only to depletion but also pollution

Flood controls
- urbanisation increases the peak of the mean annual flood, especially in moderate conditions
- a 243 per cent increase resulted from the building of Stevenage New Town
- however, during heavy prolonged rainfall, saturated soil behaves in a similar way to urban surfaces

URBAN HYDROLOGY AND THE STORM HYDROGRAPH

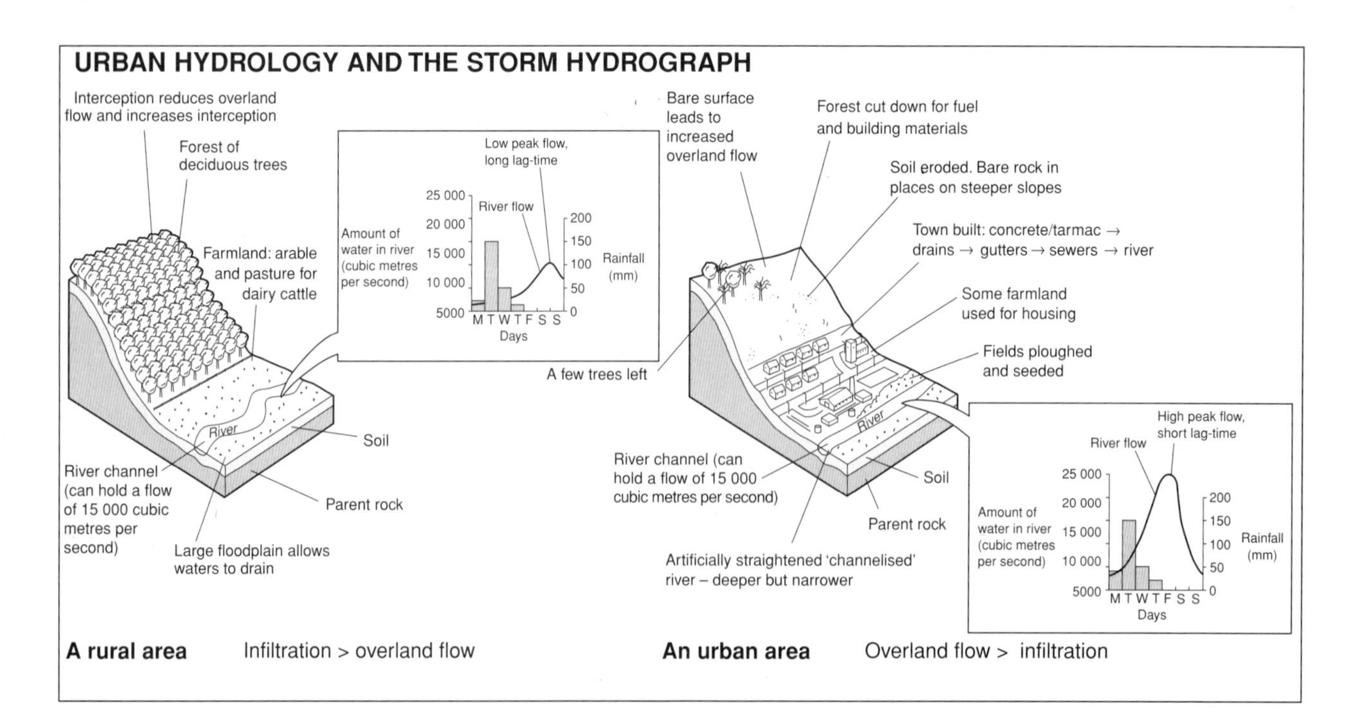

A rural area Infiltration > overland flow

An urban area Overland flow > infiltration

Long profiles

By plotting a line graph of river's height above the base level against distance from its source the long profile is revealed. As rivers evolve through time and over distance the stream passes through a series of different stages.

Upper Course

A In the initial stage a stream has lakes, waterfalls, and rapids.

- initial uplift
- overland flow is concentrated in depressions making lakes
- overflow links lake basins and initial stream system forms

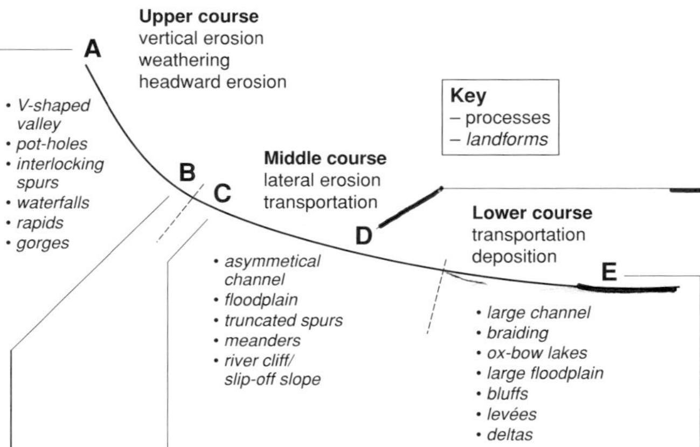

Middle Course

D Approaching the middle course, the stream has a floodplain almost wide enough to accommodate its meanders.

- river is in equilibrium
- supply of load = average rate at which stream can transport load
- floodplain gets wider through the enlargement and downstream migration of meanders

Lower Upper Course

B Further down the river the lakes are gone, but falls and rapids persist along the narrow incised gorge.

- deepening of channel
- waterfalls evolve into rapids
- headward erosion can form gorges
- exogenetic load through landslides and weathering

Upper Middle Course

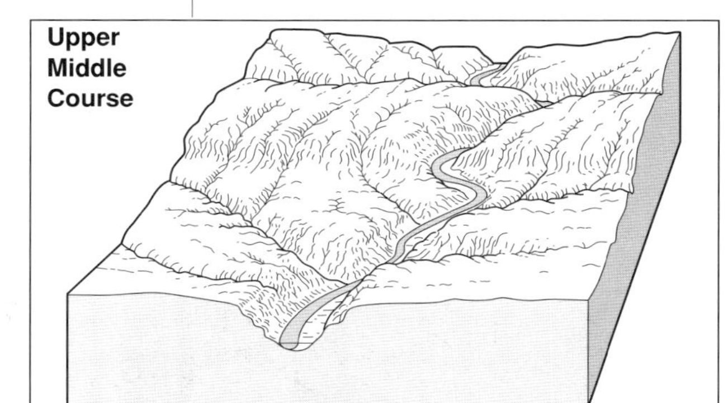

C The upper middle course brings a smoothly graduated profile without rapids or falls, but with the beginnings of a floodplain.

- smooth, even gradient
- floodplain begins to form

Lower Course

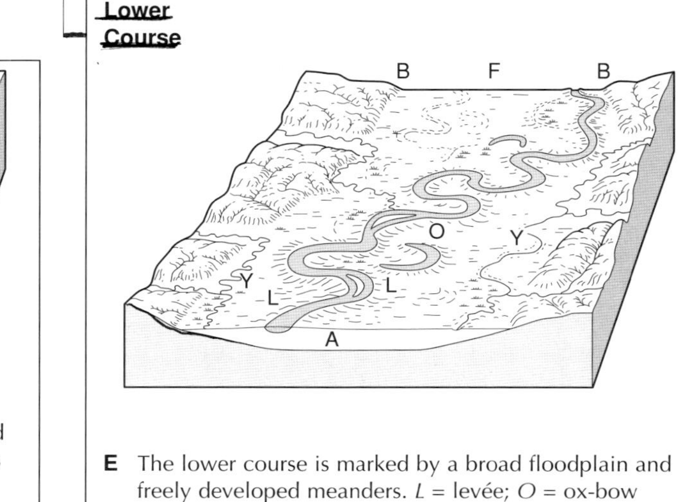

E The lower course is marked by a broad floodplain and freely developed meanders. *L* = levée; *O* = ox-bow lakes; *Y* = yazoo stream; *A* = alluvium; *B* = bluffs; *F* = floodplain.

Rivers as sediment systems

The river is a sub-system of a large unit - the drainage basin. The sediment system within a river is a further sub-system depending on many variables:

- discharge
- climate
- relief
- rock type

Sources

Sheet wash

Rill and gully erosion

Mass movement

Exogenetic (outside the channel)

Large rivers where erosive power is great

Material delivered from stream bed and banks

Relative importance dependent upon discharge and type of material

Endogenetic (within the channel)

In upland areas discharge is low and bed material is large, so endogenetic sediment yield is low.

In lowland areas discharge is high and bed and bank material is not resistant, so sediment yield will be high.

CRITICAL EROSION VELOCITY

The critical erosion velocity curve is the range of velocity needed to pick up particles of various sizes. The relationship between velocity and particle erosion, transportation, and deposition is given by the Hjulström Curve.

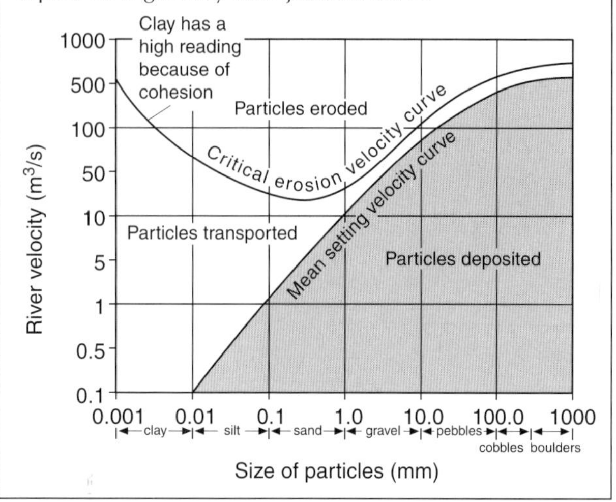

EQUILIBRIUM

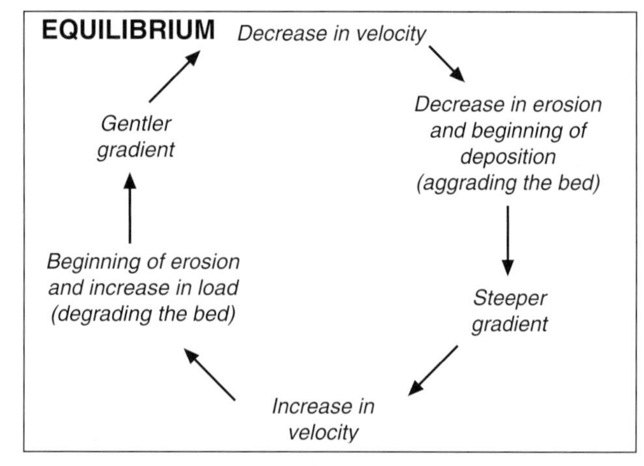

TRANSPORT LOAD

The *capacity* of a river refers to the total weight of material carried by a river. The *competence* of a river refers to the diameter of the largest particle that the river can carry.

Suspended sediment load

- carried with the body of the current
- 'wash-load' - small silt clay particles (<0.0625 mm)
- 'suspended bed material' - larger fine-medium sands which derive from channel bed

Bedload

- moves by sliding, rolling, or saltating
- maybe exogenetic or endogenetic

Dissolved load

- derives from precipitation, chemical weathering, erosion, atmospheric fallout, mineral springs, pollution

Transportation

Light material, e.g. silts and clays held in suspension by turbulence

Dissolved material held in solution

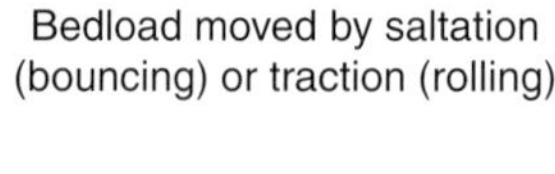

Bedrock

The river channel

TYPES OF FLOW

Streamflow is very complex. The velocity and therefore the energy is controlled by:

- *gradient* of channel bed
- *volume of water* within the channel
- the *shape* of the channel
- *channel roughness*/friction

LAMINAR FLOW

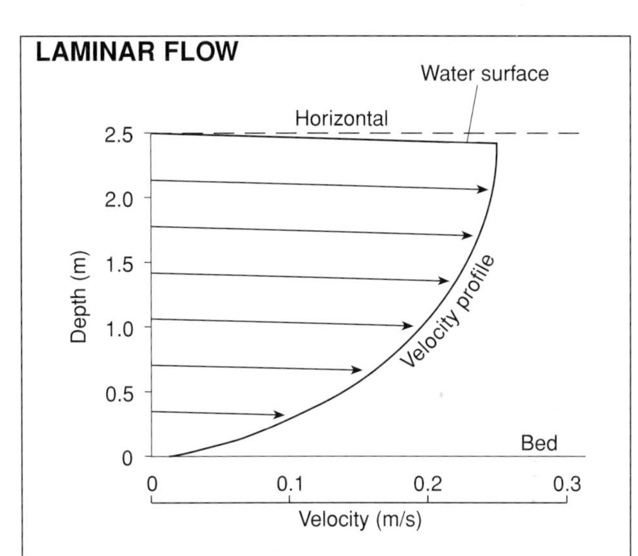

Necessary conditions for laminar flow include:

- smooth, straight channel
- shallow water
- low, non-uniform velocity allowing water to flow in sheets parallel to channel bed. Rare in reality. Most common in lower reaches

TURBULENT FLOW

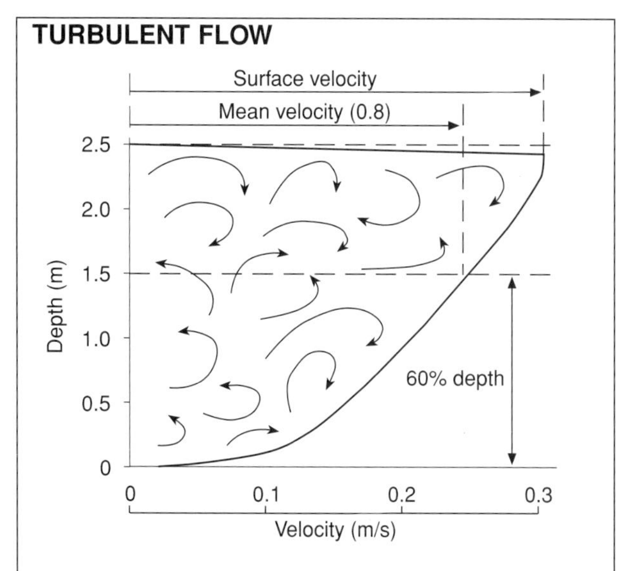

Necessary conditions for turbulent flow include:

- complex channel shape, e.g. winding channels, riffles and pools
- high velocity turbulence is associated with *cavitation* as *eddies* trap air in pores, cracks, crevices which is then released under great pressure

CHANNEL ROUGHNESS

Channel roughness causes friction which slows the flow of the river. Friction is caused by boulders, vegetation, sinuosity, and bedform. It is measured using *Manning's 'n'* which expresses the relationship between channel roughness and velocity in an equation:

$$v = \frac{R^{\frac{2}{3}} S^{\frac{1}{2}}}{'n'}$$

v = velocity
R = hydraulic radius
S = slope
n = roughness

The higher the value the rougher the bed, e.g.

Bed profile	Sand and gravel	Course gravel	Boulders
Uniform	0.02	0.03	0.05
Undulating	0.05	0.06	0.07
Irregular	0.08	0.09	0.10

CHANNEL SHAPE

Channel efficiency is measured by the *hydraulic radius*, i.e. *cross-sectional area*. This is wetted perimeter and is affected by river level and channel shape.

River level

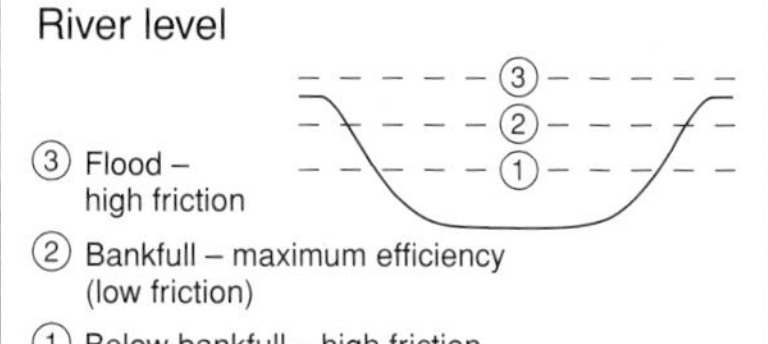

Shape

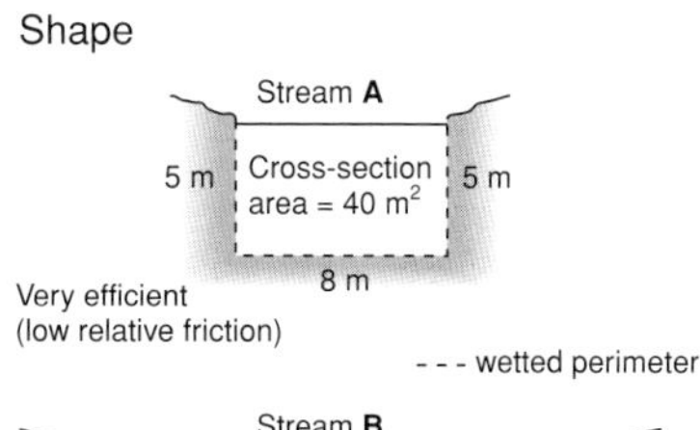

Wetted perimeters	Hydraulic radius
Stream **A**:	Stream **A**:
5 + 5 + 8 = 18 m	$\frac{40}{18}$ = 2.22 m
Stream **B**:	Stream **B**:
2 + 2 + 20 = 24 m	$\frac{40}{24}$ = 1.66 m

CHANNEL SLOPE

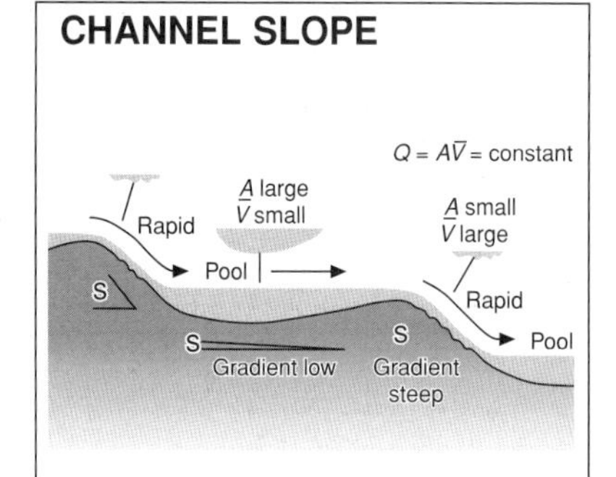

Discharge (Q) - the volume of water passing through a given cross-section in a given unit of time; velocity (v); cross-sectional area (A).

Steeper gradients should lead to higher velocities because of gravity. Velocity increases quickly where a stream passes from a pool of low gradient to a steep stretch of rapids.

As v increases cross-sectional area decreases; as v decreases cross-sectional area increases.

Meanders

WHAT CAUSES MEANDERS?

There is no single explanation, but a number of factors have been suggested:

- **sandbars** - flume tank experimentation suggests that sinuosity can be triggered by sandbars

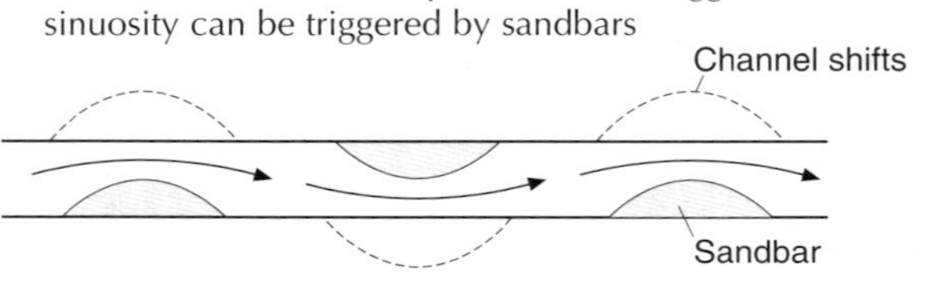

- **slope thresholds**

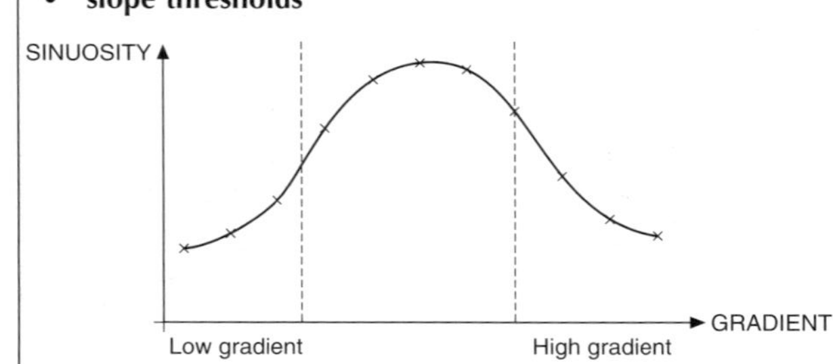

- **helicoidal flow**

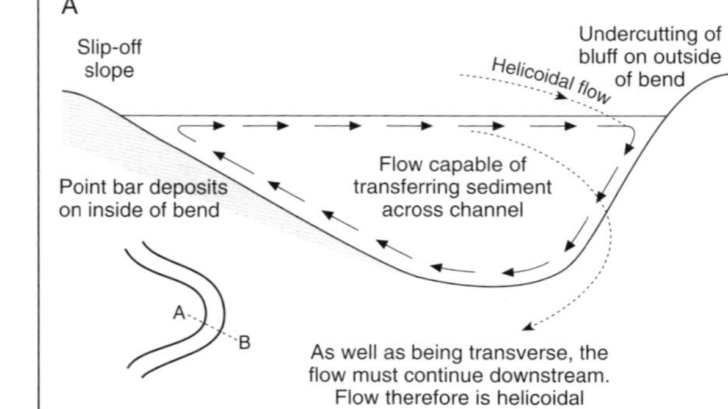

The thalweg is the line tracing the deepest and fastest water. The thalweg moves from side to side within the channel, and also corkscrews in cross-section. This is helicoidal flow and increases the amplitude of the meander.

SINUOSITY

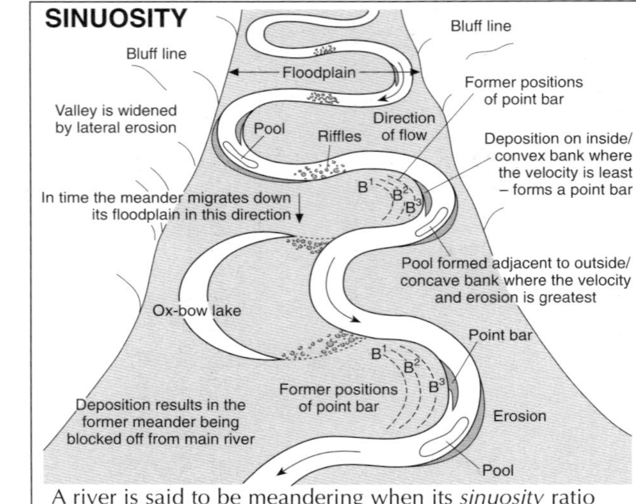

A river is said to be meandering when its *sinuosity* ratio exceeds 1.5. The *wavelength* of meanders is dependent on three major factors: channel width, discharge, and the nature of the bed and banks.

Development of a meander through time

(a) (b) (c)

Pool
Riffle
5 times the bedwidth
Original course
Sinuosity is: actual channel length / straight line distance
one wavelength – usually 10 times the bedwidth
Line of main current (THALWEG)

POOLS AND RIFFLES

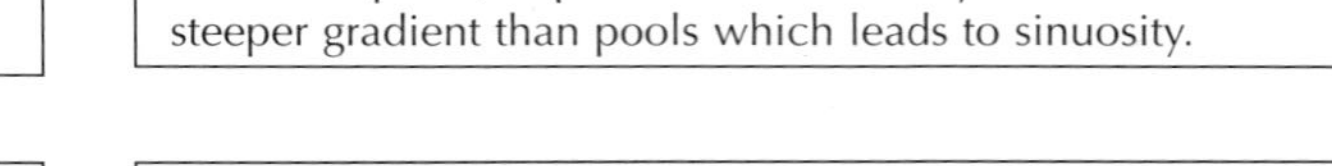

Meanders have strong links with pools and riffles. They are caused by turbulence. Roller eddies cause disposition of coarse sediment (riffles) at high velocity points and fine sediment (pools) at points of low velocity. Riffles have a steeper gradient than pools which leads to sinuosity.

IN WHAT WAYS DO MEANDERS MIGRATE?

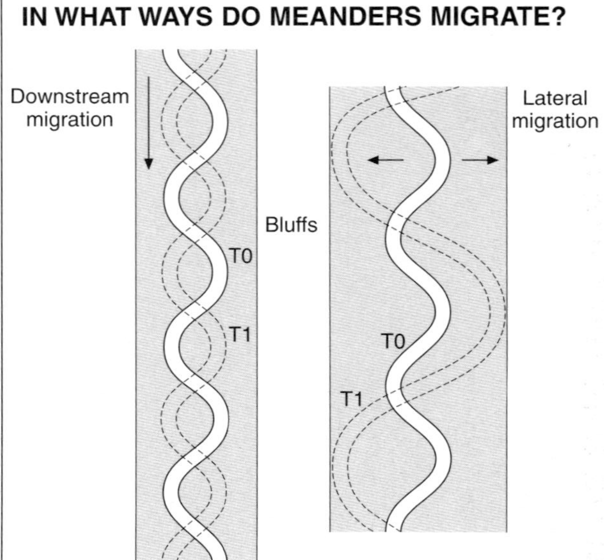

This results in a greatly widened *floodplain*.

Deltas and estuaries

DELTAS

Deltas are formed when river sediments are deposited when a river enters a standing body of water such as a lake, lagoon, or ocean. Deposition occurs because velocity is checked. A number of factors affect the formation of deltas:

- amount and calibre (size) of load - rivers must be heavily laden, coarse sediment will be dumped first
- salinity - salt-water causes charged particles in freshwater to flocculate or adhere together
- gradient of coastline - delta formation will be more likely on gentle coastline where turbidity is less
- vegetation - plant life will slow waters and so increase deposition, and also provide a service on which deposition can occur
- low energy river discharge and/or low energy wave or tidal energy.

Structure of a simple delta

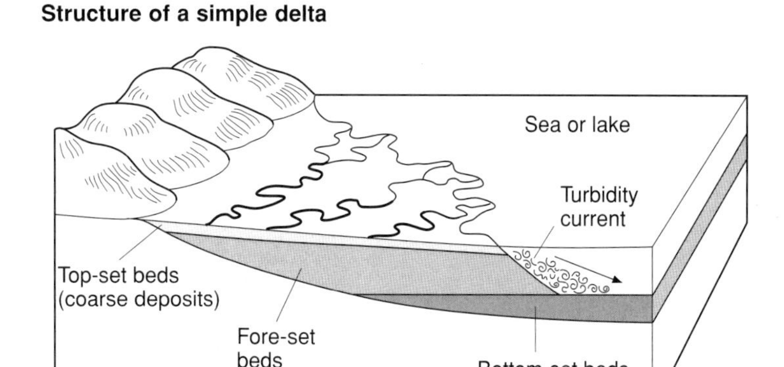

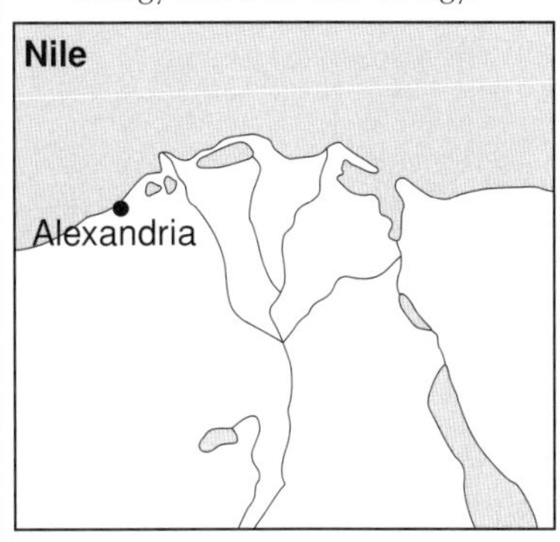

Deltas occur in three forms:

- arcuate - many distributaries which branch out radially (e.g. the Nile Delta)
- cuspate - a pointed delta formed by a dominant channel
- bird's foot - long, projecting fingers which grow at the ends of distributaries (e.g. the Mississippi Delta)

ESTUARIES

Estuaries occur where a coastal area has recently subsided or the ocean level has risen, causing the lower part of the river to be drowned. Unlike a *ria*, which is also a drowned river valley, estuaries form traps for sediment which may be exposed some or all of the time.

Landforms, sediments, and water movements in estuaries

Floodplain
Salt marsh
Inter-tidal sand flats
Reclaimed flat

Tidal limit
Ebb
Flood
Flood

Vegetation

- vegetation like *Spartina towsendii* stabilises the loose surface of the sediment preventing erosion
- it also reduces velocity and so increases deposition.

Tidal range

- rising *flood* tides and falling *ebb* tides form channels
- velocity of water flow may be great in both directions
- the result is a shifting set of channels based on erosion and deposition.

Sedimentation

- estuaries are sheltered which enhances sedimentation due to reduced velocity
- freshwater meeting salt-water may lead to *flocculation* where charged particles cluster and sink.

Rivers and people

CAUSES OF RIVER 'PROBLEMS'

INHERENT	River channel changes, e.g. migration cut-offs
EXTERNAL	Climatic change, anthropogenic causes
NATURAL	Storm and flood occurrence - frequency, timing, magnitude Climatic change
HUMAN ACTIVITIES	
Direct	River regulation Channelisation Water abstraction Waste disposal Irrigation Drainage, especially agricultural Dams
Indirect	Land-use change, especially deforestation, afforestation Urbanisation and roads Mining Agricultural practices

PROBLEMS RELATED TO RIVERS

Problem	Effects
FLOODING	→ Destruction of structures and communications Danger to life and property Destruction of crops Interruption of activities Drainage difficulties
EROSION	→ Destruction of structures Loss of land and property Boundary disputes
SEDIMENTATION	→ Flooding Drainage pattern alteration Change in ecology Water quality change Navigation difficulties
CHANNEL AND CHANGES IN STABILITY	→ Destruction of structures Loss of amenity value Navigation difficulties Boundary disputes
ECOLOGICAL CHANGES	→ Loss of amenity value Decrease in fish stocks
LAND DRAINAGE	→ Groundwater alterations Vegetation change Agricultural change

CASE STUDY: THE THREE GORGES DAM

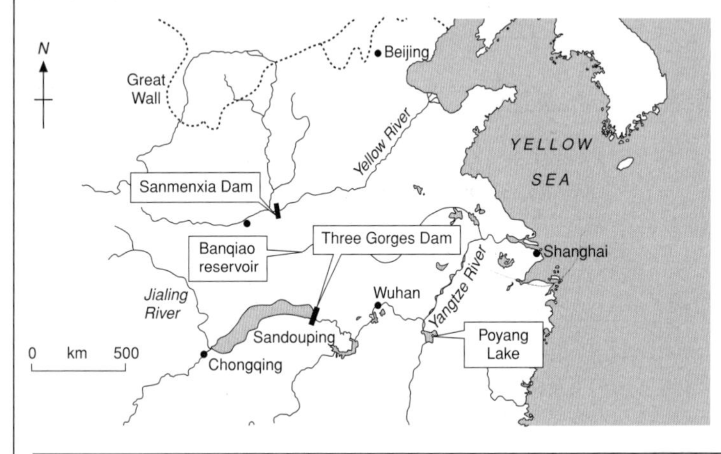

The decision to build the Three Gorges Dam on the Yangtze River highlights some of the conflicts apparent in the way people use the river. The dam will enable China to:

- generate up to 18 000 megawatts of power, reducing the country's dependence on coal
- supply Shanghai's 13 million people with water
- protect 10 million people from flooding (over 300 000 people died in China as a result of flooding last century)
- raise water levels to allow shipping above the Three Gorges (formally rapids).

Protest against the Three Gorges Dam

- Most floods in recent years have come from rivers which join the Yangtze below the Three Gorges Dam.
- The region is seismically active and landslides are frequent.
- The port at the head of the lake may become silted up as a result of increased desposition and the development of a delta at the head of the lake.
- Up to 1.2 million people will have to be moved to make way for the dam.
- Much of the land available for resettlement is over 800 m above sea-level, and is colder with infertile thin soils on relatively steep slopes.
- Dozens of towns, for example Wanxian and Fuling with 140 000 and 80 000 people respectively, will be flooded.
- Up to 530 million tonnes of silt are carried through the Gorge annually - the first dam on the river lost its capacity within seven years and one on the Yellow River filled with silt within four years.
- To reduce the silt load afforestation is needed, but resettlement of people will cause greater pressure on the slopes above the dam.
- The dam will interfere with aquatic life - the Siberian Crane and the White Flag Dolphin are threatened with extinction.
- Archaeological treasures will be drowned, including the Zhang Fei temple.

Drainage patterns and order

River systems can be analysed as networks that vary according to shape and geometry. The shape includes **accordant** and **discordant networks**, whereas geometry involves length, area, relief, and frequency. Several quantitative studies of the geometric properties of rivers and basins have been made and these have given rise to **fluvial morphometry**.

THE BIFURCATION RATIO

This describes the ratio of streams of one order to the next higher order e.g. first order to second order, second order to third order, and so on. Generally, the ratio of between 3 and 5. For example, if there are 160 first order streams and these join to form 40 second order streams, the bifurcation ratio is 4:1. The bifurcation ratio is a useful way of comparing river systems, and is an important aspect of stream networks.

STREAM ORDER

Stream order refers to stream hierarchy. Two systems are commonly used, that of Strahler and Shreve. According to **Strahler**, a stream of with no tributaries (a 'fingertip' stream) is first-order. When two first-order streams meet, they form a second-order stream. When two second-order streams meet, they form a third-order stream, and so on. To form the next higher order, two streams of a similar order must meet.

Shreve's stream order refers to the number of fingertip tributaries. This system may be useful for small basins but is unworkable for large basins.

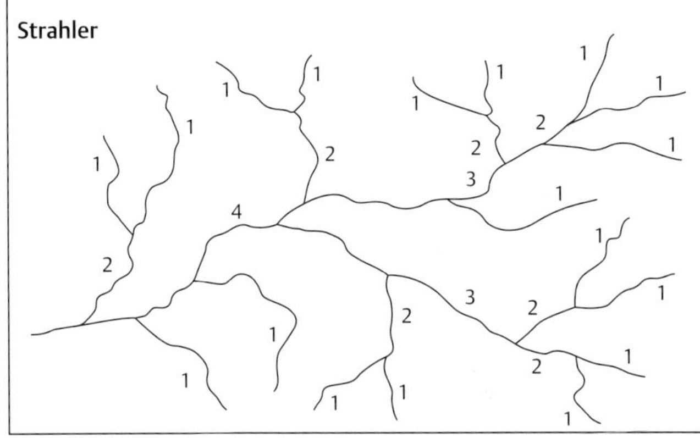

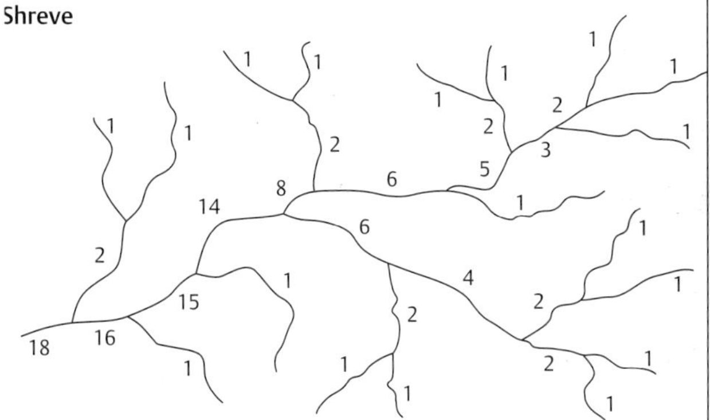

SOME MORPHOMETRIC PROPERTIES OF DRAINAGE BASINS

Network properties
Drainage density
Mean length of stream channel per unit area
Stream frequency
Number of stream segments of all orders per unit area
Length of overland flow
Mean distance from channels up maximum valley-side slope to drainage divide

Areal properties
Circulatory ratio
Total drainage basin area divided by the area of a circle having the same perimeter as the basin
Elongation ration
The diameter of a circle of the same area as the drainage basin divided by the maximum length of the basin measured from its mouth.

Relief properties
Basin relief
Difference in elevation between the highest and lowest point in a drainage basin
Relief ratio
Basin relief divided by the maximum length of the basin
Ruggedness number
Basin relief multiplied by drainage density

LAWS OF MORPHOMETRY

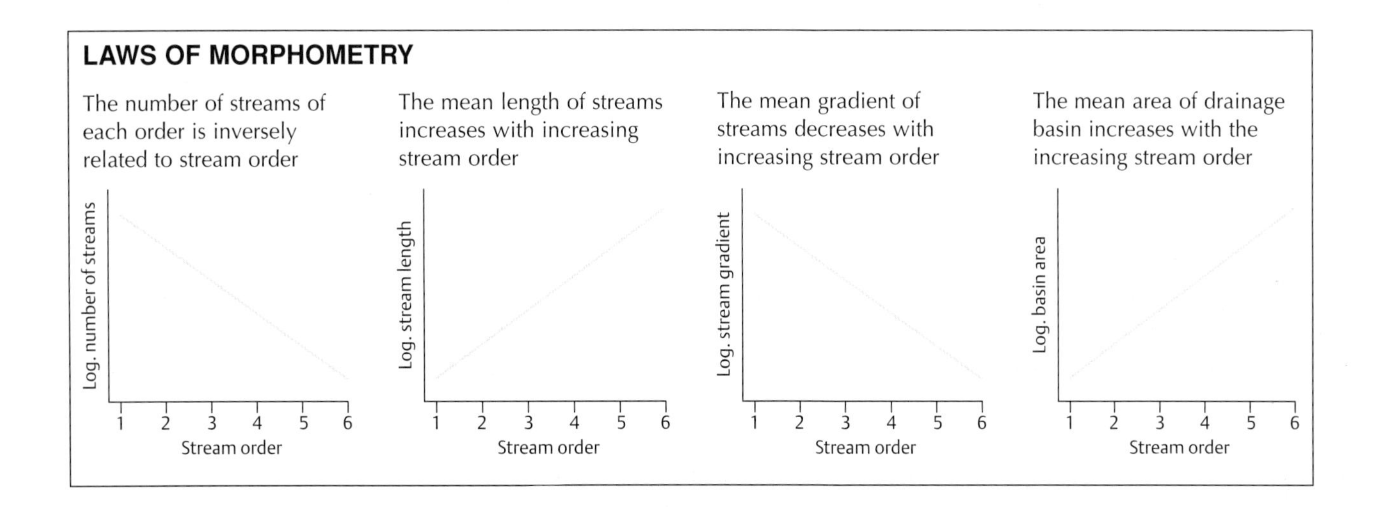

Floods and flooding

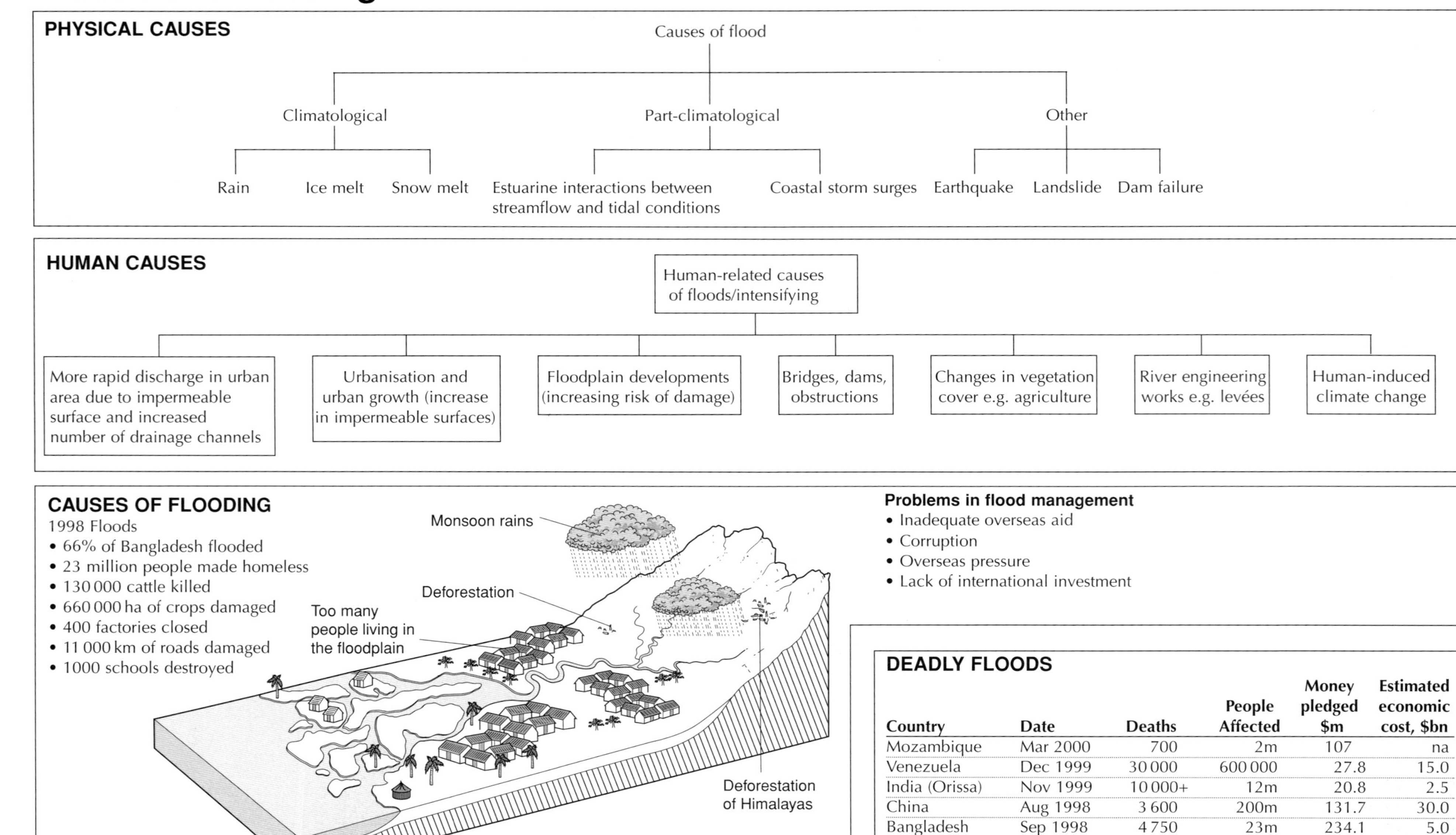

CAUSES OF FLOODING

1998 Floods

- 66% of Bangladesh flooded
- 23 million people made homeless
- 130 000 cattle killed
- 660 000 ha of crops damaged
- 400 factories closed
- 11 000 km of roads damaged
- 1000 schools destroyed

Problems in flood management

- Inadequate overseas aid
- Corruption
- Overseas pressure
- Lack of international investment

DEADLY FLOODS

Country	Date	Deaths	People Affected	Money pledged $m	Estimated economic cost, $bn
Mozambique	Mar 2000	700	2m	107	na
Venezuela	Dec 1999	30 000	600 000	27.8	15.0
India (Orissa)	Nov 1999	10 000+	12m	20.8	2.5
China	Aug 1998	3 600	200m	131.7	30.0
Bangladesh	Sep 1998	4 750	23m	234.1	5.0

Waves

WAVE TERMINOLOGY

Wavelength or **amplitude** is the distance between two successive crests or troughs.

Wave period is the time in seconds between two successive crests or troughs.

Wave frequency is the number of waves per minute.

Wave height is the distance between the trough and the crest.

The **fetch** is the amount of open water over which a wave has passed.

Velocity is the speed a wave travels at, and is influenced by wind, fetch, and depth of water.

Swash is the movement of water up the beach.

Backwash is the movement of water down the beach.

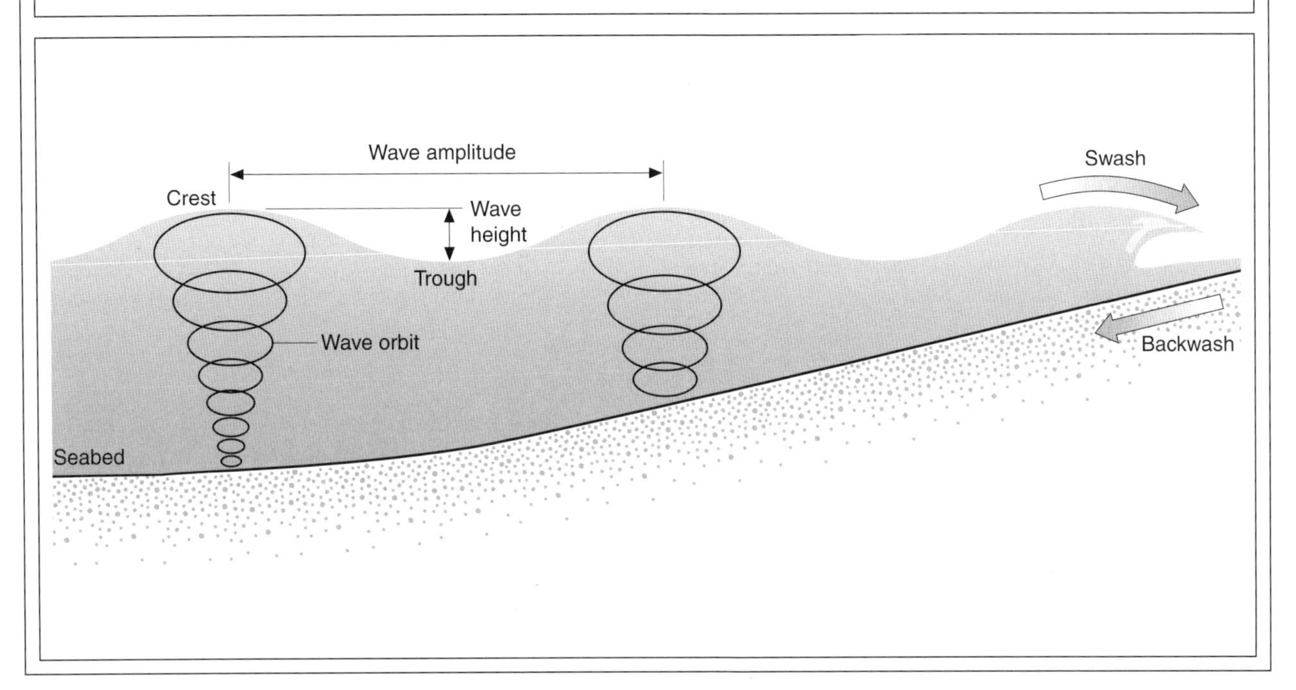

DESTRUCTIVE WAVES

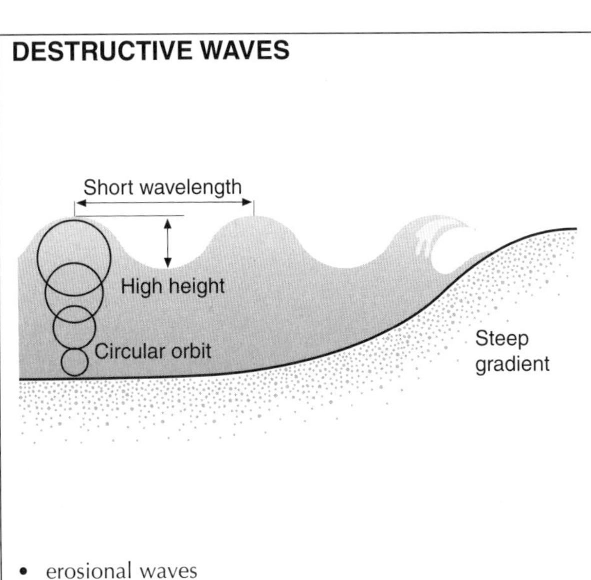

- erosional waves
- also called 'surging', 'storm', or 'plunging' waves
- short wavelength, high height
- high frequency (10-12 per minute)
- circular orbit
- low period (one every 5-6 seconds)
- backwash greater than swash
- steep gradient.

CONSTRUCTIVE WAVES

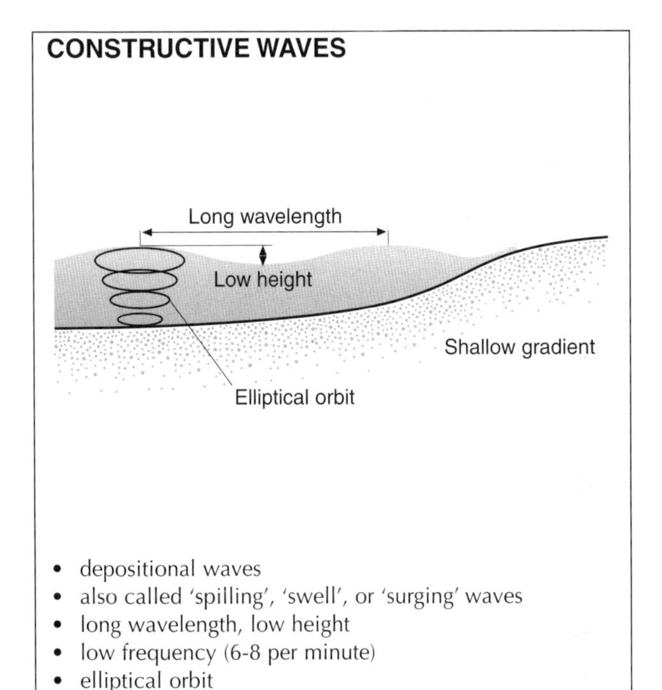

- depositional waves
- also called 'spilling', 'swell', or 'surging' waves
- long wavelength, low height
- low frequency (6-8 per minute)
- elliptical orbit
- high period (one every 8-10 seconds)
- swash greater than backwash
- low gradient.

Refraction and longshore drift

Wave **refraction** occurs when waves approach an irregular coastline or at an oblique angle (a). Refraction reduces wave velocity and, if complete, causes wave fronts to break parallel to the shore. Wave refraction concentrates energy on the flanks of headlands and dissipates energy in bays (b). However, refraction is rarely complete, and consequently **longshore drift** occurs (c).

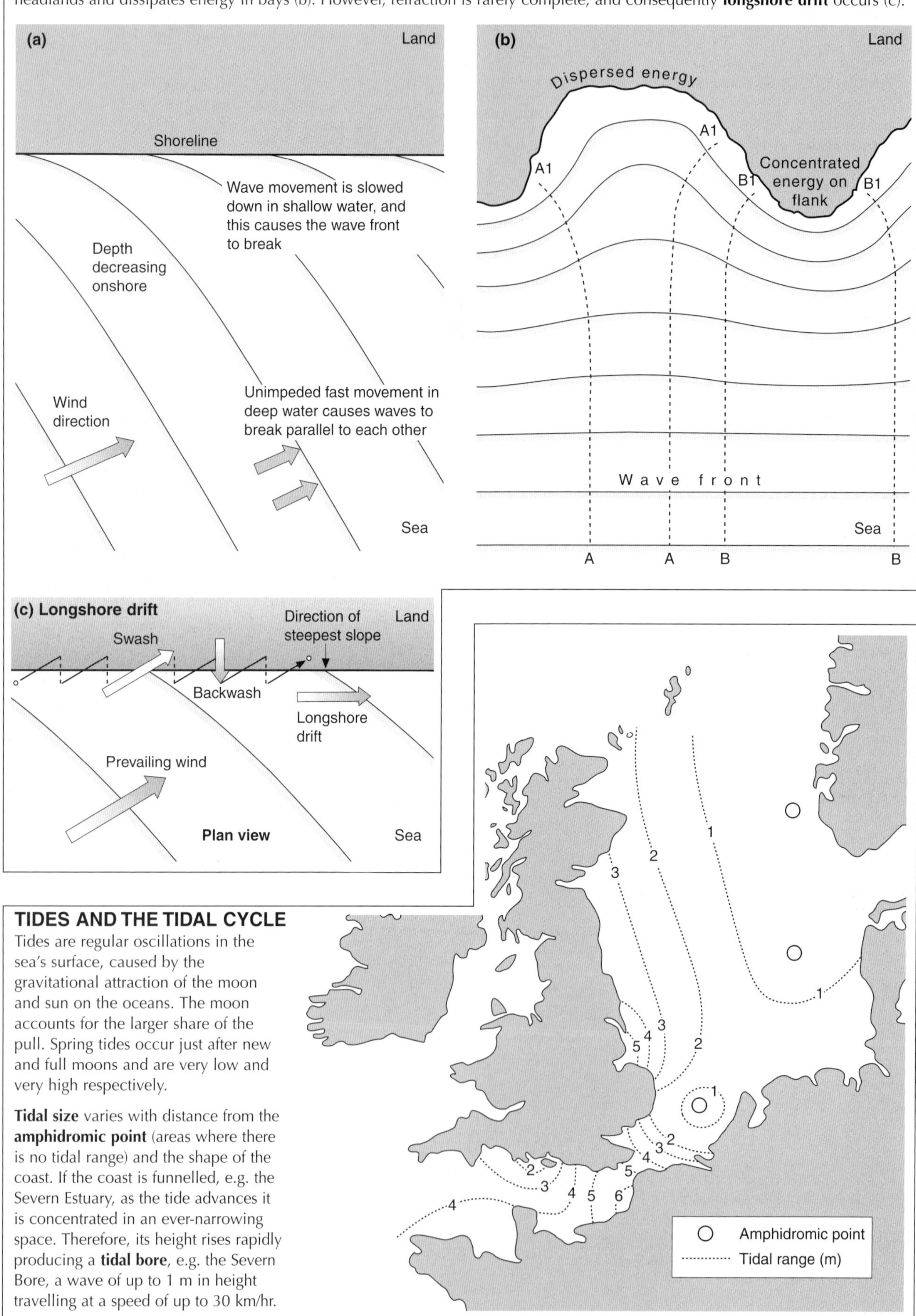

TIDES AND THE TIDAL CYCLE

Tides are regular oscillations in the sea's surface, caused by the gravitational attraction of the moon and sun on the oceans. The moon accounts for the larger share of the pull. Spring tides occur just after new and full moons and are very low and very high respectively.

Tidal size varies with distance from the **amphidromic point** (areas where there is no tidal range) and the shape of the coast. If the coast is funnelled, e.g. the Severn Estuary, as the tide advances it is concentrated in an ever-narrowing space. Therefore, its height rises rapidly producing a **tidal bore**, e.g. the Severn Bore, a wave of up to 1 m in height travelling at a speed of up to 30 km/hr.

Coastal erosion

PROCESSES

- **Abrasion** The wearing away of the shoreline by material carried by the waves.
- **Hydraulic impact** The force of water and air on rocks (up to 30 000 kg/m^2 in severe storms).
- **Solution** The wearing away of base-rich rocks, especially limestone, by acidic water. Organic acids aid the process.
- **Attrition** The rounding and reduction of particles carried by the waves.

Additionally, sub-aerial processes such as mass movements, wind erosion, and weathering are important.

CLIFF SHAPES

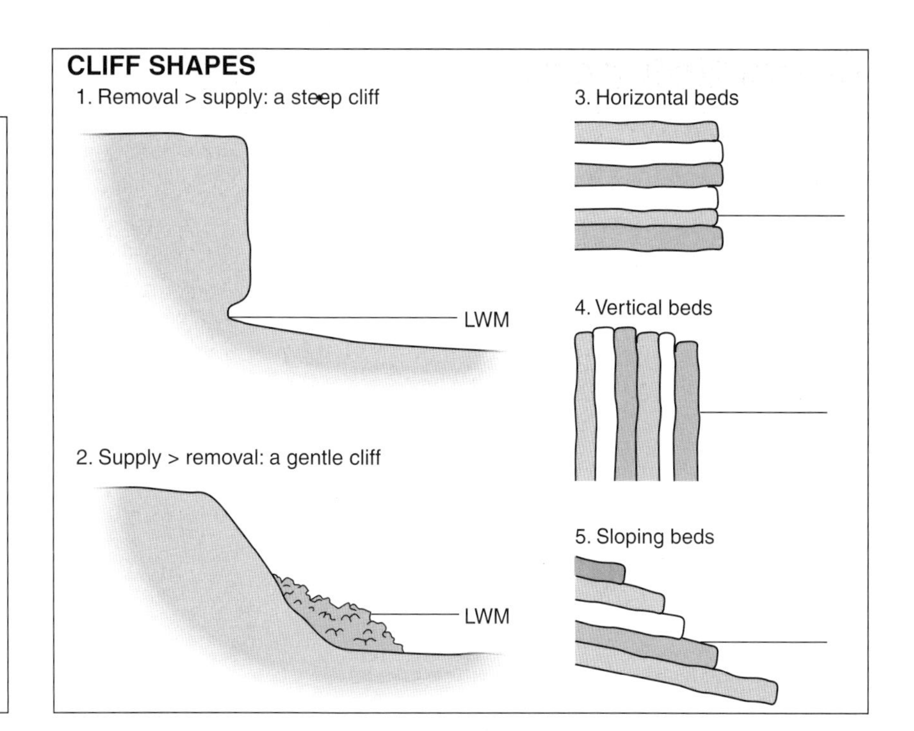

EVOLUTION OF SHORE PLATFORMS

A model of cliff and shore platform evolution shows how a steep cliff (1) is replaced by a lengthening platform and lower angle cliff (5) subjected to sub-aerial processes rather than marine forces.

Alternatively, platforms might be formed by (i) frost action, (ii) salt weathering, or (iii) biological action during lower sea-levels and different climates.

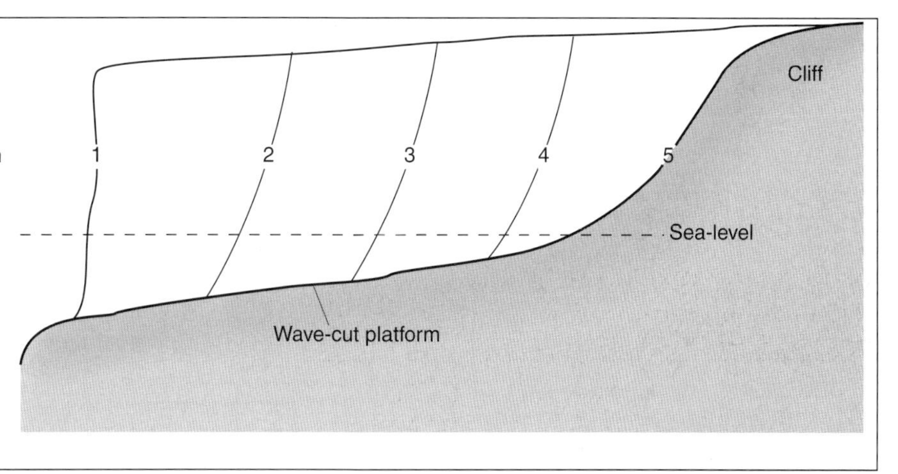

FEATURES OF COASTAL EROSION

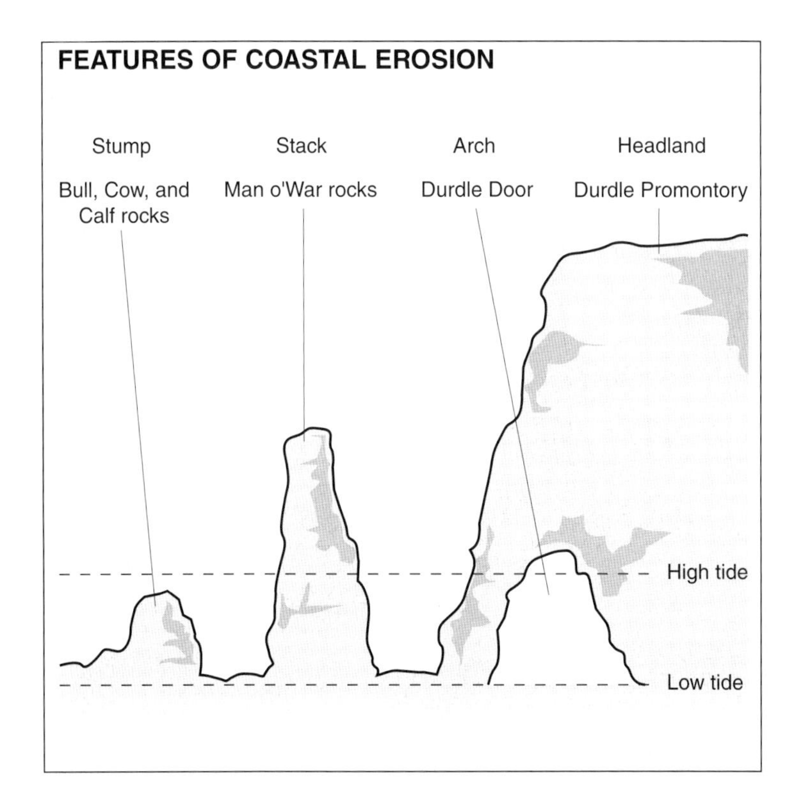

RATES OF CLIFF EROSION

Location	Geology	Erosion (metres per 100 years)
Holderness	Glacial drift	120
Cromer, Norfolk	Glacial drift	96
Folkestone	Clay	28
Isle of Thanet	Chalk	7-22
Seaford Head	Chalk	126
Beachy Head	Chalk	106
Barton, Hants	Barton Beds	58

Erosion is highest when there are:

- frequent storm waves
- easily erodable material.

Coastal deposition

BEACH PROFILE

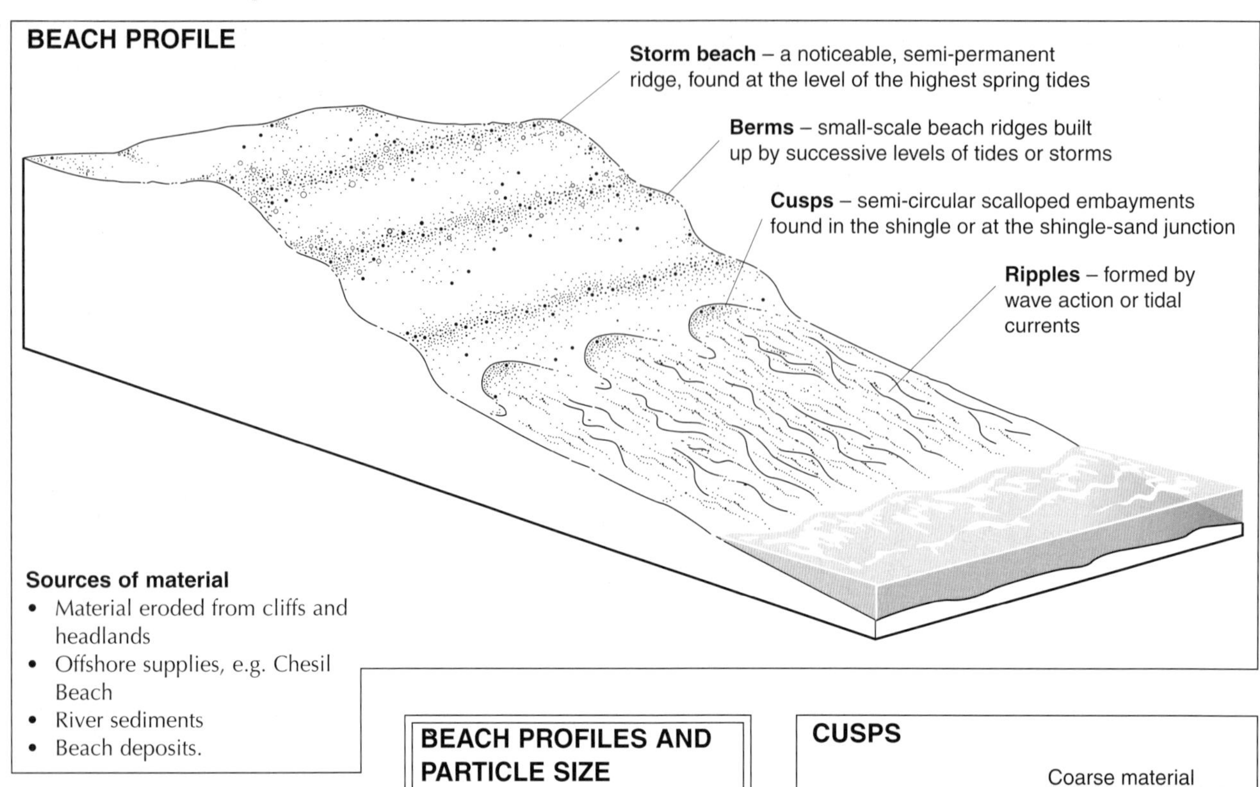

Sources of material
- Material eroded from cliffs and headlands
- Offshore supplies, e.g. Chesil Beach
- River sediments
- Beach deposits.

BEACH PROFILES AND PARTICLE SIZE

Material	Diameter (mm)	Beach angle
Cobbles	32	24°
Pebbles	4	17°
Coarse sand	2	7°
Medium sand	0.2	5°
Fine sand	0.02	3°
Very fine sand	0.002	1°

SEASONAL BEACH PROFILE CHANGES

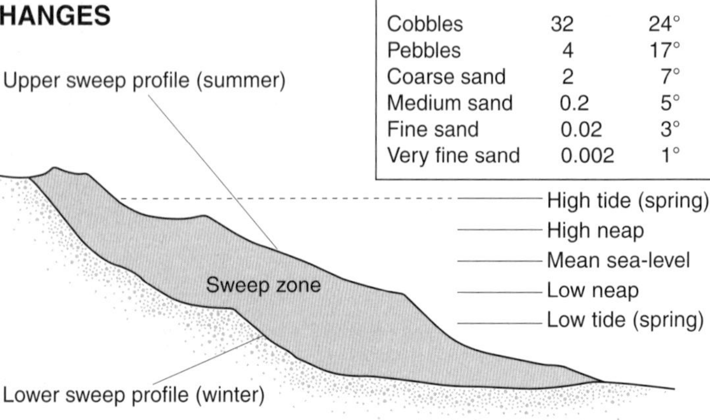

CUSPS

Coarse material
Incoming swash divides
Backwash combines to flush out cusp

FEATURES OF DEPOSITION

Essential requirements include:
- a large supply of material
- longshore drift
- an irregular, indented coastline, e.g. river mouths.

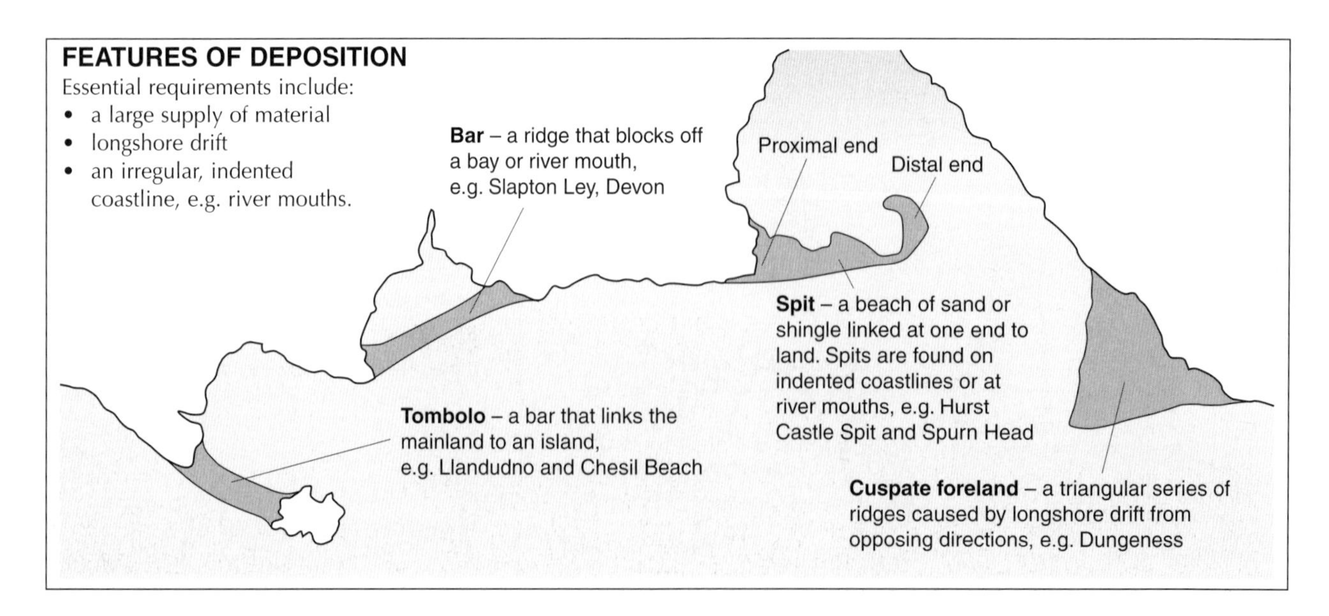

Coastal ecosystems

An ecosystem is a set of inter-related plants and animals with their non-living environment. Coastal ecosystems include sand dunes, **psammoseres**, and salt marshes, **haloseres**. These change spatially and temporally. The changes in micro-environment which allow other species to invade, compete, succeed, and dominate is termed **succession**.

SAND DUNE SUCCESSION - STUDLAND BEACH, ISLE OF PURBECK

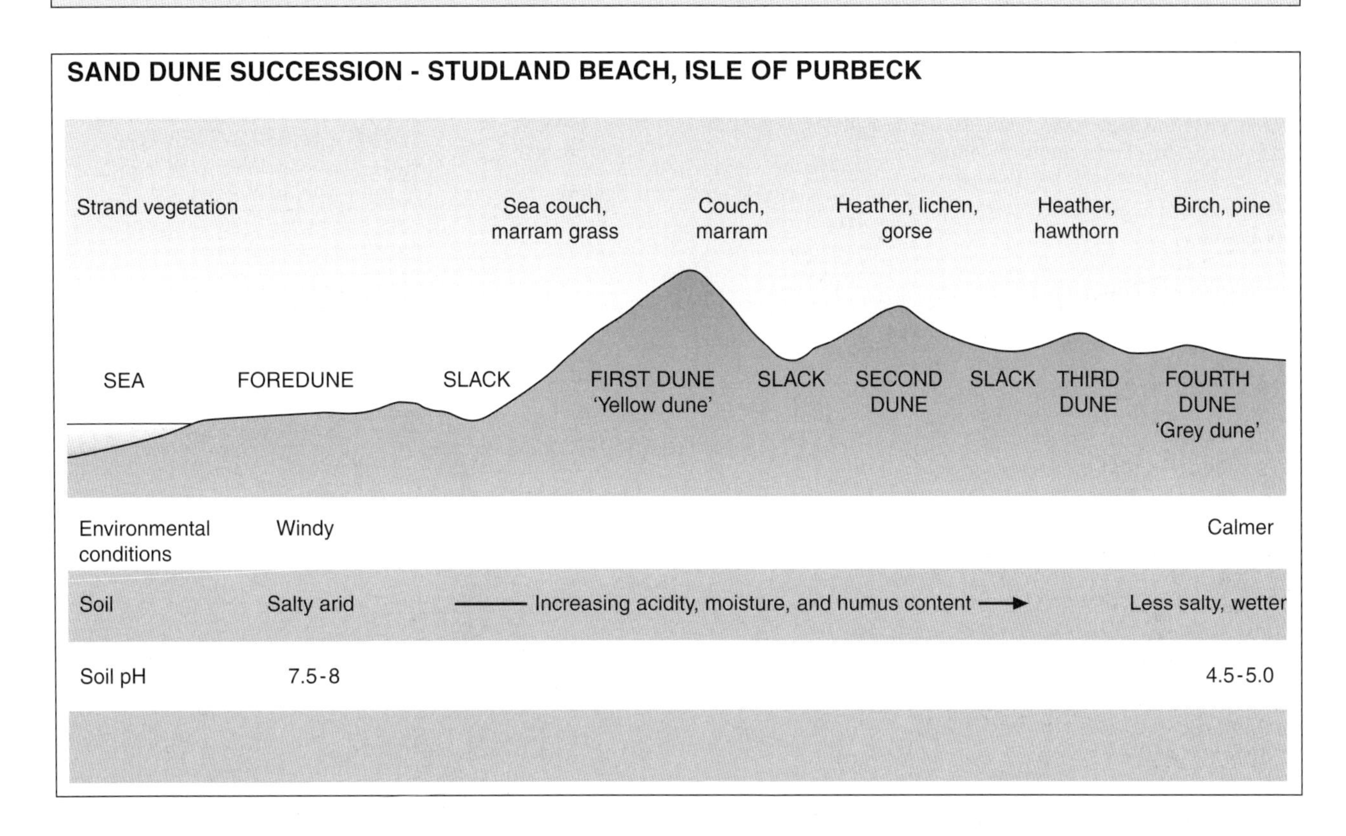

SALT MARSHES

Examples include Scolt Head Island in East Anglia and Newtown on the Isle of Wight.

Salt marshes are very productive and fertile ecosystems because of their high oxygen content, nutrient availability, and light availability, and because of the cleaning action of the tides.

I **Colonisers** on bare mud flats: algae (enteromorpha), eel grass, and marsh samphire (salicornia) increase the amount of deposition of silt. These plants can tolerate alkaline conditions and regular inundation by sea-water.

II **Halophytic vegetation** such as rice grass and spartina (cord grass) build up the salt marsh by as much as 5 cm per annum. Their roots anchor into the soft mud; the vegetation is taller and longer living than salicornia but not as salt tolerant.

III Sea lavendar **grasses**: inundated only at spring tides. Less salt tolerant.

IV A **raised salt marsh** with creeks may be formed, including turf grasses such as fescue and rushes (juncus). Inundation is rare.

V Inundation absent: ash and alder.

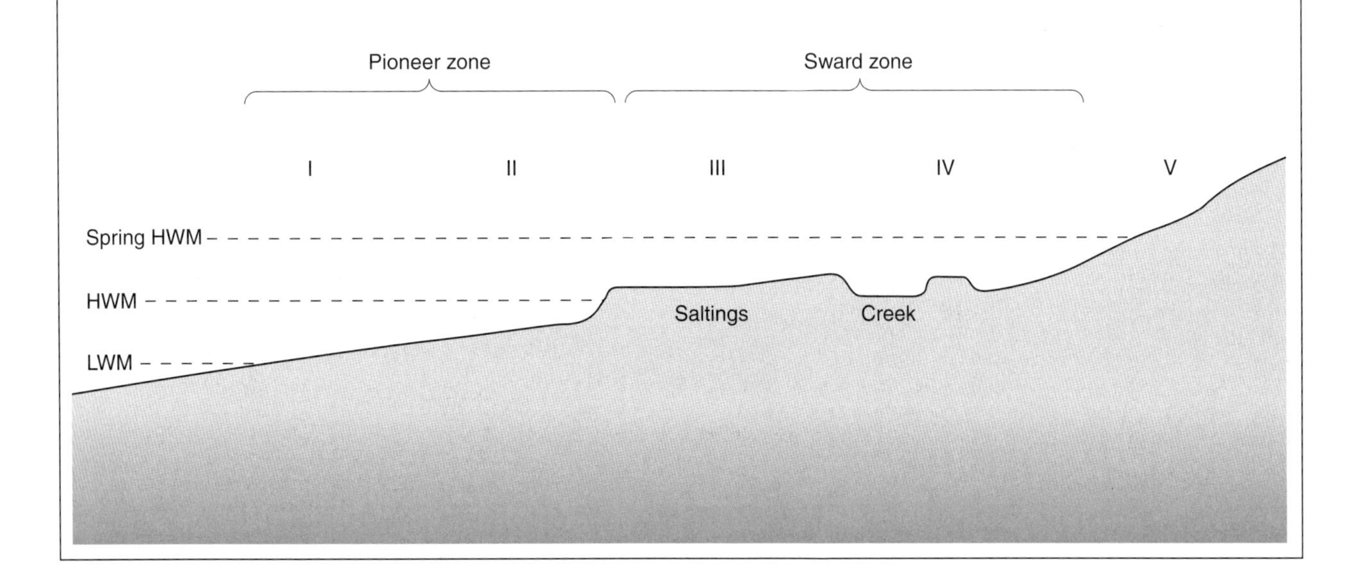

Coastal classification

SEA-LEVEL CHANGES

Sea-levels change in connection with the growth and decay of ice sheets. **Eustatic** change refers to a global change in sea-level. The level of the land also varies in relation to the sea. Land may rise as a result of tectonic uplift or following the removal of an ice sheet. The change in the level of the land relative to the level of the sea is known as **isostatic adjustment** or **isostacy**.

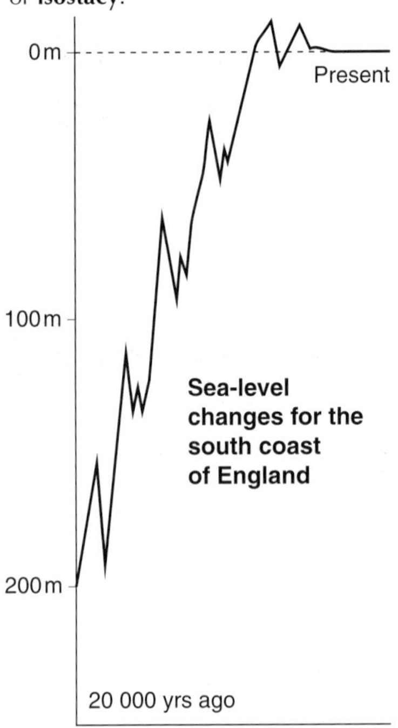

Features of emerged coastlines include:

- raised beaches, e.g. Portland raised beach
- coastal plains
- relict cliffs, e.g. the Fall Line in eastern USA.

Submerged coastlines include:

- rias, e.g. the River Fal, a drowned river valley
- fjords, e.g. Loch Torridan, a drowned U-shaped valley
- fjards or drowned glacial lowlands.

Coasts can be classified in a number of ways:

- high or low **energy** coastlines
- erosional or depositional
- submerged or emerged
- macro- meso- or micro-tidal
- storm- or swell-wave environment
- advancing or retreating coastlines.

These are mainly descriptive categories apart from Valentin's classification which links the first two factors. However, the categories overlap - high energy coasts, for example, are mainly erosional or storm-wave coasts.

TIDAL CLASSIFICATION

- Macro-tidal > 4 m
- Meso-tidal 2 - 4 m
- Micro-tidal 2 m

WAVE ENVIRONMENTS

Storm-wave environments, e.g. the British Isles and mid-latitude coastlines dominated by waves generated thousands of kilometres away.

Swell-wave environments, e.g. the tropical trade wind areas dominated by more gentle winds.

VALENTIN'S CLASSIFICATION

ADVANCING COAST

Emergence

Erosion

Deposition

RETREATING COAST

Submergence

- **Retreating coasts** include submerged coats and coasts where the rate of erosion is greater than the rate of emergence.
- **Advancing coasts** include emerged coastlines and coasts where deposition is rapid.

ATLANTIC AND PACIFIC COASTLINES

This classification is based on whether the trend (direction) of the geology is parallel to or at right angles to the coastline. In **Atlantic** coastlines, the geology is at right angles to the coast, e.g. South-West Ireland, whereas in **Pacific** coastlines the geological trend is parallel to the coastline, e.g. California. Both types of coastline are illustrated by the example of the Isle of Purbeck.

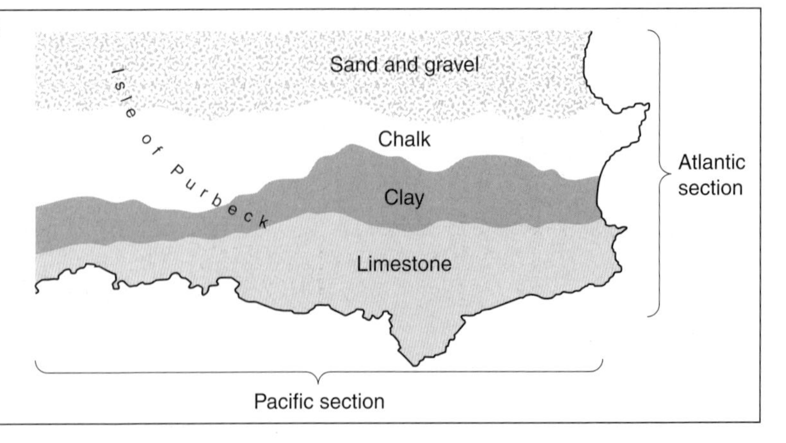

Coastal management (1)

There are a number of ways in which people affect coastal systems, e.g. by dredging and extracting sediment, and building flood protection schemes. Attempts to protect the coast in one place may increase pressure elsewhere, as is often the case with groynes. The best protection for a coast is a beach. However, many beaches are **relict** features, i.e. they have stopped forming as their sediment supply has run out.

LONGSHORE DRIFT

Groynes can prevent longshore drift from removing a beach by interrupting the natural flow of sediment. However, by trapping sediment they deprive another area, down-drift, of beach replenishment. Without its beach a coast is increasingly vulnerable to erosion, e.g. the cliffs at Barton on Sea were easily eroded following the construction of groynes along the coast at Bournemouth.

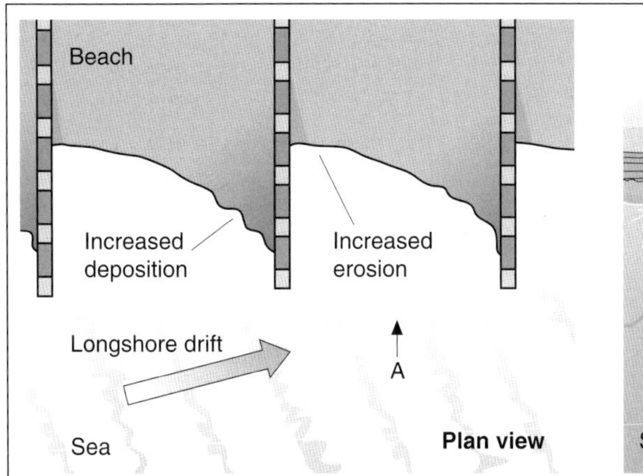

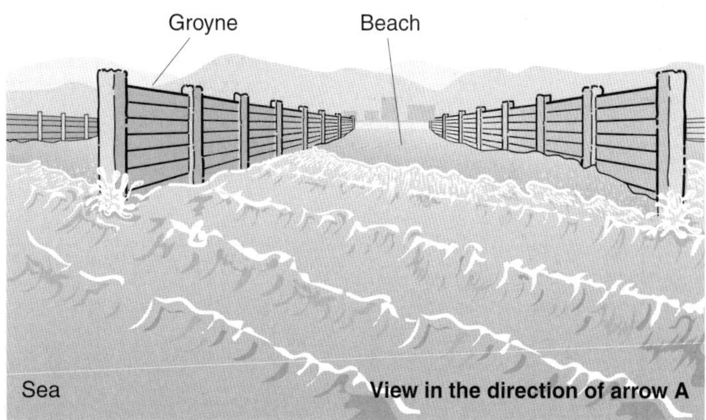

COASTAL FLOODING

This is a serious problem in the low-lying areas of eastern England. A combination of factors created severe floods in February 1953, when over 300 people were killed, over 800 km^2 was flooded, and the cost of the damage exceeded £50 million. These included:

- an intense low pressure system, 970 mb, with winds of over 100 knots
- high tides
- high river levels.

SAND DUNE EROSION

Sand dunes are extremely susceptible to erosion both by wind and sea. Trampling accelerates the process. By reducing the vegetation cover the sand or soil is not held and is open to wind erosion, which may lead to the formation of blow-out dunes.

There are a number of ways of stabilising dunes:

- planting marram grass
- building walkways or 'duckboards' to reduce trampling
- planting fences and brushwood to trap sand
- land use zoning to prevent areas from suffering pressure.

DREDGING

The fishing village of Hallsands was destroyed in 1917 by a severe storm. The shingle beach, which had previously protected the village, had been removed by contractors building the Naval Dockyard extension at Devonport. Up to 660 000 tons of shingle were removed, lowering the beach by up to 5 m. The material was never replaced, as the shingle was a relict feature deposited about 6000 years ago by rising sea-levels. Up to 6 m of cliff erosion occurred between 1907 and 1957.

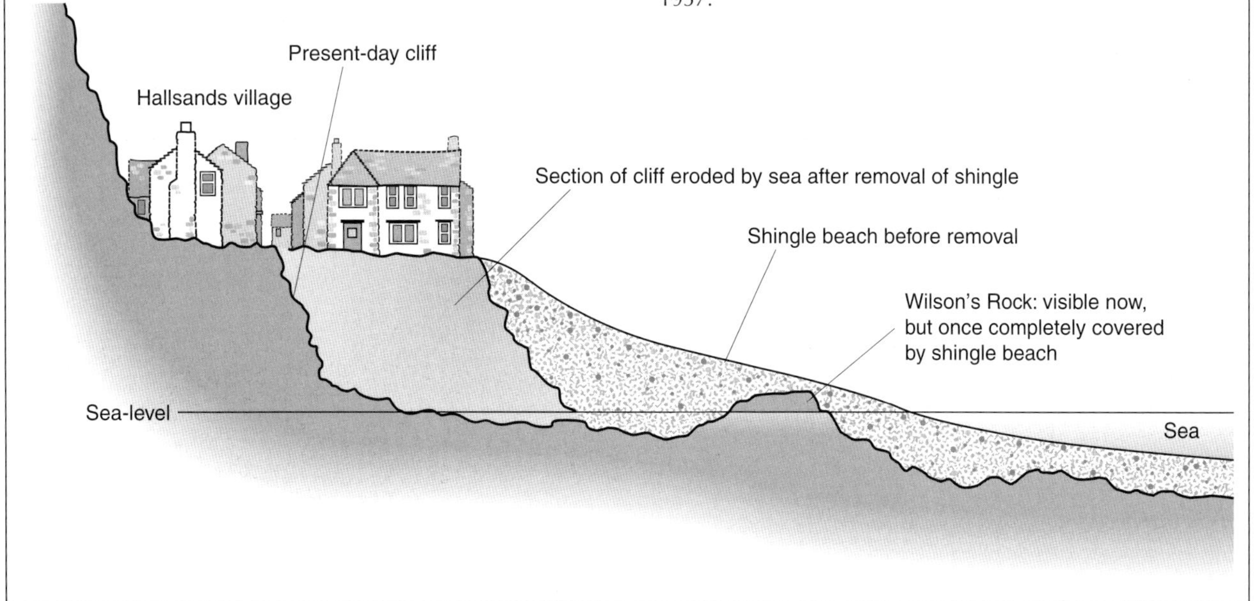

Coastal management (2)

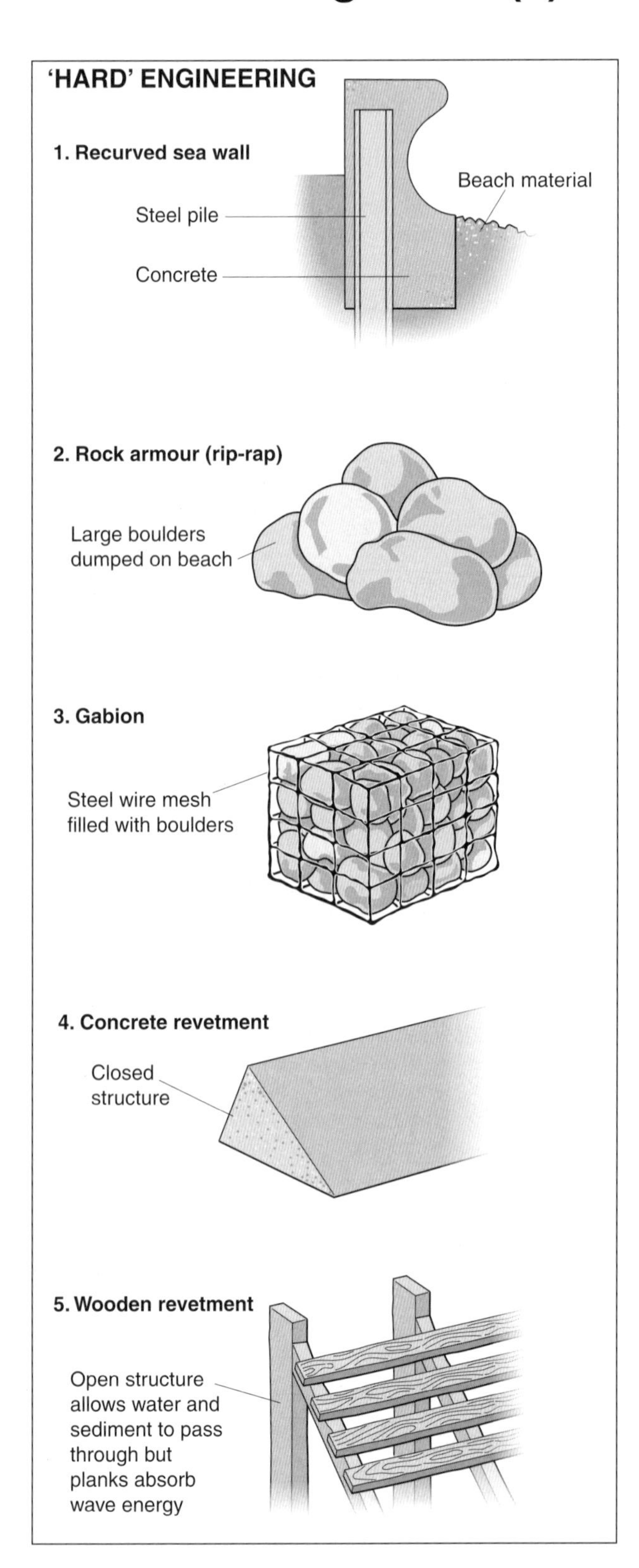

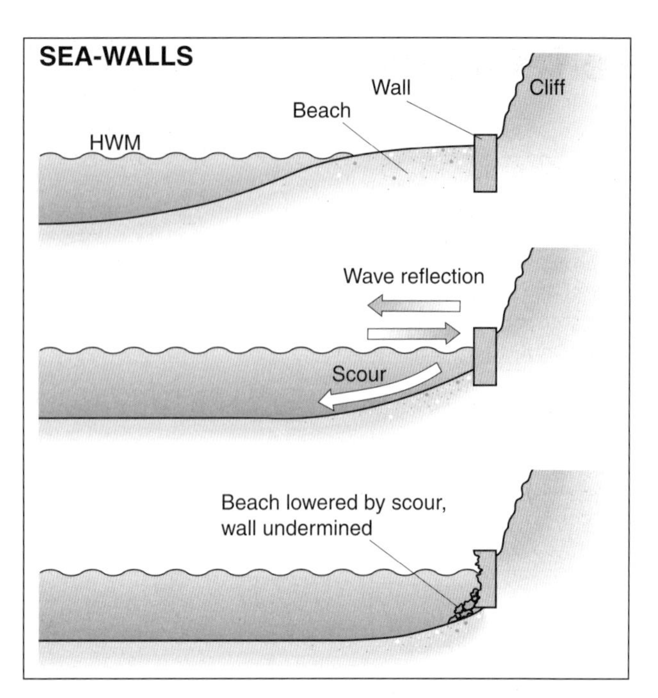

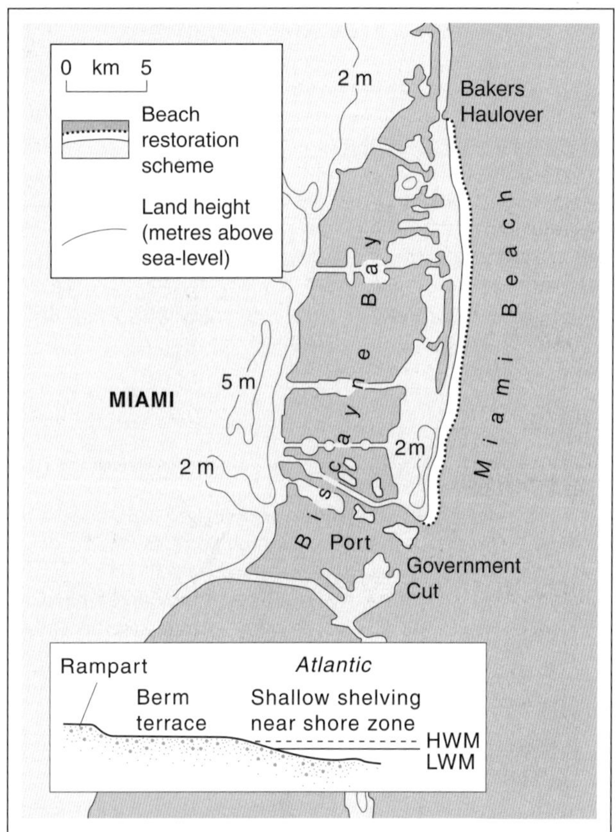

MANAGED RETREAT

The cost of protecting Britain's coastline was up to £60 million annually until the early 1990s. Since then government cuts have reduced this. Part of the problem is that southern and eastern England are slowly sinking while sea-level is rising. The risk of flooding and hence the cost of protection are increasing. 'Managed retreat' allows nature to take its course: erosion in some areas, deposition in others. Benefits include less money spent and the creation of natural environments.

BEACH REPLENISHMENT

Miami is an excellent example of beach replenishment. One of the USA's most popular tourist resorts, by the 1950s there was very little beach left. Erosion threatened the physical, economic, and social future of Miami. Between 1976 and 1982, an 18 km long, 200 m wide beach was constructed using 18 million cubic metres of sand dredged from a zone 3-4 km offshore. As far as possible it replicated a natural beach, although for the benefit of the tourists it has been kept largely free of vegetation. Access for shipping through the barrier island is still necessary, so erosion and drifting of sand still occurs. Approximately 250 000 m^3 of sand has to be replenished each year. Although expensive, the high value of tourism, industry, and residential property make this a feasible solution.

Periglaciation

PERIGLACIAL ENVIRONMENTS

Periglacial areas are found on the edge of glaciers or ice masses and are characterised by **permafrost** and **freeze-thaw** action. Summer temperatures rise above freezing so ice melts. Three types of periglacial region can be identified: Arctic continental, Alpine, and Arctic Maritime. These vary in terms of mean annual temperature and therefore the frequency and intensity with which processes operate.

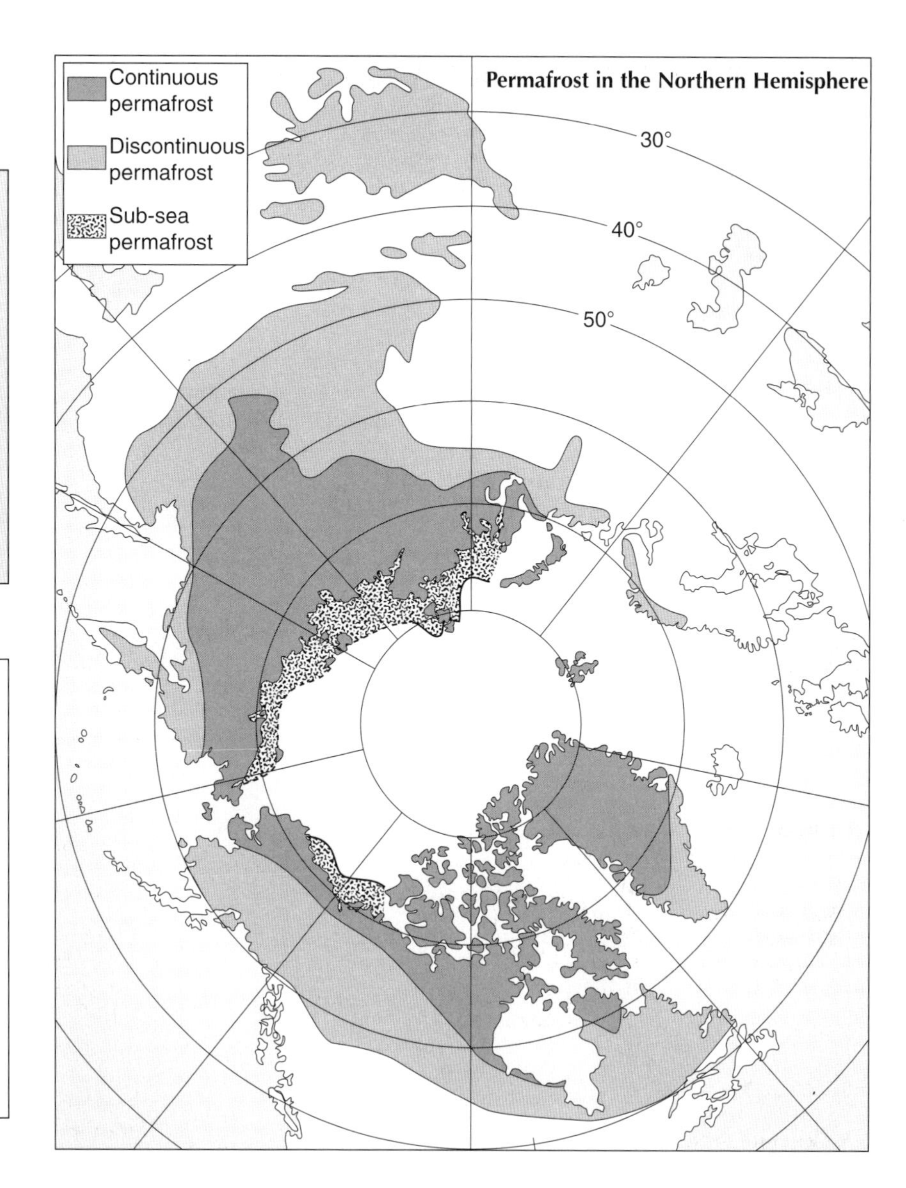

Periglacial environments extended during glacial phases. Much of southern Britain, especially those areas not covered by glaciers, was exposed to periglacial processes for considerable lengths of time. Such effects were most marked on rocks such as chalk and limestone which underwent changes in permeability during periglacial phases. Periglacial features may also be found in glaciated areas, formed during periods of glacial wasting (deglaciation). In Scotland, the periglacial zone expanded 11 000 years ago as a result of the Loch Lomond Readvance.

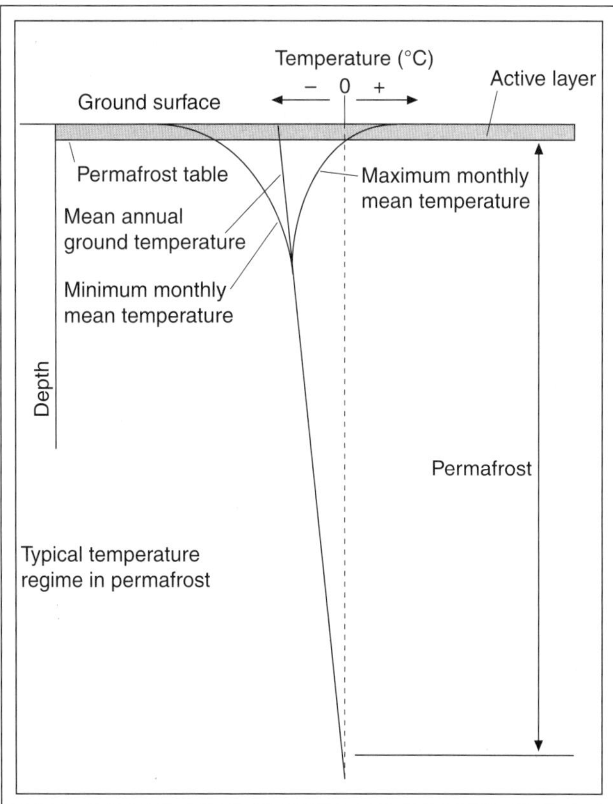

PERMAFROST

Periglacial areas are also associated with **permafrost**, impermeable permanently frozen ground. Approximately 20% of the world's surface is underlain by permafrost, in places up to 700 m deep. Three types of permafrost exist - **continuous**, **discontinuous**, and **sporadic** - and these are associated with mean annual temperatures of –5° to –50°C, –1.5° to –5°C, and 0° to –1.5°C respectively. Above the permafrost is found the **active layer**, a highly mobile layer which seasonally thaws out and is associated with intense mass movements. The depth of the active layer depends upon the amount of heat it receives and varies in Siberia from 0.2-1.6 m at 70°N and from 0.7-4 m at 50°N.

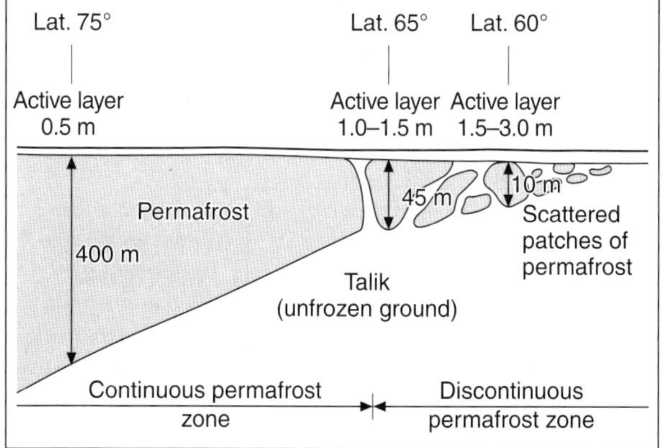

Periglacial processes

Periglacial environments are dominated by **freeze-thaw** weathering. This occurs as temperatures fluctuate above and below freezing point. As water freezes it expands by 10%, exerting pressures of up to 2100 kg/cm^2. Most rocks can only withstand up to 210 kg/cm^2. It has most effect on well jointed rocks which allow water to seep into cracks and fissures. **Congelifraction** refers to the splitting of rocks by freeze-thaw action.

Solifluction (**gelifluction** or **congelifluction**) literally means flowing soil. In winter, water freezes in the soil causing expansion of the soil and segregation of individual soil particles. In spring, the ice melts and water flows downhill. It cannot infiltrate into the soil because of the impermeable permafrost. As it moves over the permafrost it carries segregated soil particles (peds) and deposits them further downslope as a solifluction lobe or terracette.

Cambering is the process whereby segments of rock become dislodged from the main body of rock and begin to move downhill. It is aided by freeze-thaw.

Nivation (**altiplanation** or **cryoplanation**) is freeze-thaw weathering under a snow bank. The broken material is removed in spring and summer by the melted snow.

Chemical weathering is also effective in periglacial regions. **Carbonation** is an important process because of the low temperatures. Carbon dioxide is more soluble at low temperature hence the water becomes quite acidic. It is aided by the slowly rotting vegetation which releases organic acids. **Hydrolysis** is also important because of the presence of organic acids in the marshy soil.

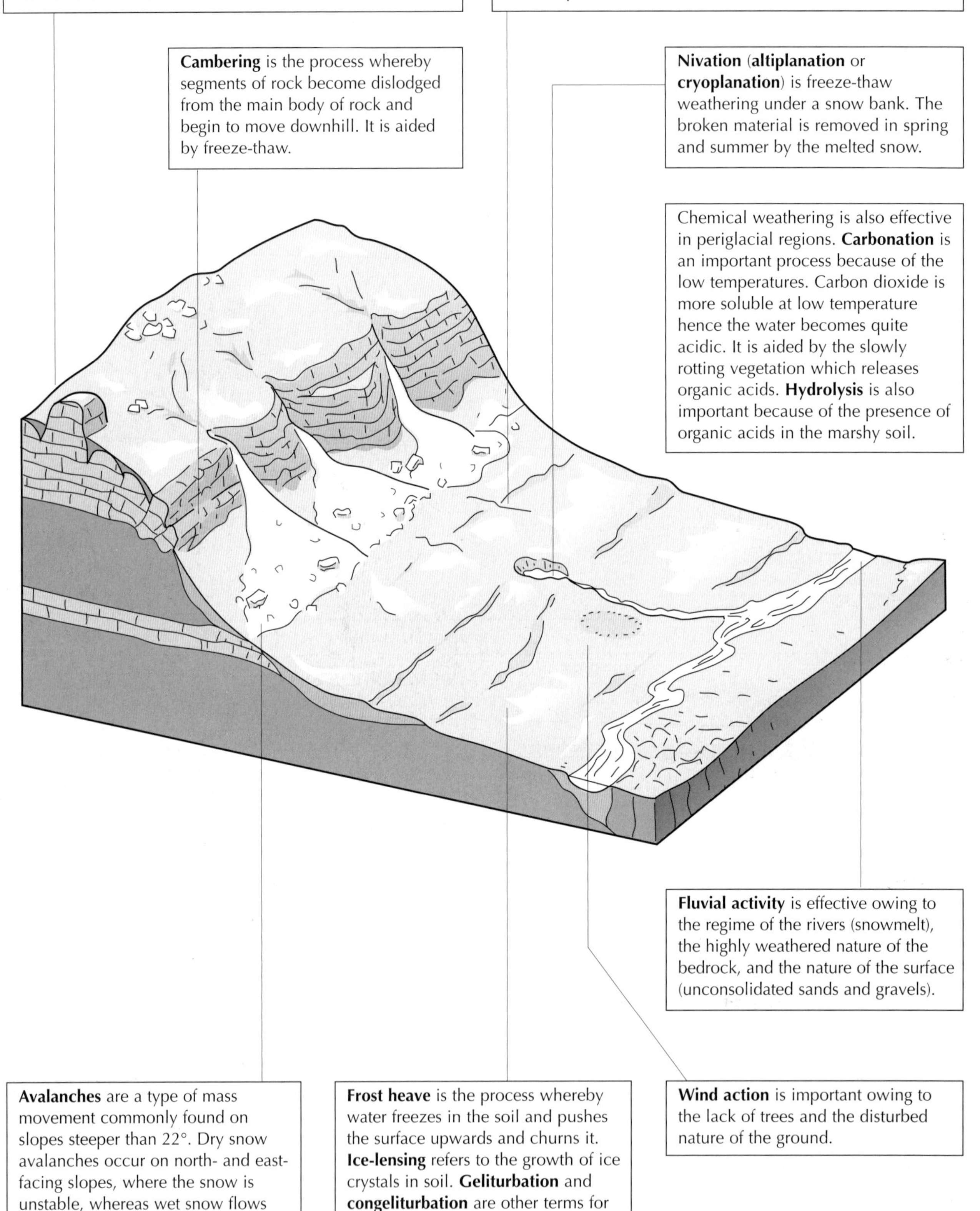

Fluvial activity is effective owing to the regime of the rivers (snowmelt), the highly weathered nature of the bedrock, and the nature of the surface (unconsolidated sands and gravels).

Avalanches are a type of mass movement commonly found on slopes steeper than 22°. Dry snow avalanches occur on north- and east-facing slopes, where the snow is unstable, whereas wet snow flows generally result from rapid snow melt.

Frost heave is the process whereby water freezes in the soil and pushes the surface upwards and churns it. **Ice-lensing** refers to the growth of ice crystals in soil. **Geliturbation** and **congeliturbation** are other terms for frost heave.

Wind action is important owing to the lack of trees and the disturbed nature of the ground.

Periglacial landforms

Tors are isolated outcrops of bare rock, e.g. Yes Tor and Hay Tor on Dartmoor. They are formed as a result of intense frost shattering under periglacial conditions and the removal of the weathered material (growan) by mass movements and fluvial activity (the Palmer-Nielson theory).

Patterned ground is a general term describing the stone circles, polygons, and stripes that are found in soils subjected to intense frost action, e.g. Grimes Graves near Thetford in Norfolk. On steeper slopes, stone stripes replace stone circles and polygons. Their exact mode of formation is unclear although ice sorting, differential frost heave, solifluction, and the effect of vegetation are widely held to be responsible.

Scree slopes are slopes composed of large quantities of angular fragments of rock, e.g. the slopes at Wastwater in the Lake District. Typically they have an angle of rest of about 35°. Extensive upland surfaces of angular rocks are known as **blockfields.**

Dry valleys are river valleys without rivers. They are most commonly found on chalk and limestone such as The Manger at Uffington (Vale of the White Horse) and the Devil's Dyke near Brighton. During the periglacial period, limestone and chalk became impermeable owing to permafrost, and therefore rivers flowed over their surfaces. High rates of fluvial erosion occurred because of spring melt, the highly weathered nature of the surface, and high rates of carbonation. At the end of the periglacial period normal permeability returned, waters sank into the permeable rocks, and the valleys were left dry.

Rivers in periglacial areas are typically **braided**, with numerous small channels separated by small linear islands, e.g. the River Eyra in Iceland. Braiding occurs because the river is carrying too much sediment as a result of the highly erosive nature of streams fed by spring melt, and is forced to deposit some.

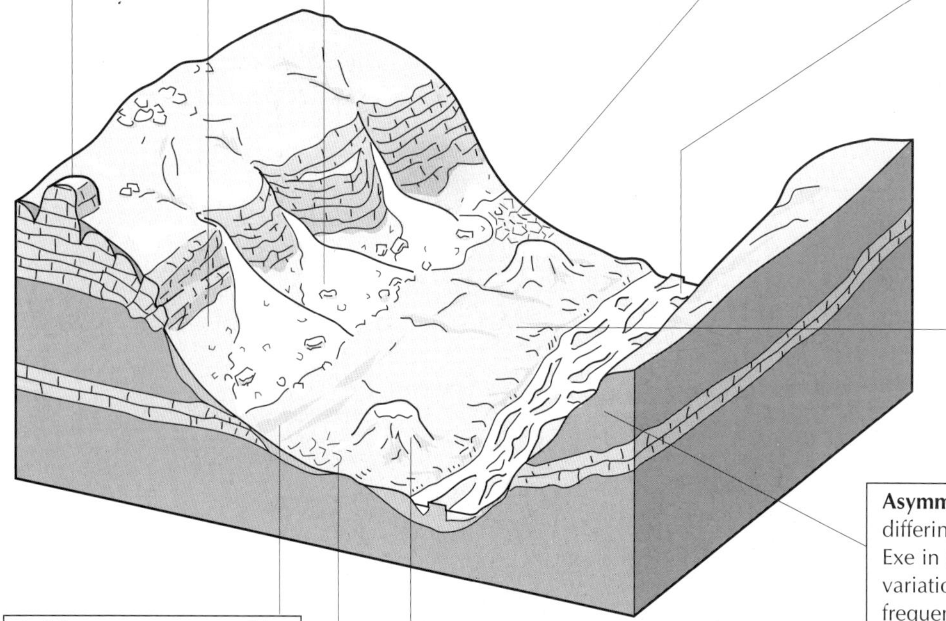

Loess refers to deposits laid down by the wind. They consist mostly of unstratified, structureless silt, and cover extensive areas in China and northern Europe and produce smaller deposits in Britain such as the Brickearth of East Anglia, e.g. Wangford Warren.

Asymmetric slopes are valleys with differing slope angles, e.g. the River Exe in Devon. They are caused by variations in **aspect** which affects the frequency and intensity of weathering. South-facing slopes receive more insolation and are subjected to more mass wasting, and hence have lower slope angles. By contrast, north-facing slopes remain frozen for longer periods, are not as highly weathered, and consequently remain steeper.

Solifluction **terracettes**, like those at Maiden Castle in Dorset, are step-like features ('sheep-walks') caused by mass movement in the active layer. **Solifluction lobes** are elongated versions typically 20 m long and over 1 m high.

Coombe rock or **head** is a periglacial deposit comprising of chalk, mud, and/or clay, compacted with angular fragments of frost shattered rock, e.g. at Scratchey Bottom near Durdle Door in Dorset.

A **pingo** is an isolated, conical hill up to 90 m high and 800 m wide, which can only develop in periglacial areas. They form as a result of the movement and freezing of water under pressure. Two types are generally identified - **open-system** and **closed-system** pingos. Where the water is from a distant elevated source, open system pingos are formed, whereas if the supply of water is local, and the pingo occurs as a result of the expansion of permafrost, closed system pingos are formed. Nearly 1500 pingos are found in the Mackenzie Delta of Canada, and examples of relict pingos can be found in the Vale of Llanberris in Wales. When a pingo collapses ramparts and ponds are left.

Problems in the use of periglacial areas

The hazards associated with the use of periglacial areas are diverse and may be intensified by human impact. Problems include mass movements such as avalanches, solifluction, rockfalls, frost heave, and icings as well as flooding, thermokarst subsidence, low temperatures, poor soils, a short growing season, and a lack of light.

For example, the Nyenski tribe in the Yamal Peninsula of Siberia have suffered as a result of the exploitation of oil and gas. Oil leaks, subsidence of railway lines, destruction of vegetation, decreased fish stocks, pollution of breeding grounds, and reduced caribou numbers have all happened directly or indirectly as a result of man's attempt to exploit this remote and inhospitable environment.

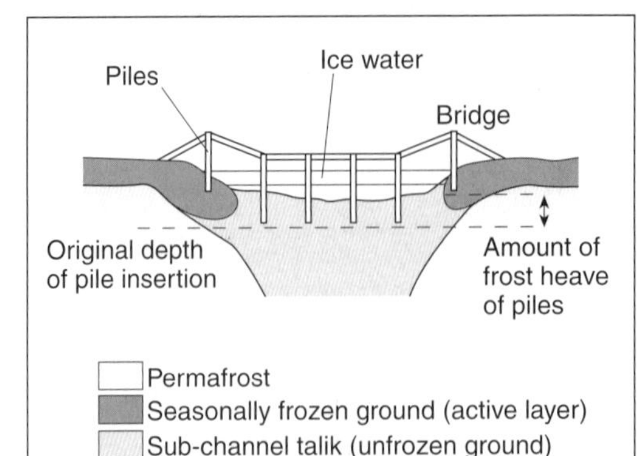

Close to rivers, **frost heave** is very significant (owing to an abundant supply of water) and can lift piles and structures out of the ground. **Piles** for carrying oil pipelines therefore need to be embedded deep in the permafrost to overcome mass movement in the active layer. In Prudhoe Bay, Alaska, they are 11 m deep. However, this is extremely expensive: each one cost over $3000 in the early 1970s.

Services are difficult to provide in periglacial environments. It is impossible to lay underground networks and so **utilidors**, insulated water and sewage pipes, are provided above ground. Waste disposal is also difficult owing to the low temperatures.

Alpine periglacial areas also suffer environmental pressures. Here the concerns are more than damage to the physical environment, as traditional economies have declined at the expense of electro-chemical and services industries, especially tourism. An elaborate infrastructure is required to cope with the demands of an affluent tourist population, and this may undermine the natural environment and traditional societies.

Traditionally, periglacial pastures have been used by Inuits for herding or hunting caribou. The abundance of lakes allows travel by float plane and the frozen winter rivers and lack of trees enables overland travel. Periglacial areas are **fragile** for two reasons. First, the ecosystem is highly susceptible to interference because of the limited number and diversity of species involved. The extremely low temperatures limit decomposition, and hence **pollution**, especially oil spills, have a very long-lasting effect on periglacial ecosystems. Secondly, **permafrost** is easily disrupted. The disruption of permafrost poses significant problems. Heat from buildings and pipelines, and changes in the vegetation cover, rapidly destroy it. Thawing of the permafrost increases the active layer and the subsequent settling of the soil causes subsidence. Consequently, engineers have either built structures on a bed of gravel, up to 1m thick for roads, or used stilts.

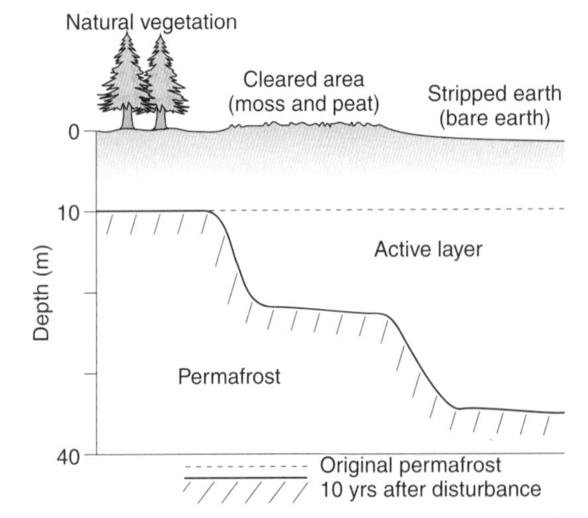

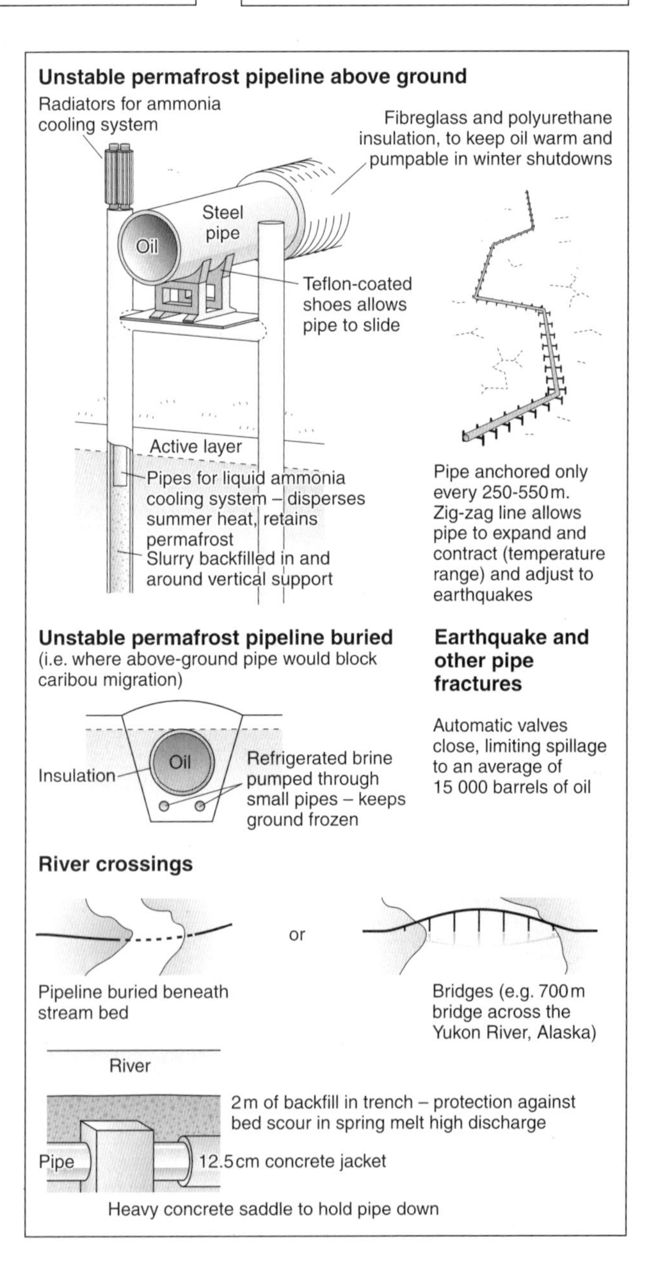

Glaciation

GLACIAL SYSTEMS

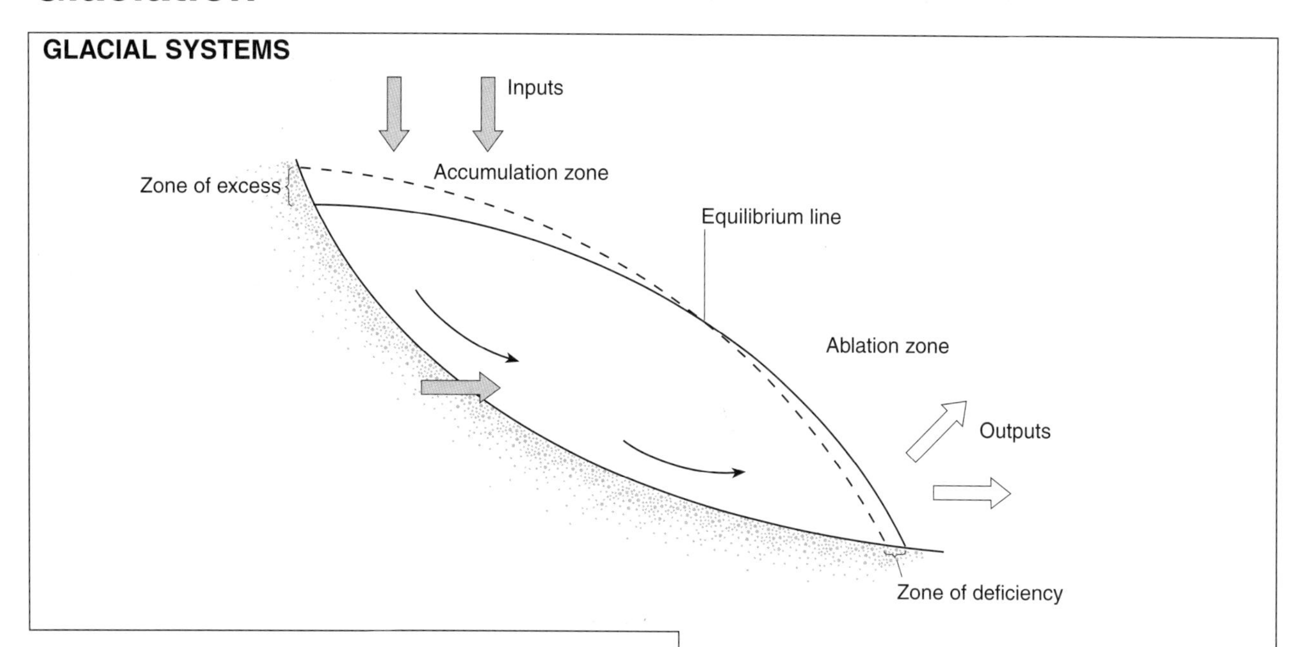

A glacial system is the balance between inputs, storage, and outputs. Inputs include **accumulation** of snow, avalanches, debris, heat, and meltwater. The main store is that of ice, but the glacier also carries debris, called **moraine**, and meltwater. The outputs are the losses due to **ablation**, the melting of snow and ice, and sublimation of ice to vapour, as well as sediment.

The **regime** of the glacier refers to whether the glacier is advancing or retreating:

if accumulation > ablation, the glacier advances
if accumulation < ablation, the glacier retreats
if accumulation = ablation, the glacier is steady

Glacial systems can be studied on an annual basis or on a much longer time scale. The size of a glacier depends on its regime, i.e. the balance between the rate and amount of supply of ice and the amount and rate of ice loss. The glacier will have a **positive regime** when the supply is greater than loss by ablation (melting, evaporation, calving, wind erosion, avalanche, and so on) and so the glacier will thicken and advance. A **negative regime** will occur when the wasting is greater than the supply (e.g. the Rhone glacier today) and thus the glacier will thin and retreat. Any glacier can be divided into two sections: an area of accumulation at high altitudes generally, and an area of ablation at the snout.

HOW SNOW BECOMES ICE

Initially snow falls as flakes and the accumulation is known as **alimentation**. With continued alimentation the lower snowflakes are compressed under more snow and gradually change into a collection/aggregate of granular ice pellets called **neve** or firn. Increased pressure causes the flakes to melt. Meltwater seeps into the gaps between the pellets and freezes. Some places still have air, which gives the neve a white colour. With continued **accumulation** and pressure the neve becomes tightly packed, and any remaining air is expelled, its place filled with freezing water. Some molecules of water vapour may condense straight into ice, again filling any spaces. Thus, the neve changes into glacier ice, by **compaction** and **crystallisation**, with a characteristic bluish colour. The change from neve to ice occurs, typically, when the neve is 30 m thick. Glacier ice in bulk is a 'granular aggregate of interlocking grains', each grain being an ice crystal. Between each crystal there remains an extremely thin 'intergranular film', consisting of a water-like solution containing chlorides and other salts. The presence of these salts lowers the freezing point around the crystals and so keeps the solution in a liquid state. This film thus acts as a lubricant, aiding ice movement.

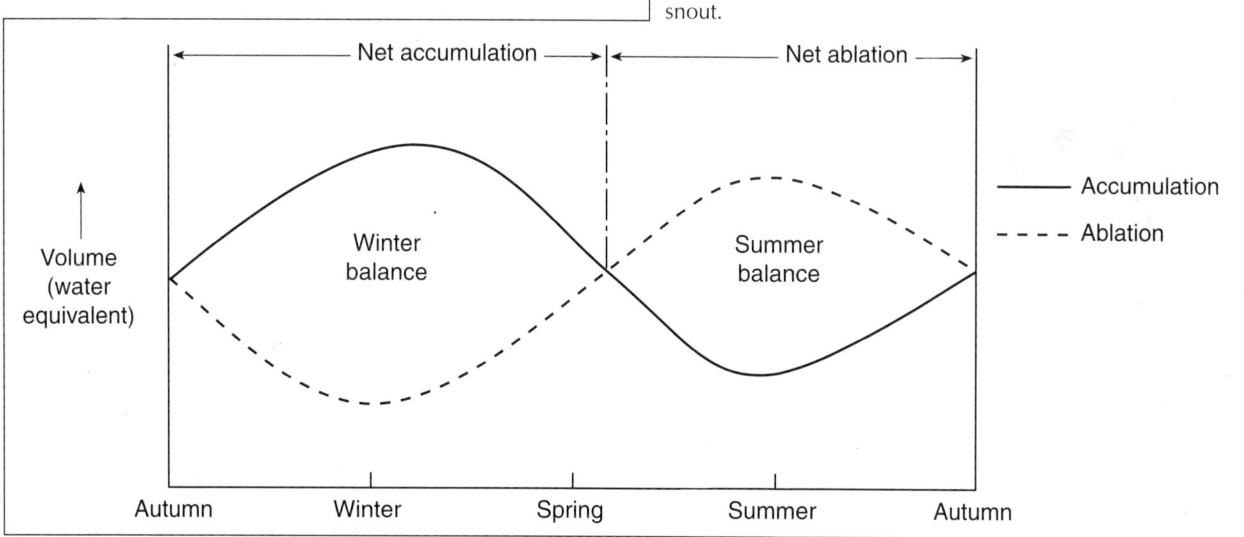

Past glaciations

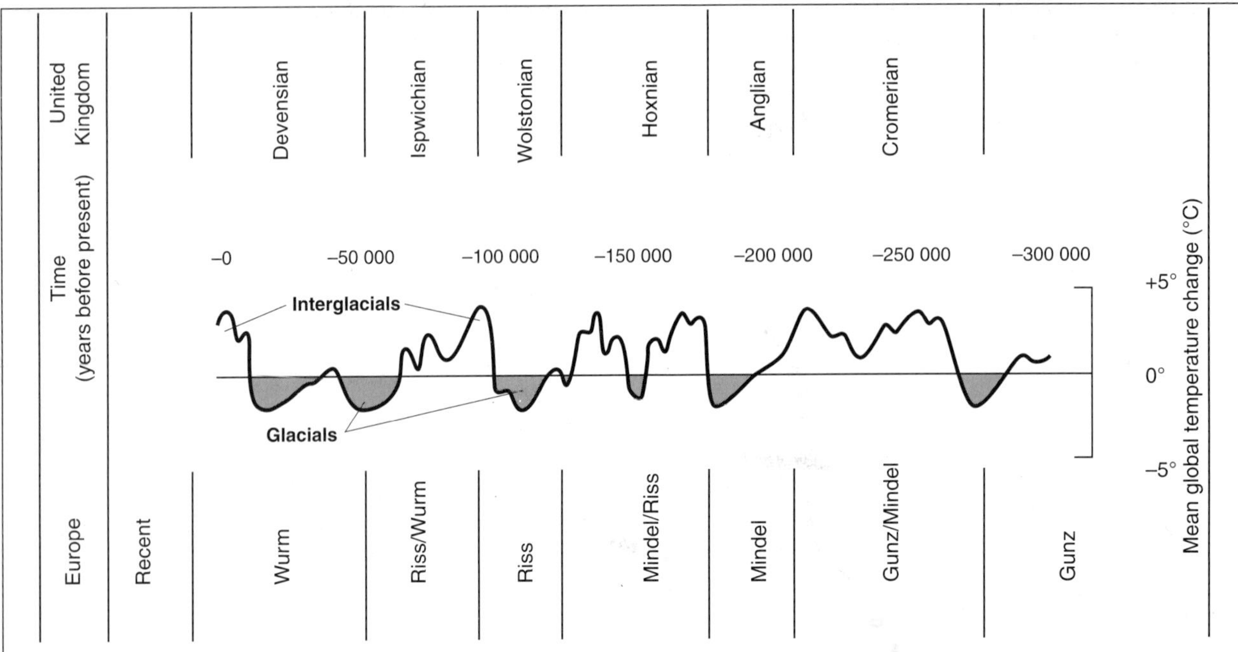

Glacial regimes are extremely sensitive to climatic change. Over the last two million years - the **Quaternary Era** - there have been cold **glacial** phases in which glaciers were present or advancing, and warm **interglacial** periods when ice retreated. This is important in explaining the effectiveness of glacial erosion and transport. Each glacial period is preceded and followed by a **periglacial** period, characterised by intense freeze-thaw activity, nivation, and snowmelt. This breaks up the landscape and the glacier is then able to erode and transport the prepared material. Moreover, at the end of each glacial period, there is pressure release, or dilation, whereby the underlying rocks expand and break as the weight of the glacier is removed. This exposes an even greater area of bedrock to be attacked by freeze-thaw. Hence, glacial activity is in part dependent upon processes that operate in the periglacial stages.

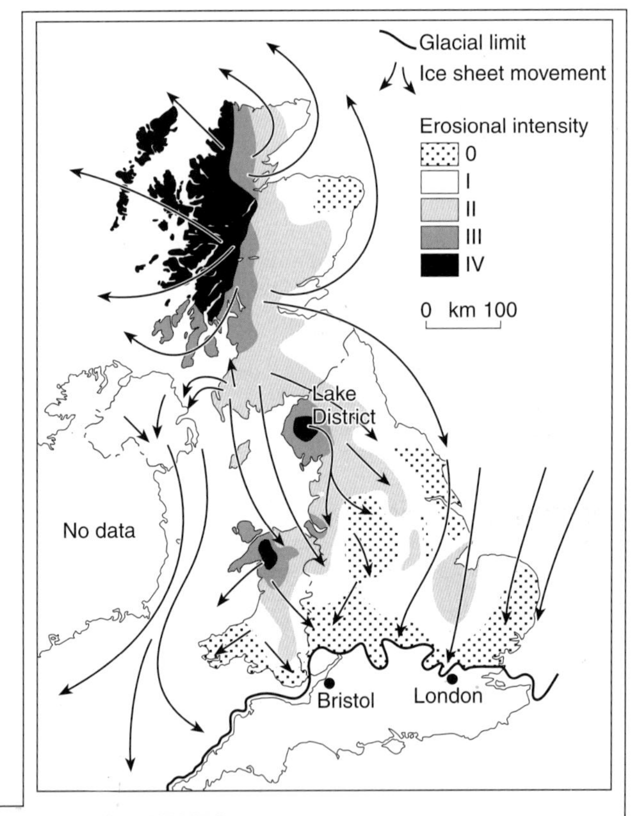

CAUSES OF GLACIAL PERIODS AND ICE ADVANCES

Many different causes of these climatic changes have been proposed and include changes in the earth's orbit, the tilt of the earth's axis, solar radiation, the position of the poles, the amount of water vapour in the atmosphere, the distribution of land and sea, the relative levels of land and sea, the nature and direction of ocean currents, and a reduction of CO_2 in the atmosphere.

Studying the evidence for glacial activity is complicated for a number of reasons: there have been many glaciations; the later glaciations remove the evidence of earlier ones; it is not possible or practical to study processes happening under glaciers; and many of the landforms have been subsequently modified by mass movements and fluvial activity.

PAST AND PRESENT DISTRIBUTION OF GLACIAL ENVIRONMENTS

It is possible to piece together the main actions of glacial activities during the **Pleistocene Ice Age** (2 m.y. b.p.). Most of Britain, Ireland, the North Sea, Scandinavia, and northern Europe was covered by an ice sheet, whereas the advance in Alpine areas was more limited. During glacial advances the summer temperature remained below 0°C allowing snow and ice to remain all year. In Britain, glaciers pushed as far south as the Bristol Channel. Upland areas were affected by more glacial periods and more intense activity. The main advances include the Anglian Glaciation between c.425 000 and 380 000 years ago, the Wolstonian Glaciation 175 000 to 128 000 years ago, and the Devensian advances between 26 000 and 15 000 years ago and between 12 000 and 10 000 years ago. The latter, the Loch Lomond readvance, was limited to western Scotland. Although there is much debate regarding the precise timings of these phases, and in some cases even the sequence, the effect of **multiple glaciations** on the environment is clear.

Thermal classification and glacier movement

CLASSIFICATION

i) **Temperate glaciers** are warm-based glaciers. Water is present throughout the ice mass and acts as a lubricant allowing ice to move freely and erode the rock. The heat is a result of the pressure of the ice or meltwater percolating down through the ice and/or release from the underlying bedrock.

ii) **In polar or cold-based glaciers** the ice remains frozen at the base, and consequently there is little water or movement. Very little erosion results.

Temperate glaciers generally have velocities of between 20 m and 200 m per year, but can reach speeds of up to 1000 m/yr. By contrast, polar glaciers may advance at only a few metres per year.

GLACIER VELOCITY

The velocity of a glacier is controlled by (i) the **gradient** of the rock floor, (ii) the **thickness of the ice** (which controls pressure and meltwater), and (iii) **temperature** within the ice. Velocity varies across a glacier as well as with depth and these variations cause **crevasses** to form. Glaciers move in three main ways - **basal slide**, **internal deformation**, and **compression and extension** of the ice surface.

Plan view

Glacier width

Valley walls

(i) Velocity reaches maximum in centre, minimum at edges due to friction

Side view

Ice surface

Basal slip

Bedrock

(ii) Internal flow greatest near surface, least at depth

BASAL SLIDE

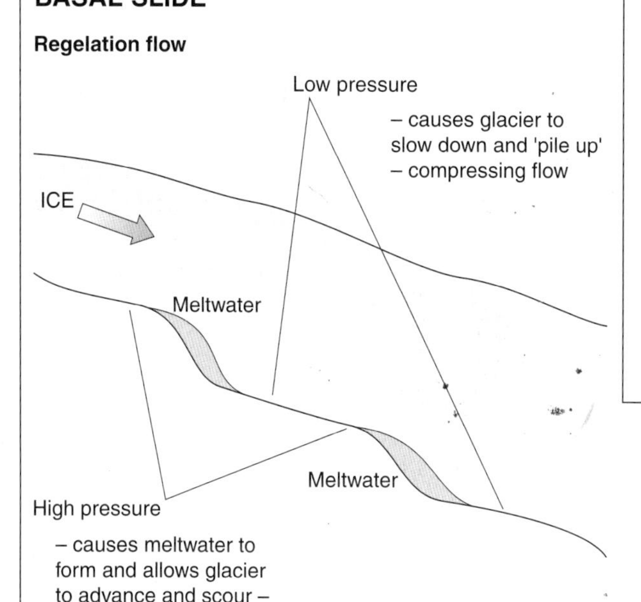

A thin film of meltwater at the base of warm glaciers allows it to slide over the bedrock. This accounts for up to 75% of glacier movement. Part of the movement is caused by **regelation**, the melting and subsequent re-freezing of water around irregularities in the valley floor. The increase of pressure upstream of the irregularity causes the melting to take place.

INTERNAL DEFORMATION

This involves the movement of ice crystals as a result of gravity. Ice acts like plasticine: on a horizontal surface it remains intact but when suspended at an angle it warps. Internal deformation is especially acute when gradients are high.

Individual crystals may move relative to other crystals (intergranular flow) or along layers - laminar flow - within the glacier.

EXTENDING AND COMPRESSING FLOW

Ice cannot deform rapidly. Consequently it fractures and movement takes place along a series of planes causing crevasses to form.

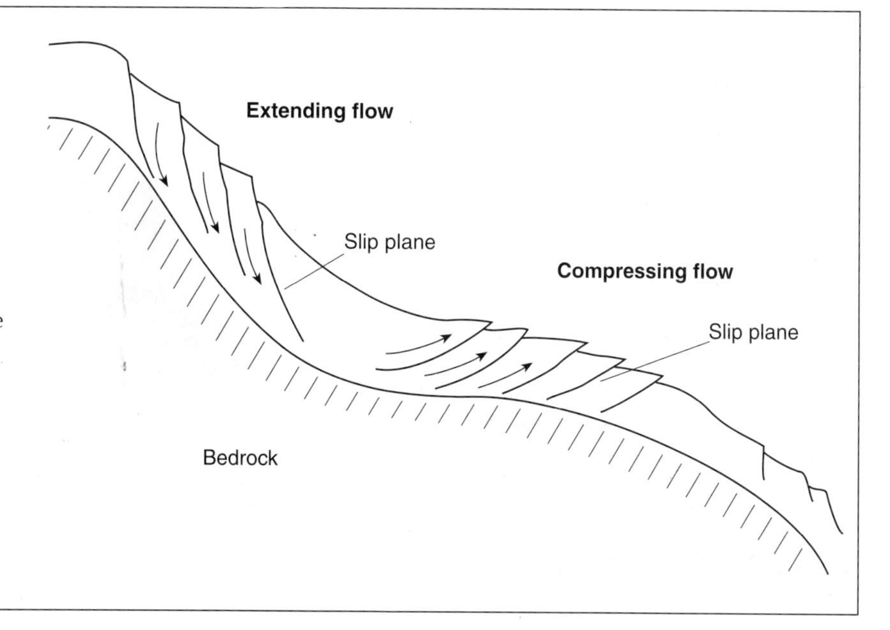

Glacial erosion

PROCESSES AND CONTROLS

The amount and rate of erosion depends on (a) the local geology, (b) the velocity of the glacier, (c) the weight and thickness of the ice, and (d) the amount and character of the load carried. The methods of glacial erosion include plucking and abrasion.

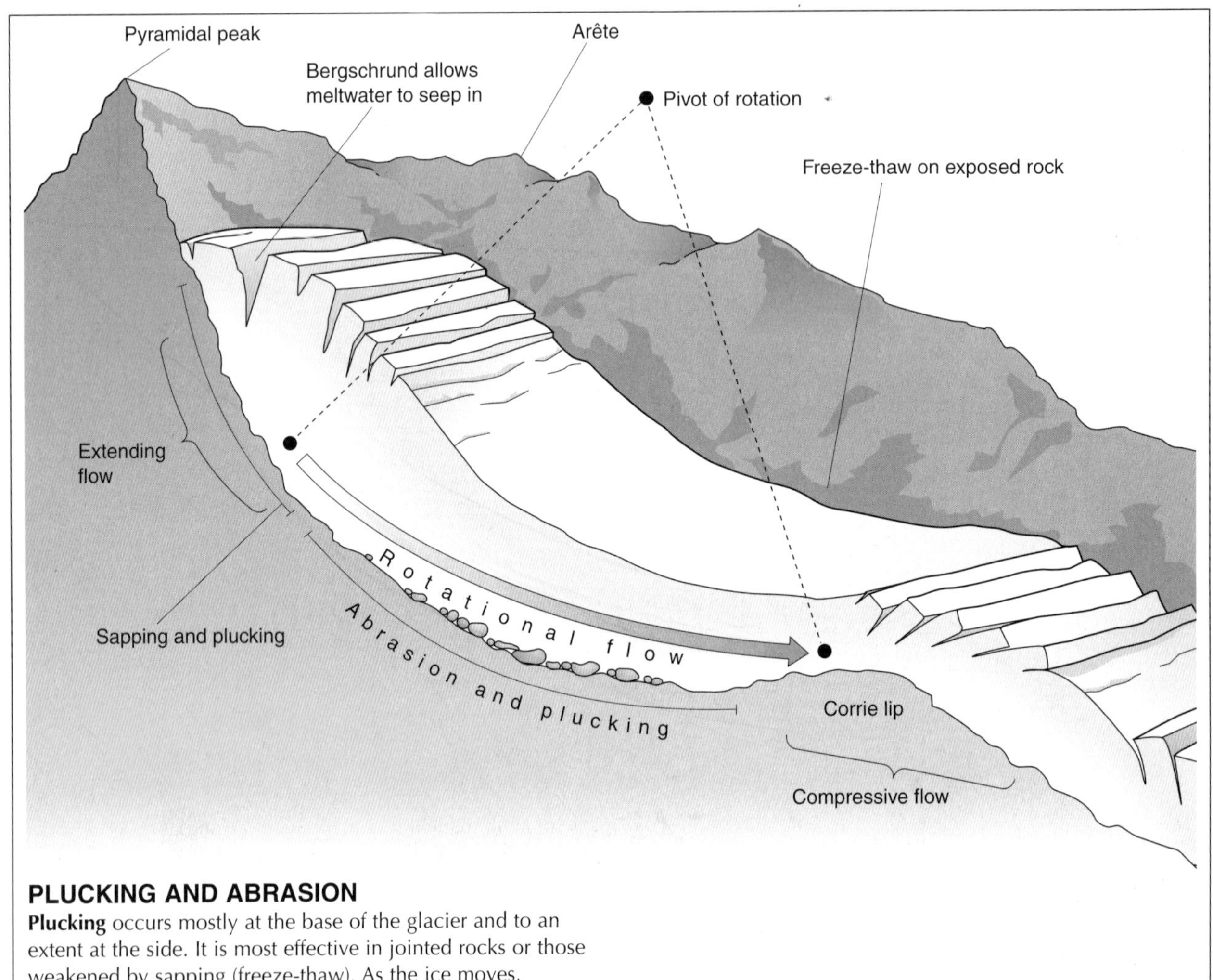

PLUCKING AND ABRASION

Plucking occurs mostly at the base of the glacier and to an extent at the side. It is most effective in jointed rocks or those weakened by sapping (freeze-thaw). As the ice moves, meltwater seeps into the joints and freezes onto the rock, which is then ripped out by the moving glacier.

Abrasion is where the debris carried by the glacier scrapes and scratches the rock leaving **striations**.

Other mechanisms include meltwater, freeze-thaw weathering, and pressure release. Although not strictly glacial nor erosional, these processes are crucial in the development of glacial scenery.

CIRQUES

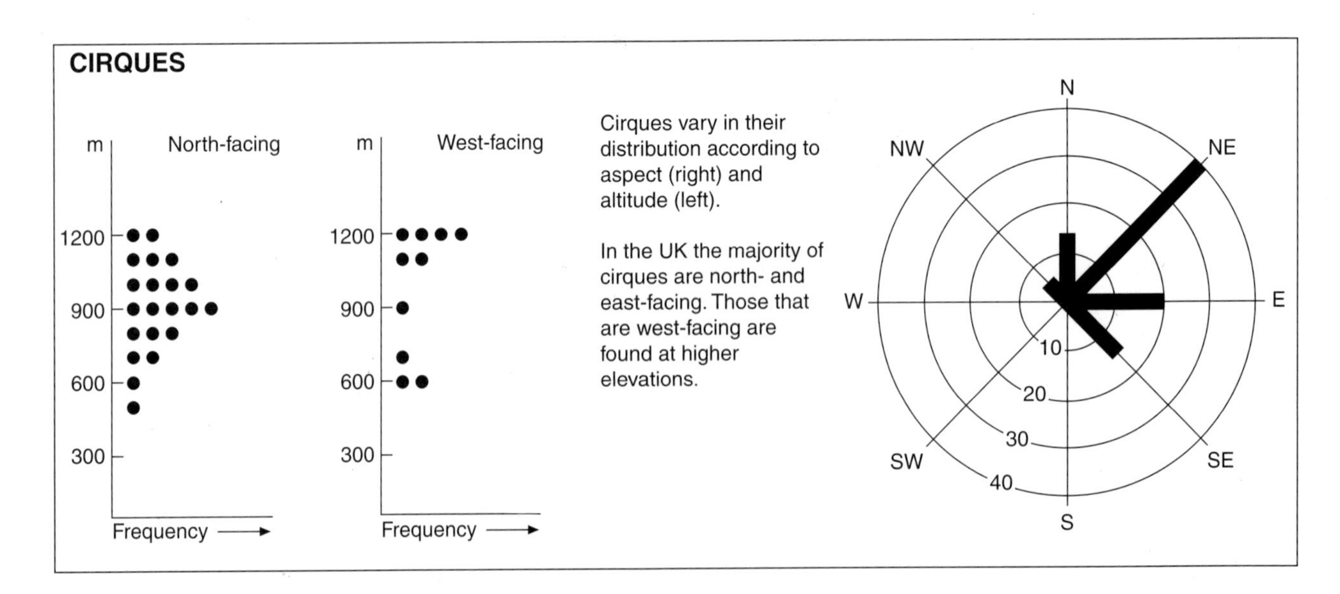

Cirques vary in their distribution according to aspect (right) and altitude (left).

In the UK the majority of cirques are north- and east-facing. Those that are west-facing are found at higher elevations.

Landforms produced by glacial erosion

CIRQUES, ARÊTES, HORNS, AND U-SHAPED VALLEYS

In the UK **cirques** are generally found on north- or east-facing slopes where accumulation is highest and ablation is lowest. They are formed in stages: (i) a pre-glacial hollow is enlarged by **nivation** (freeze-thaw and removal by snow melt); (ii) ice accumulates in the hollow; (iii) having reached a critical weight and depth, the ice moves out in a rotational manner, eroding the floor by plucking and abrasion; (iv) meltwater trickles down the bergschrund allowing the cirque to grow by freeze-thaw. After glaciation, an armchair-shaped hollow remains, frequently filled with a lake, e.g. Red Tarn in the Lake District.

Other features of glacial erosion include **arêtes** and **pyramidal peaks (horns)** caused by the headward recession (cutting back) of two or more cirques. Glacial troughs (or **U-shaped valleys**) have steep sides and flat floors. In plan view they are straight since they have truncated the interlocking spurs of the pre-glacial valley. The ice may also carve deep rock basins frequently filled with **ribbon lakes**. **Hanging valleys** are formed by tributary glaciers which, unlike rivers, do not cut down to the level of the main valley, but are left suspended above, e.g. Stickle Beck. They are usually marked by waterfalls.

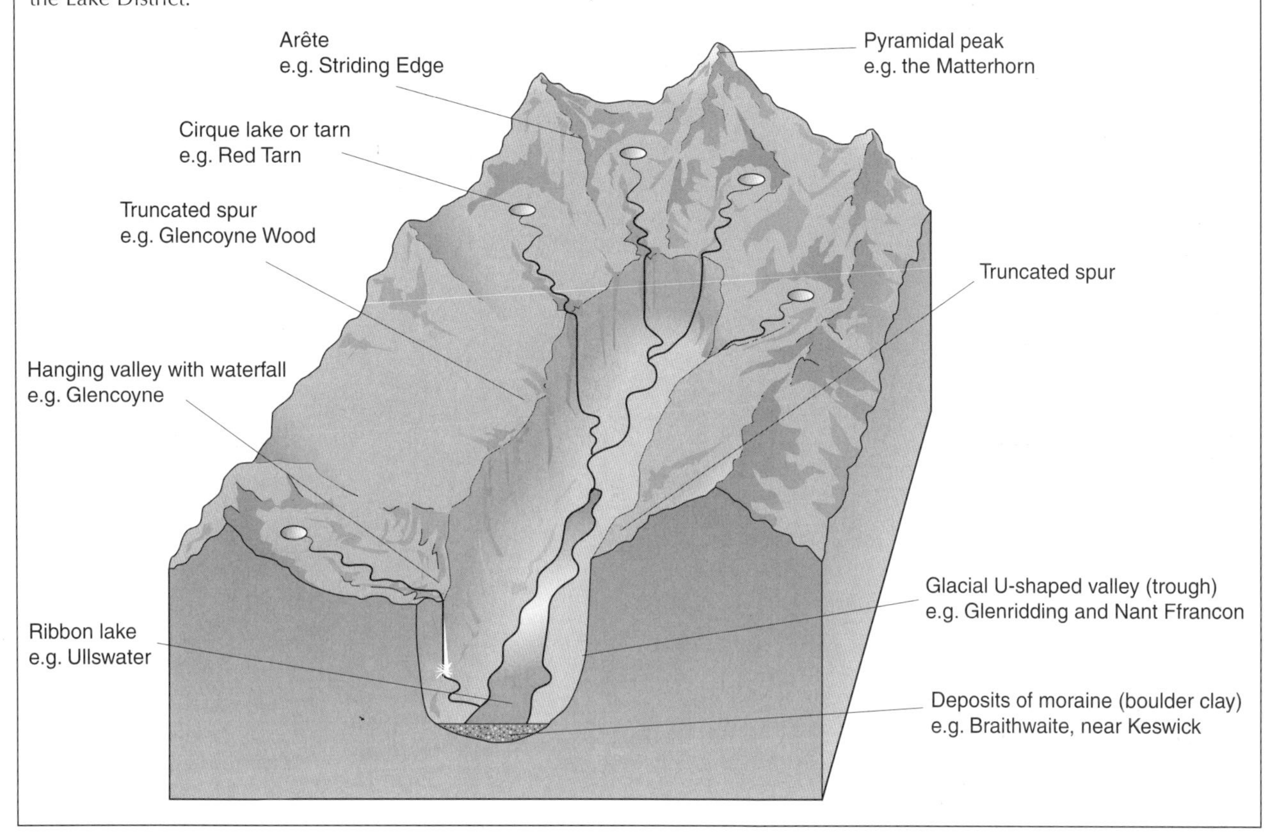

CRAG AND TAIL

A **crag and tail** is formed when a very large resistant object obstructs ice flow. The ice is forced around the obstruction, eroding weaker rock. Material immediately in the lee of the obstruction is protected by the crag and forms a tail. Edinburgh Castle rock is an ancient volcanic plug whereas its tail is formed of limestone.

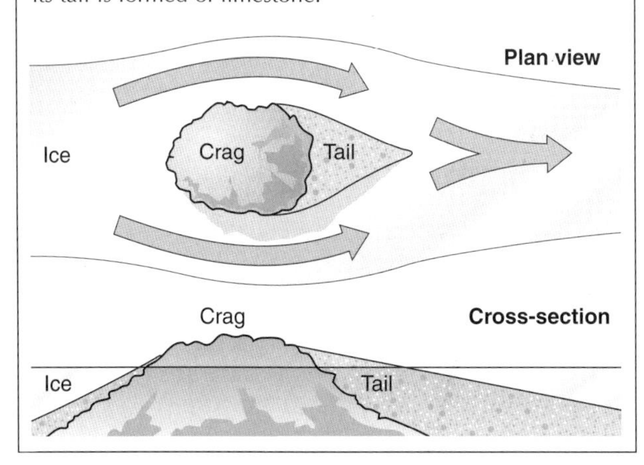

ROCHES MOUTONNÉES

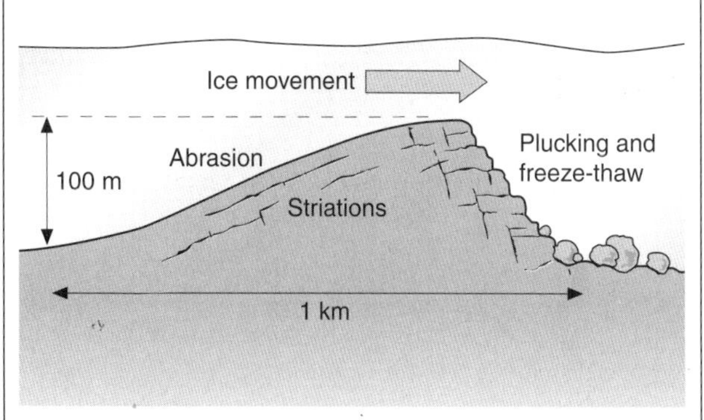

Roches moutonnées vary in size from a few metres to hundreds of metres. They are smoothed and polished on the up-valley side (stoss) by abrasion, but plucked on the lee side (down valley) as the ice accelerates. They can be over 100 m in height and several kilometres long. In Scotland, the rocky ridges are interspersed with small basins filled with water giving a **cnoc and lochan** landscape (hillock and lake).

Glacial deposition

The term **drift** refers to all glacial and fluvioglacial deposits left after the ice has melted. Glacial deposits or **till** are angular and unsorted, and their long axes are orientated in the direction of glacier flow. These include erratics, drumlins, and moraines. Till is often subdivided into **lodgement till**, material dropped by actively moving glaciers, and **ablation till**, deposits dropped by stagnant or retreating ice.

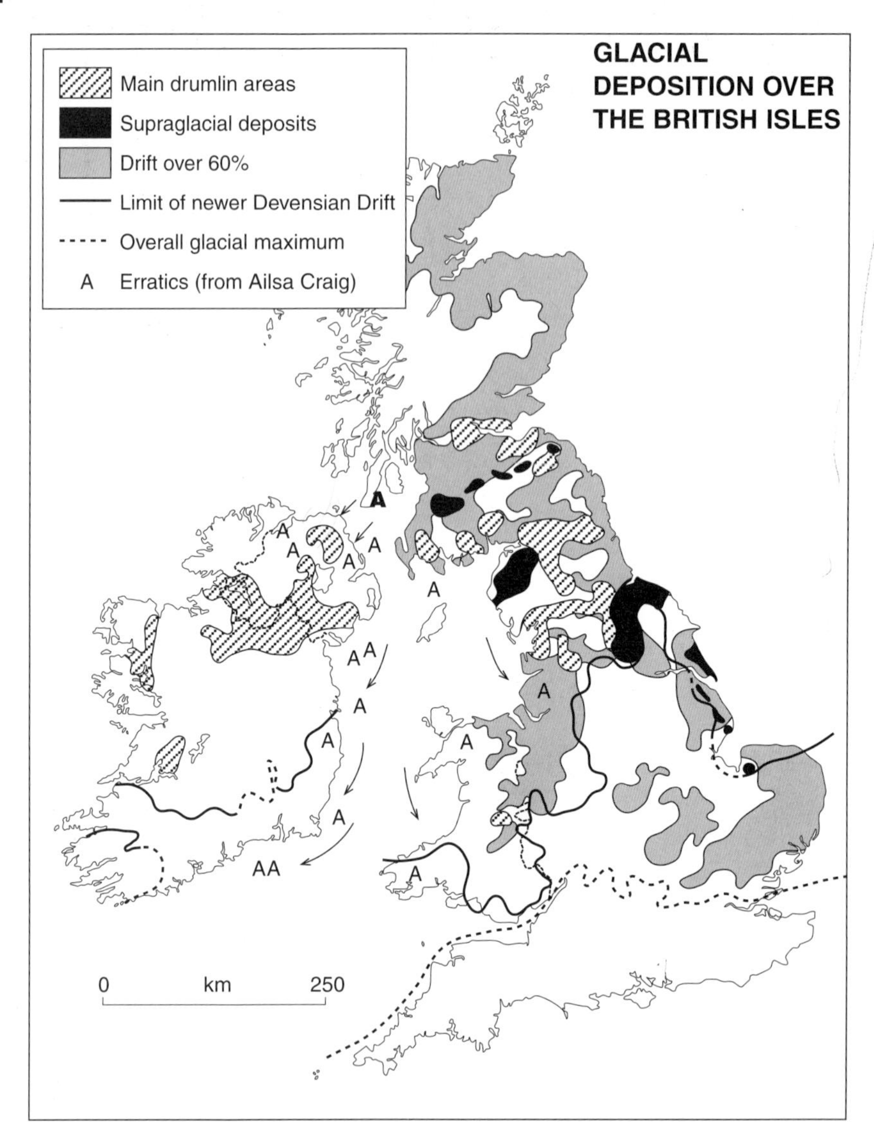

DEPOSITIONAL FEATURES

Erratics

Erratics are large boulders foreign to the local geology, e.g. the Bowder Stone in Borrowdale and the Norber Stone on the North Yorks Moors.

Moraines

Moraines are lines of loose rocks, weathered from the valley sides and carried by the glaciers. At the snout of the glacier is a crescent-shaped mound of **terminal moraine**. Its character is determined by the amount of load the glacier was carrying, the speed of movement, and the rate of retreat. The ice-contact slope (up-valley) is always steeper than the down-valley slope. The finest example in Britain is the Cromer Ridge, up to 90 m high and 8 km wide.

Drumlins

Drumlins are small oval mounds up to 1.5 km long and 100 m high, e.g those in the Ribble Valley or the drowned drumlins of Clew Bay in Co. Mayo, Ireland. They are deposited due to friction between the ice and the underlying geology, causing the glacier to drop its load. As the glacier continues to advance it streamlines the mounds.

100 m

Long axis

Ice flow

Plan view

100 m

1 km

These features can be used to determine the **direction of glacier movement**. Erratics pinpoint the origin of the material, and drumlins and the long axes of pebbles in glacial till are orientated in the direction of glacier movement.

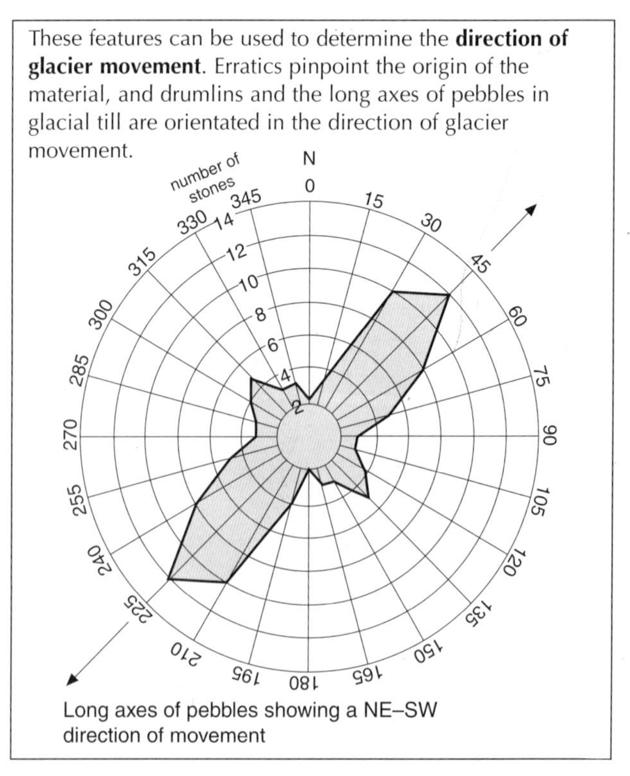

Long axes of pebbles showing a NE–SW direction of movement

Fluvioglacial deposition

Fluvioglacial erosion also has a pronounced impact on the landscape. **Proglacial** lakes are formed adjacent to the ice mass. They may be trapped by high ground or between a retreating ice mass and terminal moraine. **Cols** or spillways are channels formed by meltwater in highland areas when it drains through the lowest part of a watershed. A series of proglacial lakes in the North Yorks Moors were envisaged, where meltwater and rivers were trapped between high ground and the advancing ice mass, thereby creating massive lakes such as Lake Pickering. These grew until they reached a low point in the surrounding landscape, and then drained through in huge meltwater channels, such as Newtondale. However, this theory has been challenged by others claiming the lakes were not as large and that the channels were either marginal meltwater channels or subglacial channels.

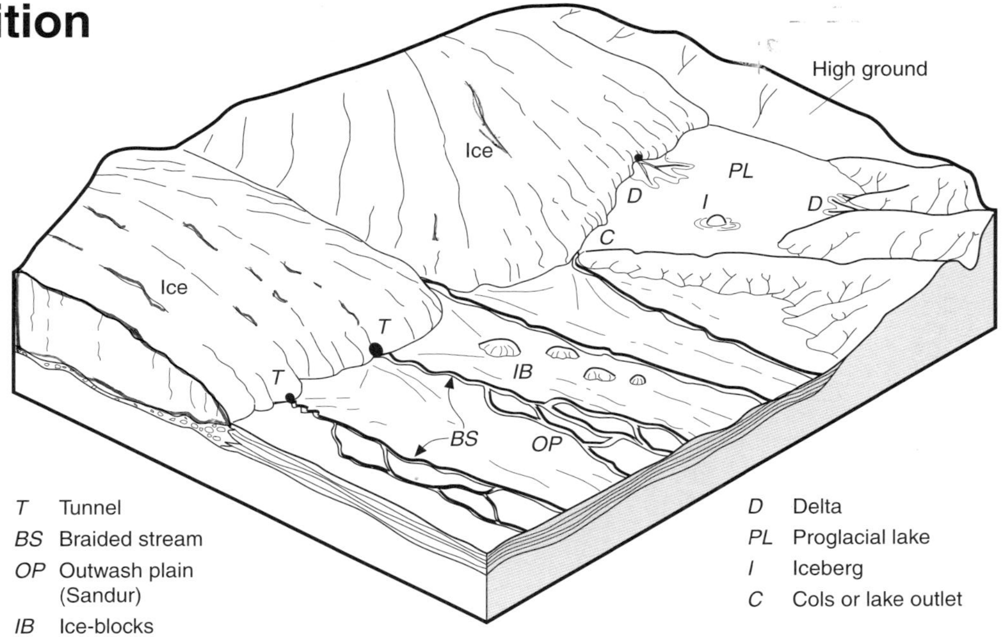

Ice
Ice
High ground
PL
D
I
D
C
T
T
IB
BS
OP

T Tunnel
BS Braided stream
OP Outwash plain (Sandur)
IB Ice-blocks

D Delta
PL Proglacial lake
I Iceberg
C Cols or lake outlet

Fluvioglacial or **meltwater** deposits can be subdivided into **prolonged drift**, in which the material is very well sorted, e.g. varves and outwash plains, and **ice-contact stratified drift**, e.g. kames and eskers, which are more varied in character.

Kames

Kames are irregular mounds of sorted sands and gravels, formed by supraglacial streams on stagnating ice sheets. Often they contain **kettle holes**, caused by the deposition of material around broken blocks of ice. **Kame terraces** are found at the side of the valley, laid down by streams occupying the site between the valley wall and the glacier, e.g. the Lammermuir Hills in eastern Scotland

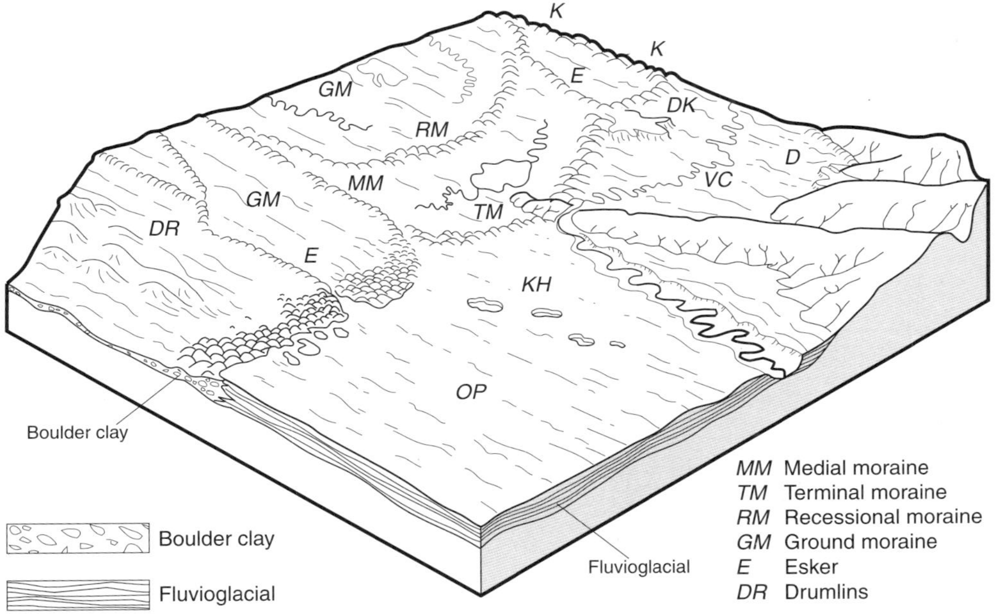

Eskers

Eskers are elongated ridges of coarse, stratified, fluvioglacial sands and gravels. Two explanations are given for their formation: (i) material is deposited in subglacial meltwater tunnels or (ii) eskers may represent a rapidly retreating delta, formed as the ice melts and subglacial streams are suddenly released of pressure.

FLUVIOGLACIAL LANDFORMS

Outwash plains

Sandur or **outwash plains** are gently sloping plains comprised of sands and gravels that are sorted and stratified. The coarser gravels are deposited first owing to a reduction of meltwater competence, and closer to the ice margin, whereas the sands are carried further down the plain. There is also vertical layering as well as horizontal stratification. These are characterised by braided rivers, heavily laden with sand and gravel, varying enormously in seasonal discharge, causing numerous islets and channels to be formed, e.g. the River Eyra in Iceland.

K
K
E
DK
GM
RM
D
VC
GM
MM
TM
DR
E
KH
OP
Boulder clay
Fluvioglacial

Boulder clay
Fluvioglacial

MM Medial moraine
TM Terminal moraine
RM Recessional moraine
GM Ground moraine
E Esker
DR Drumlins

K Kame
KH Kettle hole
VC Varved clay
DK Delta kame

Varves

Varves or **varved clays** are layered deposits of alternating coarse and fine material. Each varve is a year's deposit: coarser material is carried in the lakes by spring meltwaters whereas finer material does not settle until later in the autumn.

The structure of the atmosphere

VERTICAL STRATIFICATION

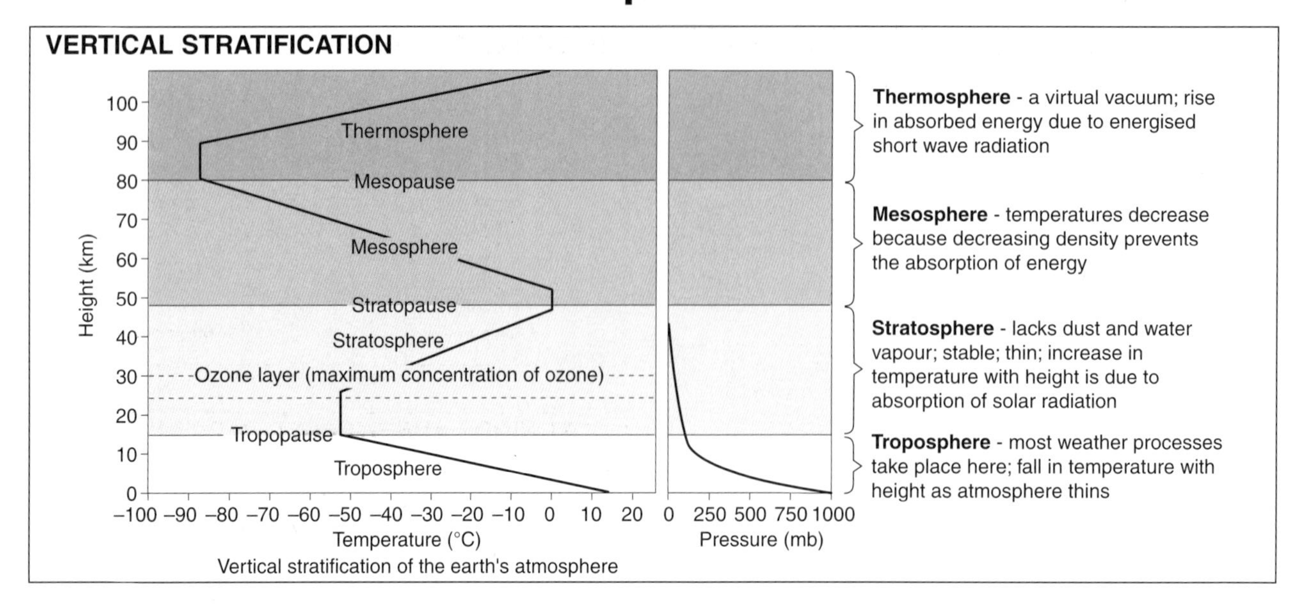

Vertical stratification of the earth's atmosphere

THE EXCHANGE OF ENERGY BETWEEN EARTH AND ATMOSPHERE

Long-wave re-radiation
Short-wave energy is converted to long-wave energy which can be absorbed by the atmosphere; some is re-radiated back. Water vapour acts as an insulator.

Latent heat transfers
Energy is used to convert water into water vapour (evaporation). The heat is retained as latent heat. When vapour turns back into water the heat is released.

Compression heating
When air contracts its temperature increases.

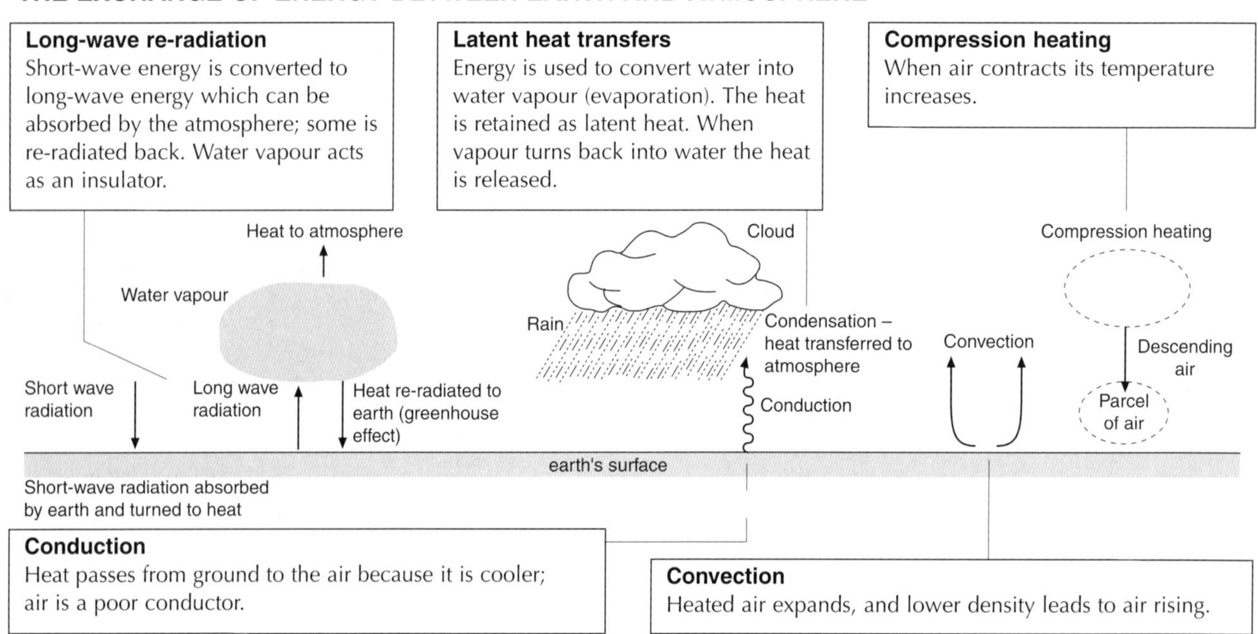

Conduction
Heat passes from ground to the air because it is cooler; air is a poor conductor.

Convection
Heated air expands, and lower density leads to air rising.

TEMPERATURE AND SOLAR RADIATION

The amount of solar radiation received and lost depends on a number of factors:

- **Distance from the sun according to the earth's elliptical orbit**

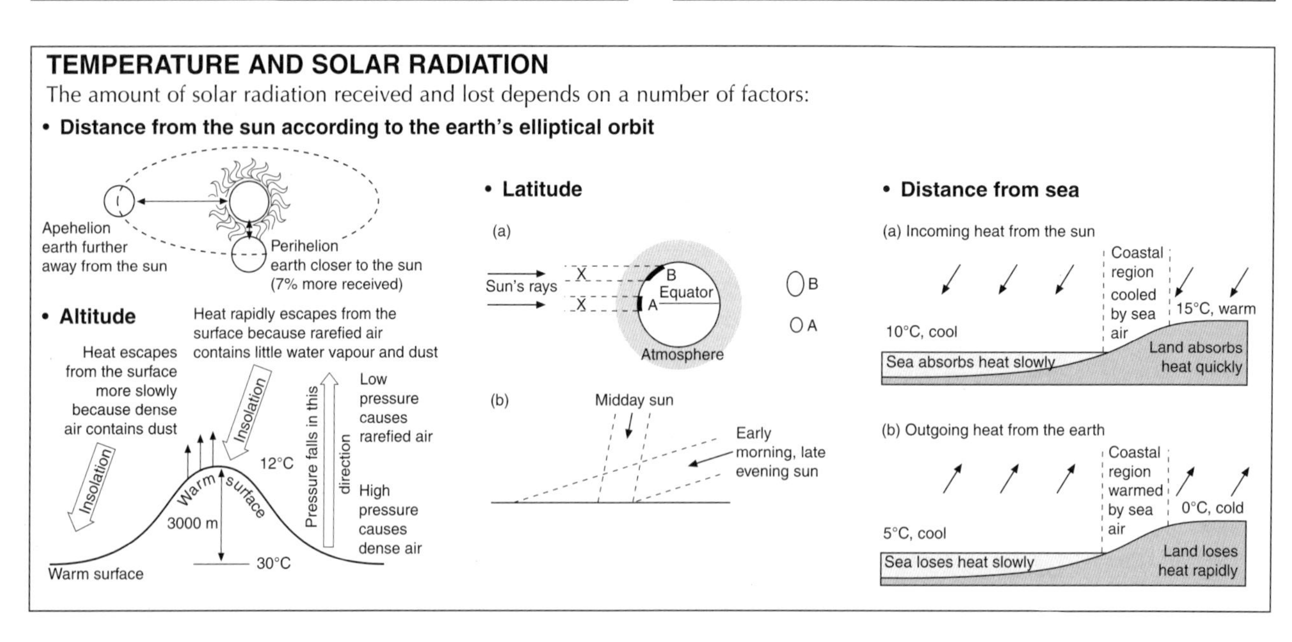

Atmospheric motion

Atmospheric motion is controlled by the combination of the following forces:

- pressure-gradient force
- Coriolis force
- the geostrophic wind
- the gradient wind
- friction.

THE PRESSURE-GRADIENT FORCE

Pressure gradient wind is the movement of air that occurs along pressure gradients from high to low pressure.

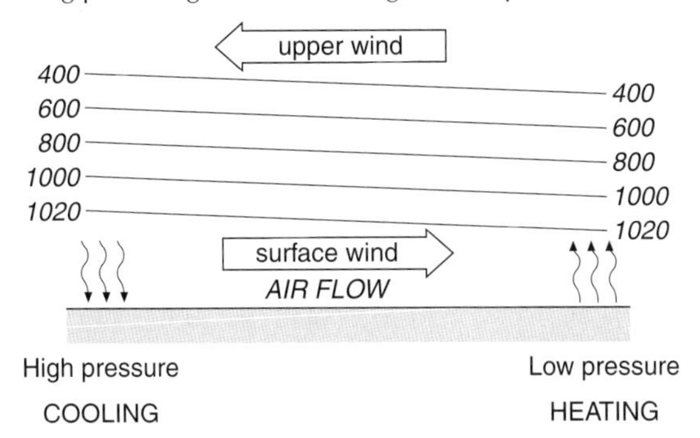

THE CORIOLIS FORCE

The Coriolis force is the deflection of winds due to the earth's rotation. Deflection is to the right in the northern hemisphere, and to the left of a path in the southern hemisphere.

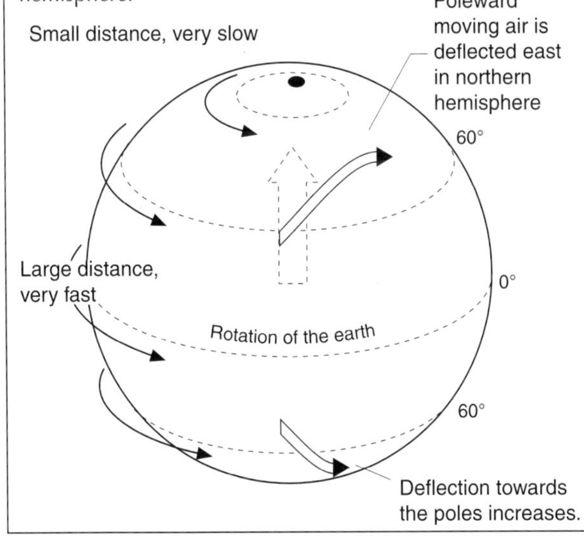

THE GRADIENT WIND

Where isobars are curved, centrifugal and centripetal forces act upon the wind to maintain a flow parallel to the isobars. The curved path is called a gradient wind.

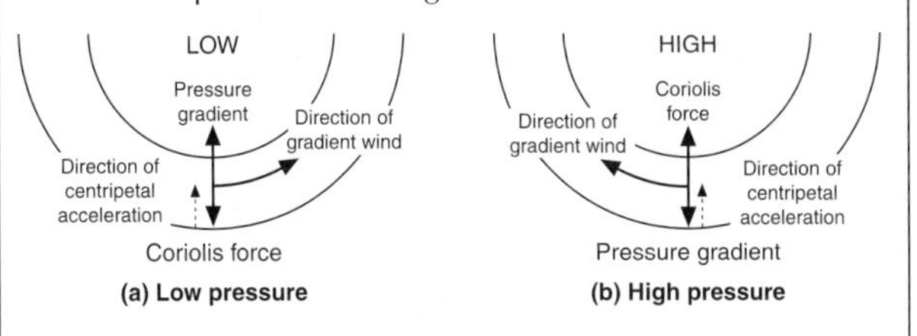

HOW WINDS ARE STEERED IN THE UPPER ATMOSPHERE

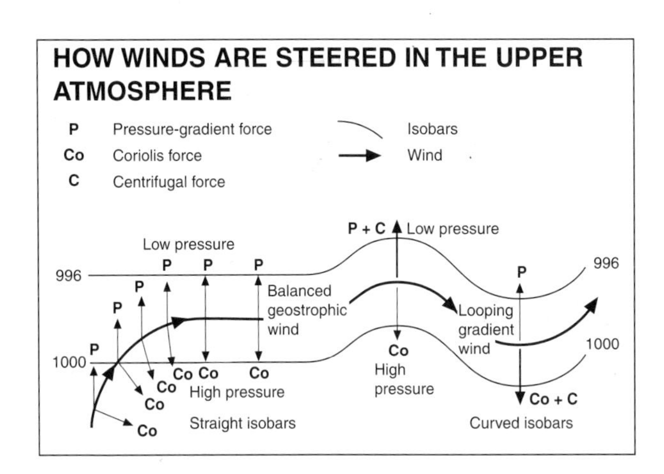

THE EFFECT OF FRICTION

Frictional drag from the earth's surface modifies the balance between horizontal gradient force and the Coriolis force. Friction decreases wind speed but also changes wind direction. Friction causes the wind to cross isobars at an angle. With increasing height the effect of friction is reduced. This means wind changes direction with height. The change in pattern is known as Ekman spirals.

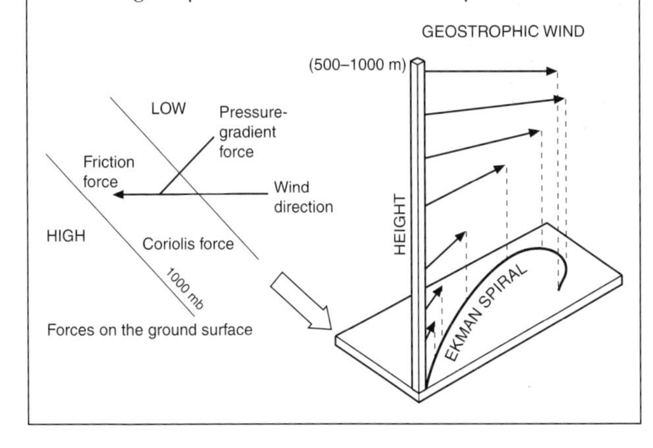

THE GEOSTROPHIC WIND

In mid-latitudes, the pressure-gradient force and the Coriolis force are directly balanced. This leads to air moving not from high to low pressure but between the two, parallel to the isobars. This is called a geostrophic wind.

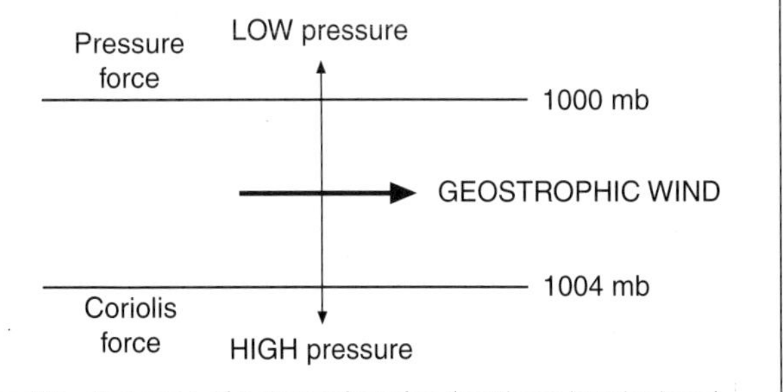

The geostrophic wind case of balanced motion (northern hemisphere)

Global circulation models

Low latitudes are warmer than higher latitudes. This energy deficit should result in a large convection cell:

- air rises over the equator due to strong heating
- air then moves polewards to sink
- this is then drawn back to the low pressure.

This simple model can be modified in a number of ways.

The Hadley Cell

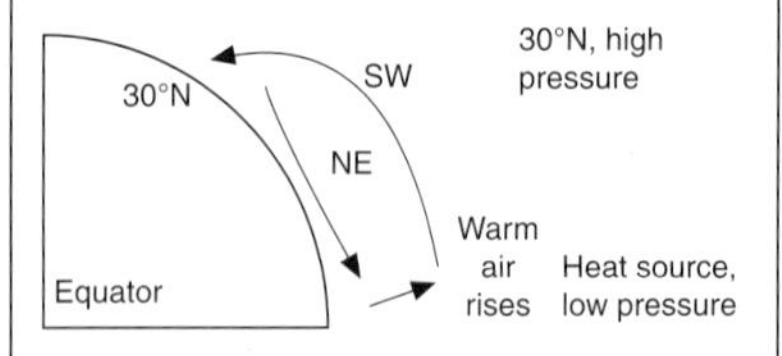

Poleward-flowing currents deflected to the right in the northern hemisphere to become south-westerlies.

The three-cell model

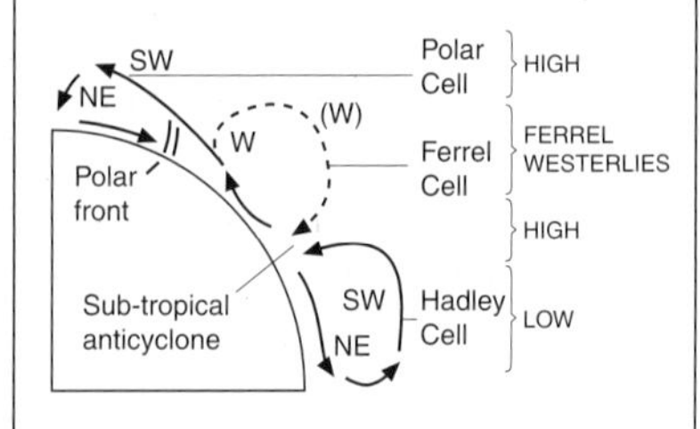

The Hadley Cell model was modified to include three cells in each hemisphere: the Hadley Cell, the Ferrel Cell, and the Polar Cell.

A GENERALISED MODEL OF GLOBAL CIRCULATION

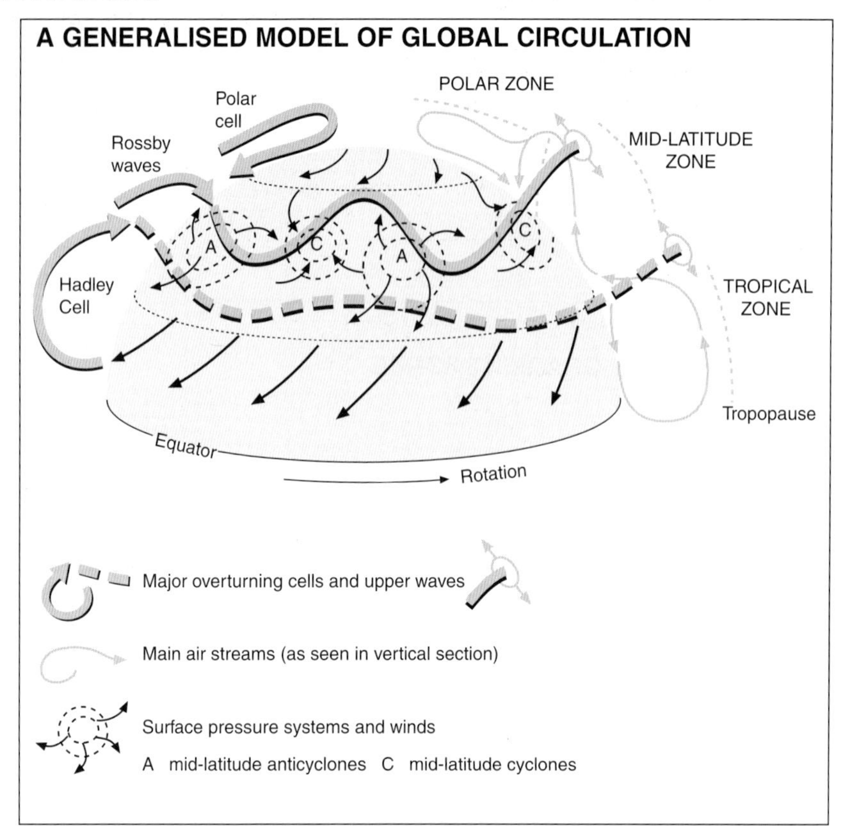

NEW CIRCULATION MODELS

New models change the relative importance of the three convection cells in each hemisphere. These changes are influenced by:

- **jet streams** - strong and regular winds which blow in the upper atmosphere about 10 km above the surface; they blow between the poles and tropics (100-300 km/h)
- two streams that occur in each hemisphere – one between 30° and 50°, the other between 20° and 30°. In the northern hemisphere the polar jet flows eastwards and the subtropical jet flows westwards.
- **Rossby waves** - 'meandering rivers of air' formed by westerly winds; three to six waves in each hemisphere; formed by major relief barriers, thermal differences; uneven land-sea-land interface

The diagram below shows a flow model relating summer convection, the easterly jet stream, and high pressure subsidence over northern Africa and the eastern North Atlantic. It shows how convection cells can be modified by upper air movements.

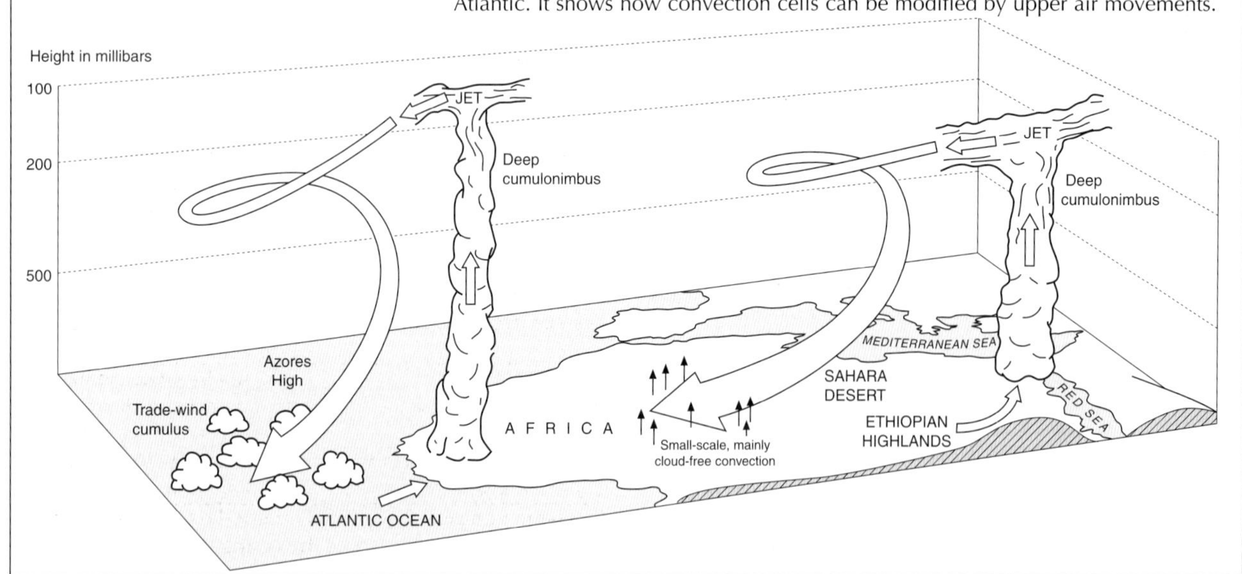

Clouds and fog

CLOUD TYPES Clouds are classified according to their appearance, form, and height. There are four main types of cloud:

A High clouds (6000 – 12 000 m)
- Cirrus - composed of small ice crystals (wispy).
- Cirrocumulus - ice crystals (globular or rippled).
- Cirrostratus - ice crystals (thin, white, almost transparent sheet which causes the sun and moon to have 'haloes').

B Middle clouds (2100 – 6000 m)
- Altocumulus - water droplets in layers or patches (globular or bumpy-looking).
- Altostratus - water droplets in sheets (grey or watery-looking clouds).

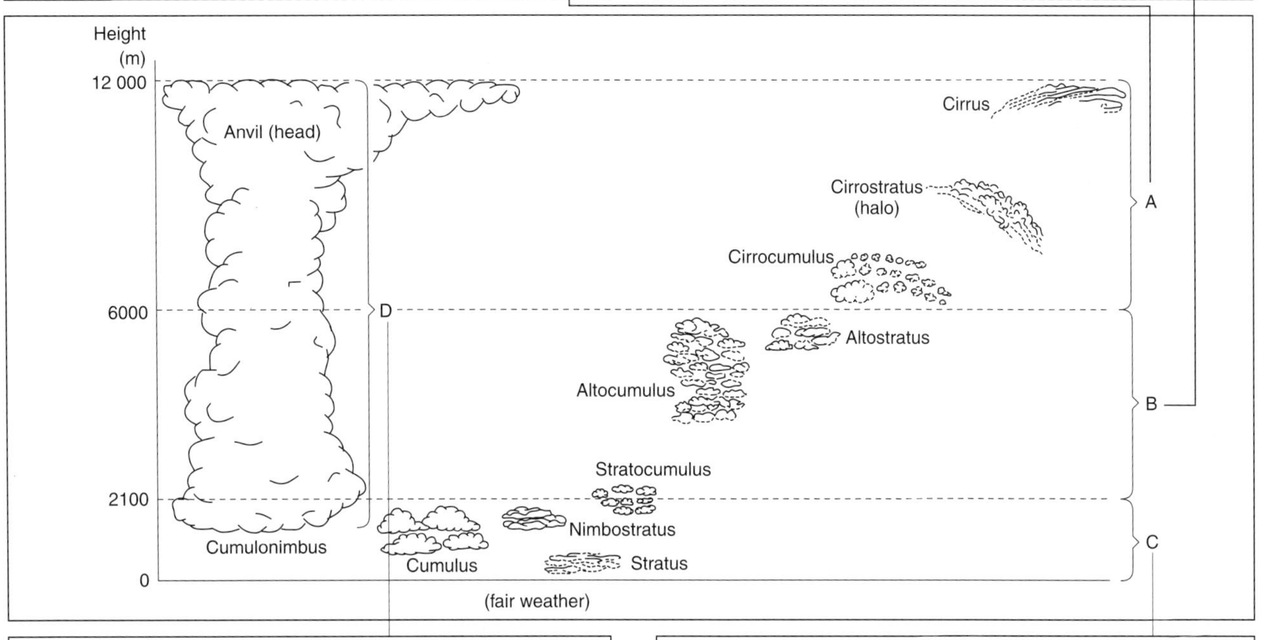

D Clouds with vertical development (1500 – 9000 m)
- Cumulus - round-topped and flat-based (whitish-grey globular mass, individual cloud units).
- Cumulonimbus - great vertical extent (white or black globular masses, rounded tops often spread out in the form of an anvil).

C Low clouds (below 2100 m)
- Stratocumulus - large globular masses (soft and grey with pronounced regular pattern).
- Nimbostratus - dark grey and rainy-looking (dense and shapeless, often leading to regular rain).
- Stratus - low, grey, and layered (brings overcast conditions).

CONDENSATION

Clouds can also be classified according to the mechanism of vertical motion which produces condensation. Four categories exist:

- gradual uplift of air associated with a low-pressure system
- thermal convection
- uplift by mechanical turbulence (forced convection)
- ascent over an orographic barrier.

FOG

Fog is cloud at ground level. Fog mostly occurs in high pressure (calm) conditions - as winds tend to mix and disperse fog. During the morning, when the ground is heated by the sun, the rising air 'lifts' the fog and it disappears.

Types of fog

Type of fog	Season	Areas affected	Mode of formation	Mode of dispersal
Radiation fog	October to March	Inland areas, especially low-lying, moist ground	Cooling due to radiation from the ground on clear nights when the wind is light	Dispersed by the sun's radiation or by increased wind
Advection fog *Over land*	Winter or spring	Often widespread inland	Cooling of warm air by passage over cold ground	Dispersed by a change in air mass or by gradual warming of the ground
Over sea and coastline	Spring and early summer	Sea and coasts; may penetrate a few miles	Cooling of warm air by passage over cold sea	Dispersed by a change in air mass and may be cleared on coast by the sun's heating
Frontal fog	All seasons	High ground	Lowering of the cloud base along the line of the front	Dispersed as the front moves and brings a change of air mass
Smoke fog (smog)	Winter	Near industrial areas and large conurbations	Similar to radiation fog	Dispersed by wind increase or by convection

Precipitation

FORMATION OF PRECIPITATION

There are four conditions needed for the formation of major precipitation (rain, snow, hail):

- air cooling
- condensation and cloud formation
- an accumulation of moisture
- growth of cloud droplets.

Two main groups of theories attempt to explain the rapid growth of raindrops:

(i) growth of ice crystals at the expense of water droplets

(ii) coalescence of small water droplets by the sweeping action of falling drops.

Collision theory: the growth of ice crystals

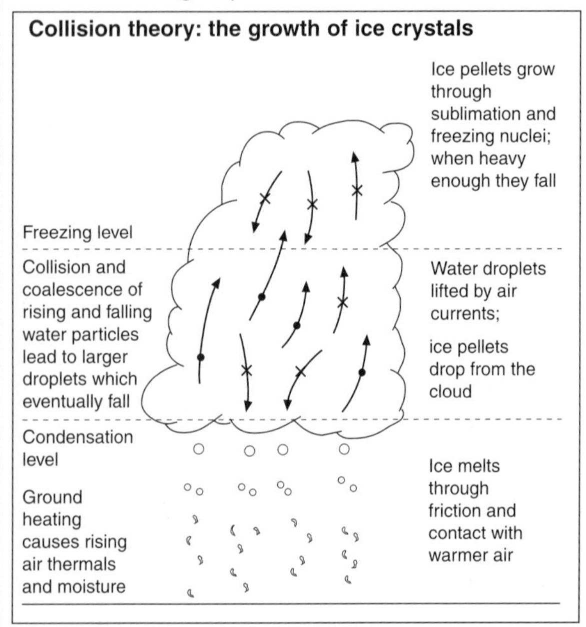

RAINFALL IN BRITAIN

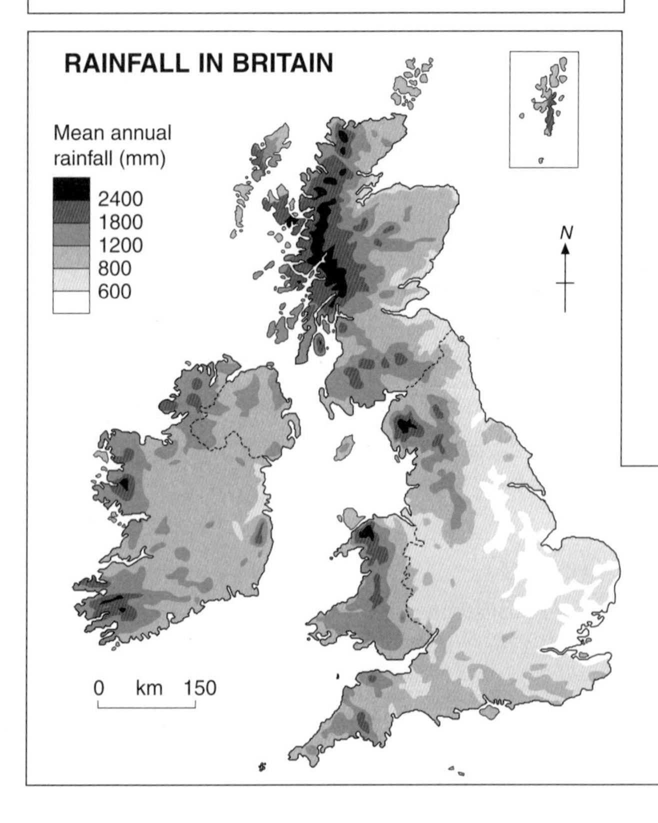

There is a great variation in the amount, type and seasonal nature of rainfall in Britain.

- The heaviest rainfall is over highland areas e.g. Wales, the Lake District, and the Scottish Highlands; much of this is relief rainfall.
- Rainfall is heavier in the west than the east; much of this is frontal rain driven across the country by prevailing winds.
- The South East experiences some convectional rainfall in Summer.

TYPES OF RAIN

1 Depression, cyclonic or frontal rain

This occurs in mid-latitudes. Warm air rises over cold air; it expands and cools; condensation occurs and cloud and rain form.

2 Convection rain

Humid tropics and the interior of continents in summer; due to strong, upward moving buoyant air; **thunderstorms.**

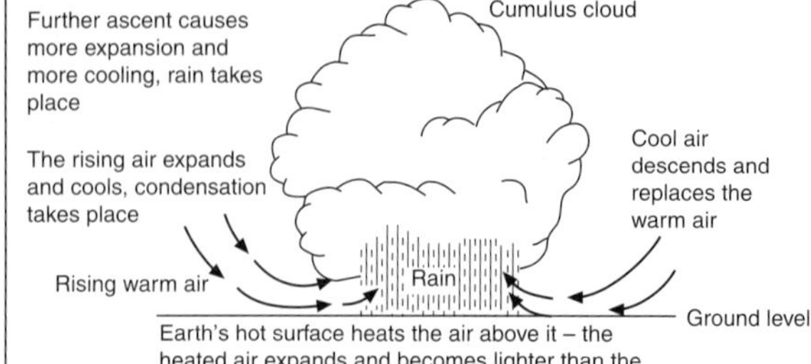

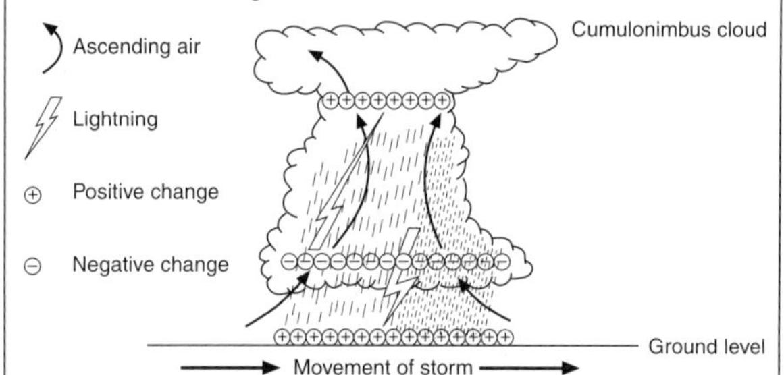

3 Orographic or relief rainfall

Upland areas that encourage warm, moist air to rise; **steady rain, drizzle.**

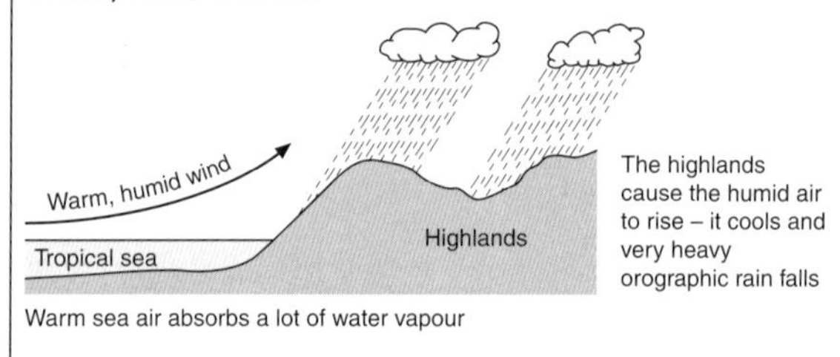

When moist air is forced to rise over a mountain range, clouds and rain, often heavy, occur

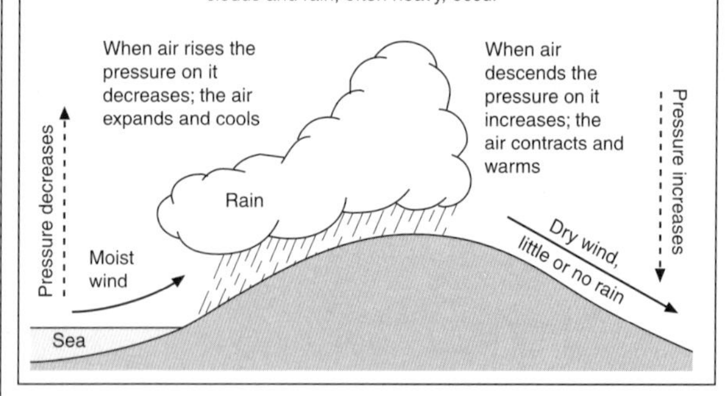

Precipitation patterns

THE ZONAL MODEL

- Abundance of rain in the equatorial zone; moderate to large amounts in the mid-latitudes; relatively low rainfall in the sub-tropics and particularly at the poles.
- Rainfall is abundant in the uplift areas of the convergence/convection zone of the equatorial trough (ITCZ) and in the polar frontal zones of the mid-latitudes.

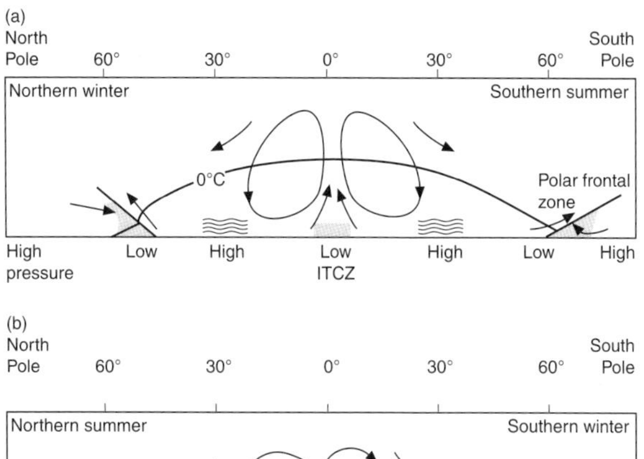

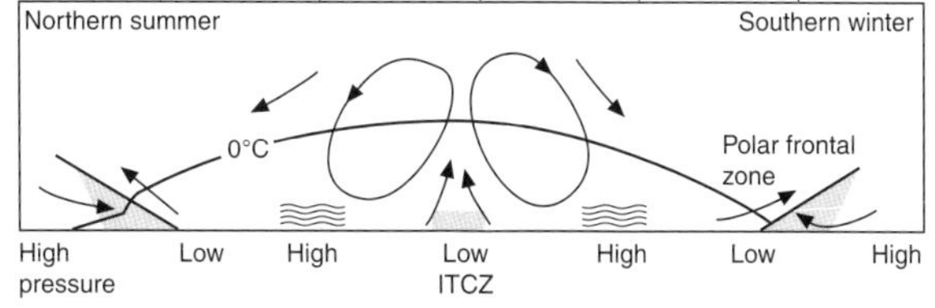

(c)

8	7	6	5	4	3	2	1	2	3	4	5	6	7	8
Sparse precipitation all seasons	Precipitation in all seasons	Winter rain, summer dryness	Slight winter rain	Dry all seasons	Slight summer rain	Summer rain, winter dryness	Rain in all seasons	Summer rain, winter dryness	Slight summer rain	Dry all seasons	Slight winter rain	Winter rain, summer dryness	Precipitation in all seasons	Sparse precipitation all seasons

Zone 1 Equatorial zone

- abundant rainfall throughout the year associated with the permanence of the ITCZ

Zone 2 Wet and dry tropics

- wet in summer, dry in winter
- summer rains due to ITCZ
- winter dry period due to sub-tropical anticyclones

Zone 3 Tropical semi-arid

- small amount of rain in summer, very dry in other seasons
- associated with equatorial margin of sub-tropical high

Zone 4 Arid zones

- permanent dryness
- year-long dominance of sub-tropical highs

Zone 5 Sub-tropical highs

- scanty winter rainfall
- dominated by sub-tropical highs, but occasional mid-latitude depressions bring rain

Zone 6 Mediterranean zone

- semi-arid/sub-humid region
- long dry summer, short wet winter
- sub-tropical highs in summer
- mid-latitude depressions in winter

Zone 7 Middle and high latitudes

- depressions and fronts
- precipitation in all seasons
- maximum precipitation in winter (cyclonic activity)

Zone 8 Polar regions

- low precipitation
- cold subsiding air
- some depressions in winter

MODIFICATIONS TO THE ZONAL MODEL

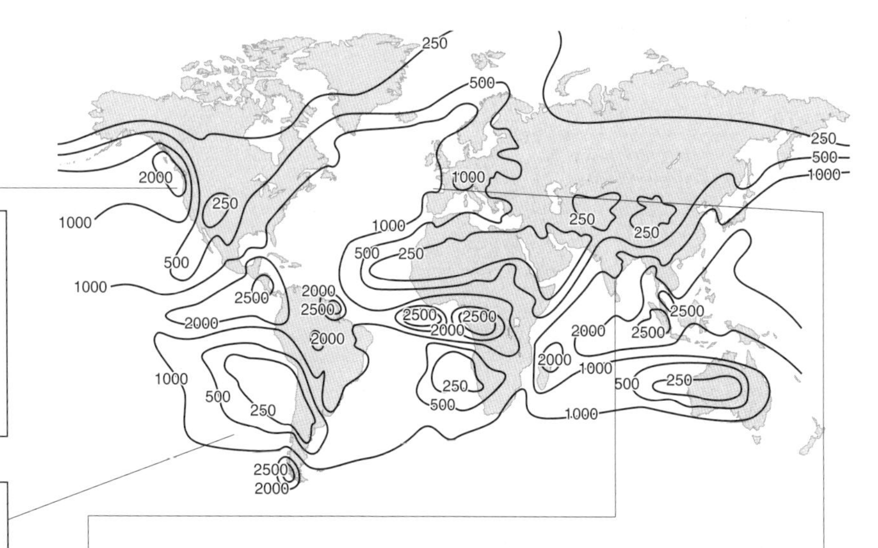

Orographic barriers

- zone of mid-latitude depressions; rise due to western Cordillera
- heavy orographic precipitation along windward sides, e.g. Olympic mountains receive 3750 mm, the leeward side only 750 mm

Ocean currents

- warm air passes over cold ocean current
- the result is temperature inversion and fog on the coast
- the stability prevents convection cell
- the result is the Atacama desert

Monsoon

- shift of ITCZ gives rise to intense but strongly seasonal rainfall, e.g. India

Mid-latitude cyclonic belt

- mid-latitude depressions moving west to east
- moist air due to evaporation over the warm Atlantic
- rain dropped on western sides of continents, e.g. the UK

Air masses

An air mass is an area of air which has similar properties of temperature and humidity.
Air masses develop over areas of similar geographical character, like the polar ice caps or hot deserts.

LAMB'S TYPES OF AIRFLOWS

Lamb has identified seven major categories of airflows (movement of pressure systems) that influence the British Isles. Each airflow type is associated with corresponding air masses.

Type	Characteristic	General weather
Westerly	Pm, Tm	Unsettled weather. Mild and stormy in winter, cool and cloudy in summer.
North-westerly	Pm, Am	Cool changeable conditions. Strong winds and showers.
Northerly	Am	Cold weather at all seasons, often associated with polar lows or troughs. Snow and sleet showers in winter, especially in the north and east.
Easterly	Ac, Pc	Cold in the winter half-year, sometimes very severe weather in the south and east with snow or sleet. Warm in summer with dry weather in the west. Occasionally thundery.
Southerly	Tm (Tc) - summer Tm or Pc in winter	Generally warm and thundery in summer. Mild, damp weather, in winter especially in the south-west.
Cyclonic	Pm, Tm	Rainy, unsettled conditions often accompanied by gales and thunderstorms.
Anticyclonic		Warm and dry in summer apart from occasional thunderstorms. Cold in winter with night frosts and fog, especially in autumn.

AVERAGE AIR MASS FREQUENCY AT KEW, LONDON

Lamb has identified seven major categories of airflows (movement of pressure systems) that influence the British Isles. Each airflow type is associated with corresponding air masses.

Polar maritime (Pm)

- an unstable air mass
- cool, showery weather, especially in winter
- gains moisture over the sea, leading to unstable air
- 'nice morning, bad day' cumuliform clouds
- in winter often about 8°C – the air mass is warmed by the north Atlantic drift
- in summer about 16° – cooled by the north Atlantic drift
- the warm north Atlantic drift encourages convection. Visibility is excellent – rising air disperses particles in the air.

Tropical maritime (Tm)

- a stable air mass brings warm air from low latitudes
- commonly forms warm sector of depressions
- in winter, air is unseasonably mild and damp –11°C
- stratus or stratocumulus cloud with drizzle; sea fog is common in coastal regions
- summer temperatures are 16 – 18°C
- visibility is poor; solid particles remain near the ground

Arctic maritime (Am)

- extreme weather
- good visibility

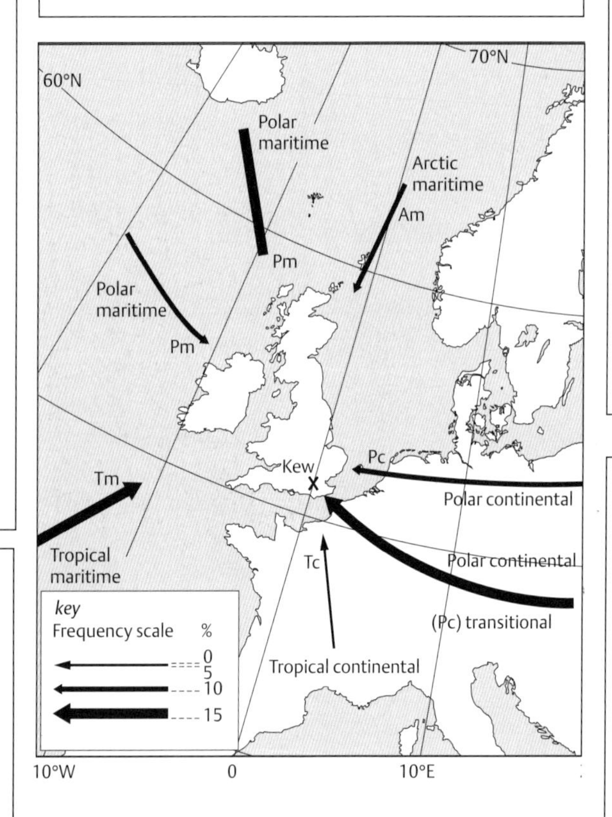

Polar continental (Pc)

- affects British Isles between December and February
- very cold, dry air from Siberia, around 0°C in January (absent in summer)
- picks up moisture from the sea and can lead to snow showers, especially on the east coast (as air becomes unstable over the North Sea)
- wind chill factor (dry air) exaggerates coldness

Tropical continental (Tc)

- warmest air entering the British Isles, –13°C in January, 25° in July
- can lead to heatwaves or late summer warming – the September 'Indian summer'
- can lead to instability and thunderstorms
- in winter it can bring fine, hazy, mild weather
- originates in North Africa
- moderate visibility; solid particles are not dispersed and air from the Sahara contains much dust.

Mid-latitude weather systems

Mid-latitude climates occur between the tropical and polar climates in both the northern and southern hemispheres.

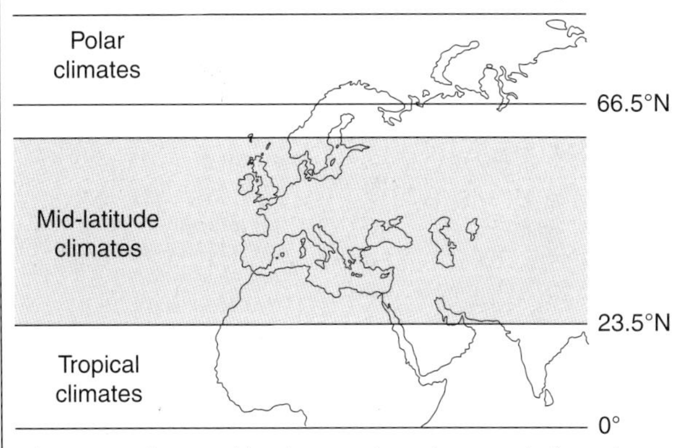

They are influenced by the meeting of warm air from the south and cold air from the north to form **fronts**. These give rise to *low pressure* systems (cyclones or depressions) and *high pressure* systems (anticyclones).

WHAT IS A FRONT?

Fronts occur when two air masses with different temperatures and densities meet. Cold air undercuts warm air.

Temperature characteristics of a frontal zone

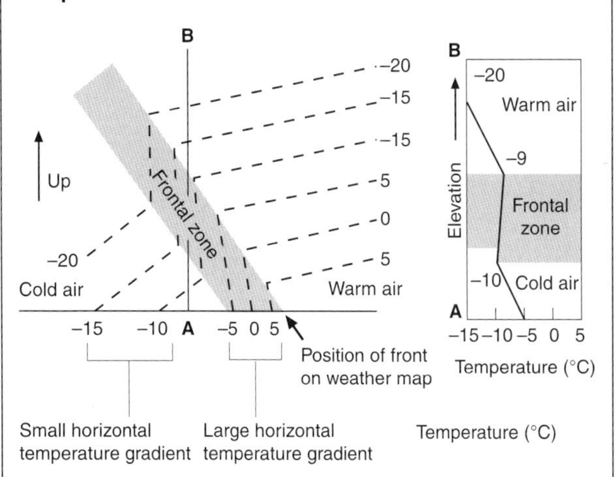

The interaction between air masses results in the formation of depressions and anticyclones.

WHAT IS A DEPRESSION?

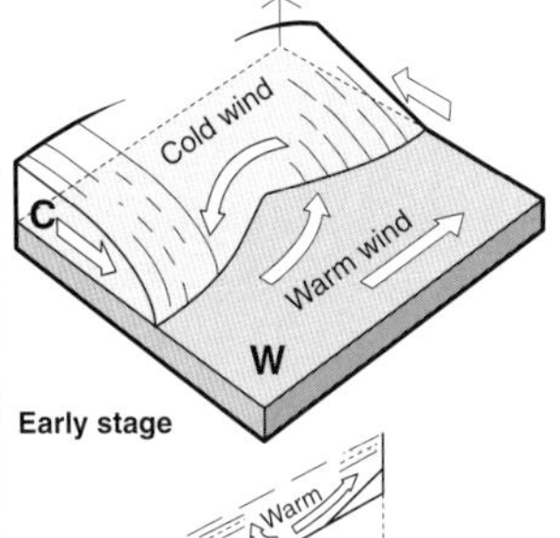

Polar front boundary between cold and warm air.

- An instability occurs on the polar front.

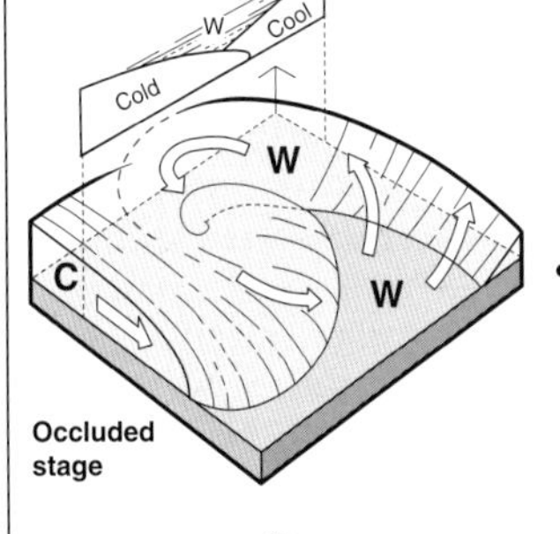

- Cold air pushes warm air north in the northern hemisphere. Warm air rises over cold air. The boundary between the two forms the *warm front*. The colder advancing air to the west is denser and undercuts the warmer air. The boundary forms the *cold front*.

- The cold front moves faster than the warm front and eventually catches it up and lifts it away from the ground, forming an *occluded* front.

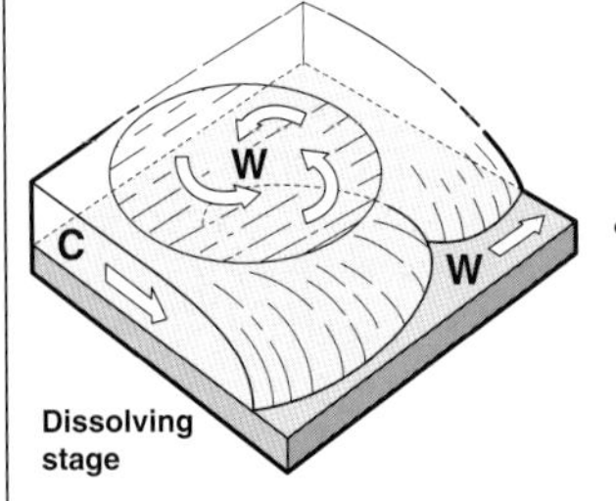

- After occlusion the depression dissolves.

WHAT IS AN ANTICYCLONE?

- Anticyclones develop in regions of descending air.
- Air moving towards the pole from equatorial regions descends, forming sub-tropical high pressures.
- The highest pressure is in the centre.
- Winds blow outwards from the centre in a clockwise direction in the northern hemisphere (due to the Coriolis effect).
- An anticyclone is a uniform air mass which gives fair weather, especially in summer.

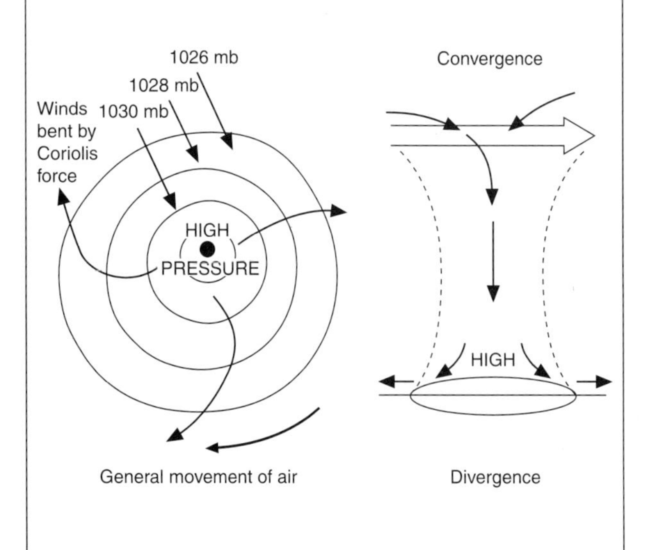

The passage of a depression

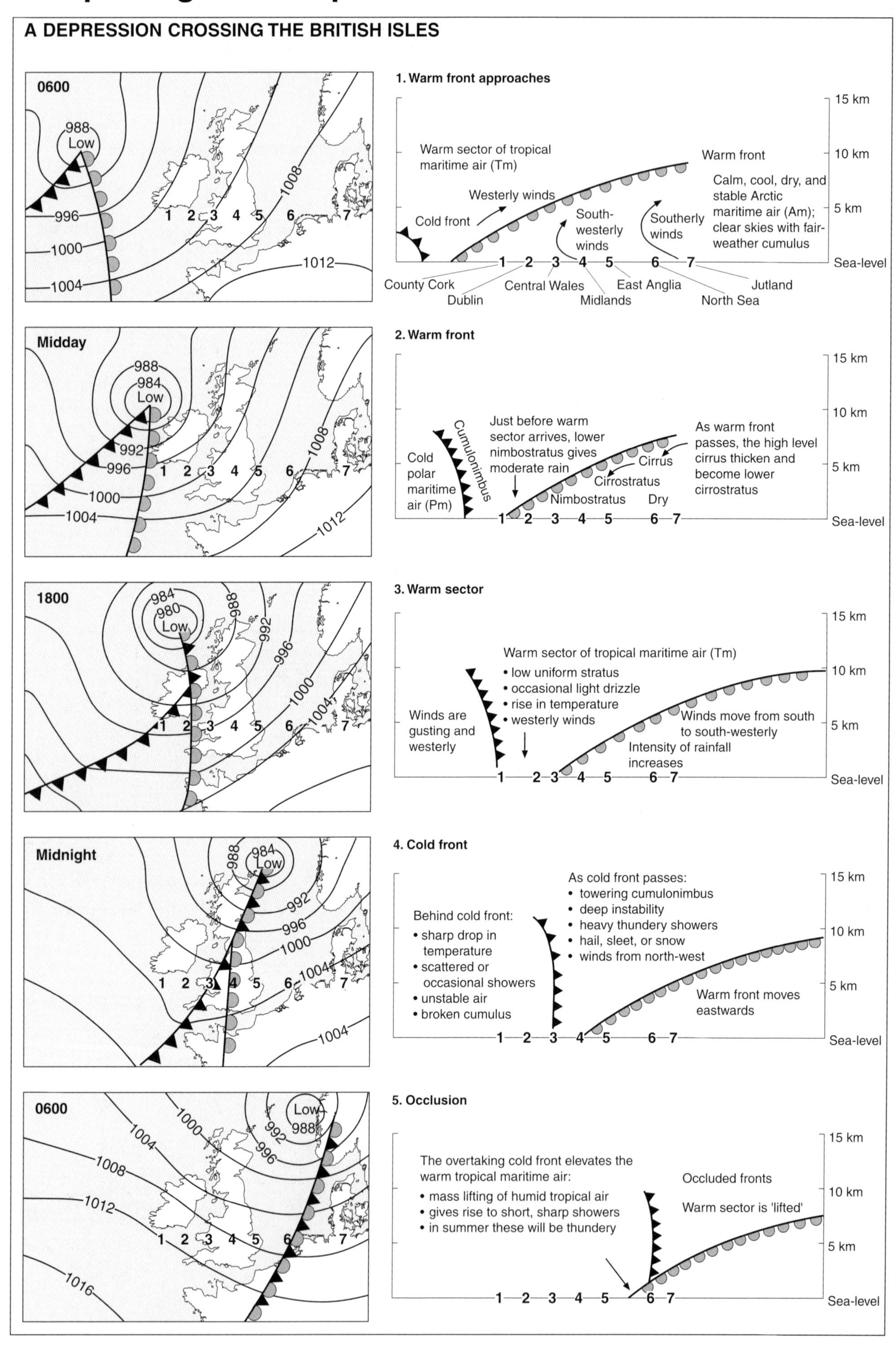

Weather associated with a depression

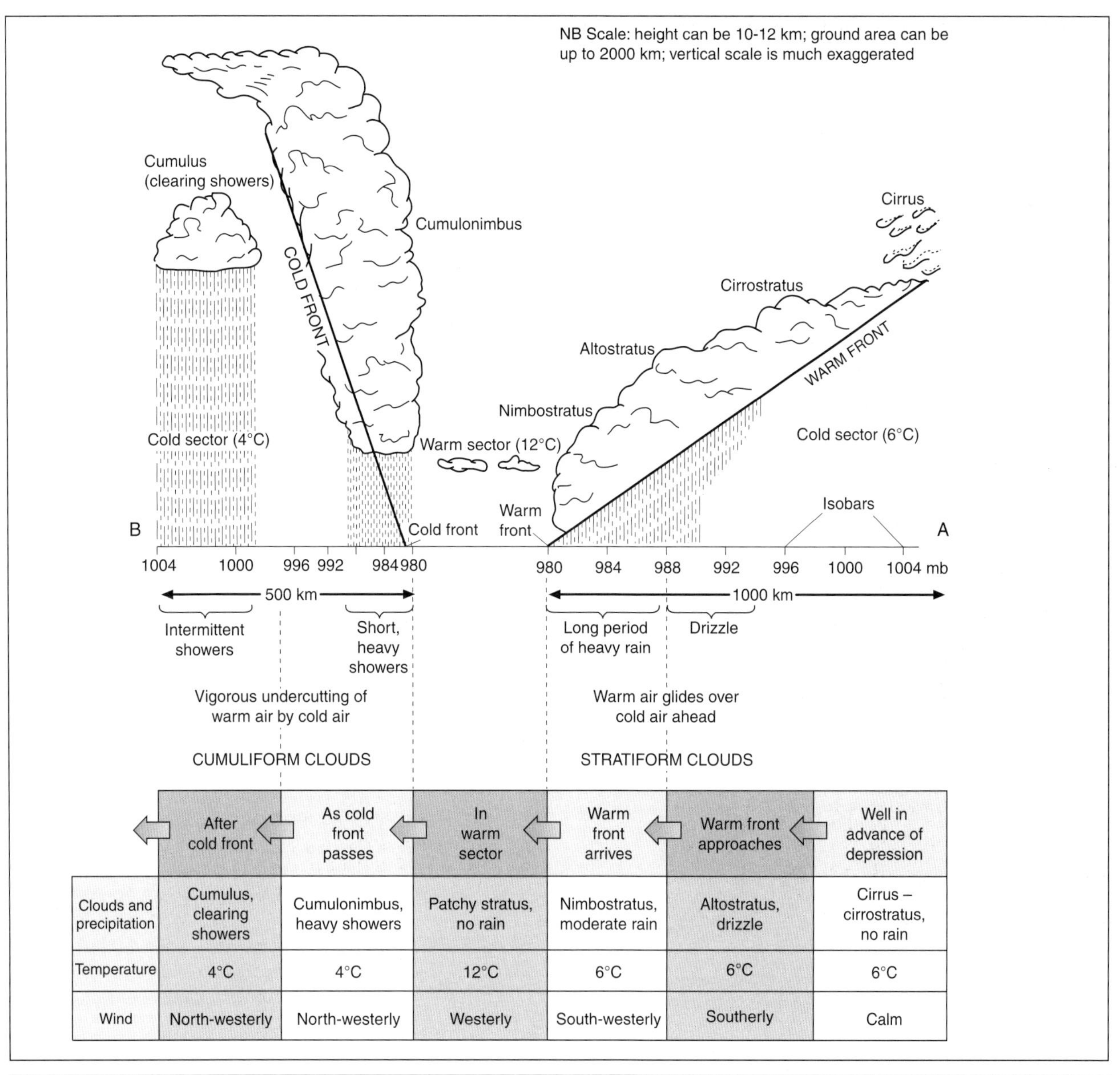

	After cold front	As cold front passes	In warm sector	Warm front arrives	Warm front approaches	Well in advance of depression
Clouds and precipitation	Cumulus, clearing showers	Cumulonimbus, heavy showers	Patchy stratus, no rain	Nimbostratus, moderate rain	Altostratus, drizzle	Cirrus – cirrostratus, no rain
Temperature	4°C	4°C	12°C	6°C	6°C	6°C
Wind	North-westerly	North-westerly	Westerly	South-westerly	Southerly	Calm

ANA AND KATA FRONTS

No two fronts are the same. In general, the greater the difference between the temperatures of the air masses involved, the greater the frontal activity. Two main types can be distinguished:

- Anafronts occur where there is a strong contrast in the air masses involved, and uplift is vigorous.
- Katafronts develop when there is little difference between the fronts involved, and uplift is limited.

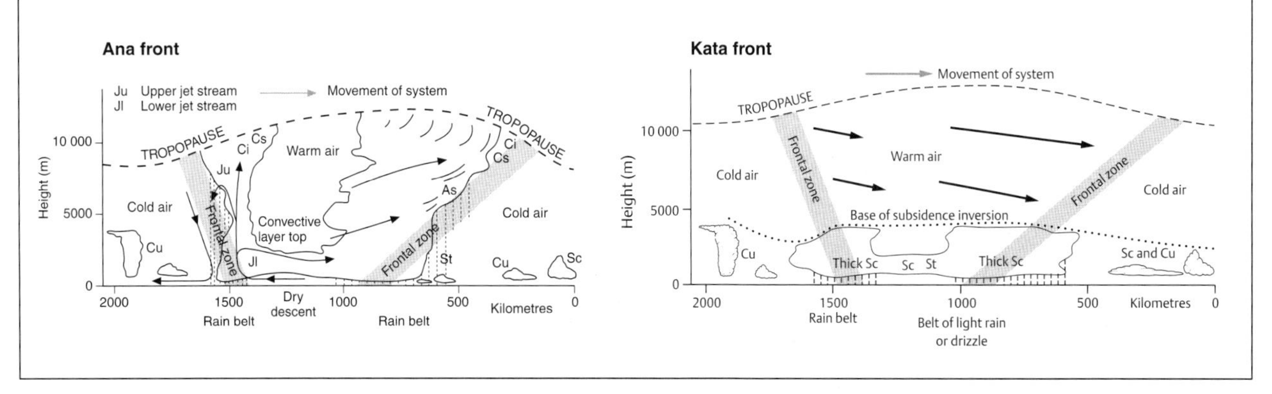

Hurricanes and tornadoes

CHARACTERISTICS OF A HURRICANE

Hurricanes are intense hazards that bring heavy rainfall, strong winds, high waves, and cause other hazards such as flooding and mudslides. Their path is erratic hence it is not always possible to give more than 12 hours notice. This is insufficient for proper evacuation measures. Hurricanes typically:

- develop as intense low pressure systems over tropical oceans.
- have a calm central area known as the eye, around which winds spiral rapidly
- have a diameter of up to 800 km, although the very strong winds that cause the most damage are found in a narrower belt up to 300 km wide.

In a mature hurricane pressure may fall to as low as 880-970 millibars. This, and the strong contrast in pressure between the eye and outer part of the hurricane lead to strong gale force winds.

Factors affecting the development of hurricanes

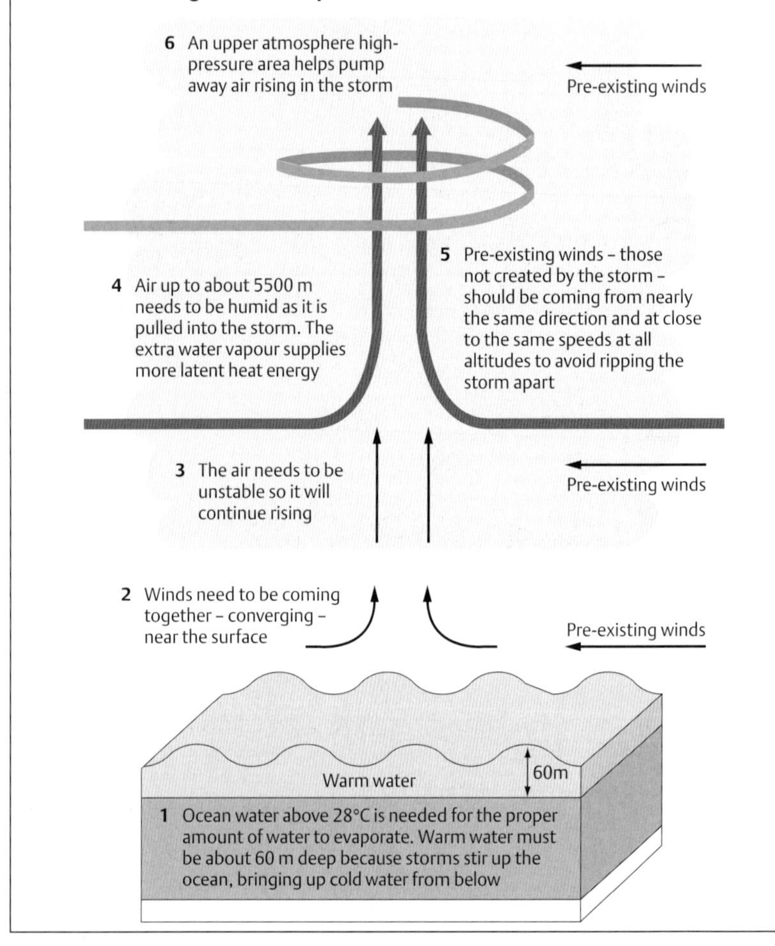

TORNADOES

Tornadoes are the most violent storms on the Earth. In the USA a number of factors need to occur simultaneously for tornadoes to form:

- a northerly flow of marine tropical air from the Gulf of Mexico that is humid and has temperatures at the ground of over 24°C
- a cold, dry air mass moving down from Canada or out from the Rocky Mountains at speeds greater than 80 kmph;
- jet-stream winds racing east at speeds greater than 380 kmph.

These three air masses, all moving in different directions, set up shearing conditions, imparting spin to a thunder-cloud.

The core of the tornado is usually less than 1 km wide and acts like a giant vacuum cleaner sucking up air and objects. When a tornado passes over a building, there can be a 10-20% drop in air pressure, causing tightly closed buildings to explode. Some of the exploded debris is sucked up with the updrafting air, which may be rising at over 160 kmph.

THE LOCATION OF THE MAIN HURRICANE TRACKS

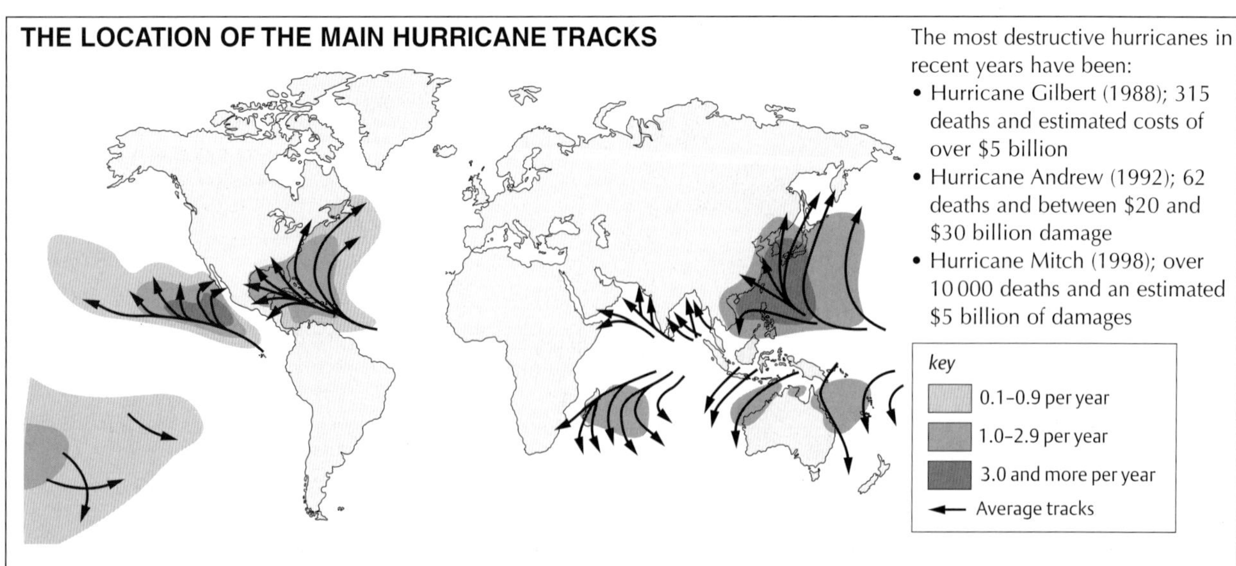

The most destructive hurricanes in recent years have been:

- Hurricane Gilbert (1988); 315 deaths and estimated costs of over $5 billion
- Hurricane Andrew (1992); 62 deaths and between $20 and $30 billion damage
- Hurricane Mitch (1998); over 10 000 deaths and an estimated $5 billion of damages

Monsoon

WHAT IS THE MONSOON?

The monsoon is the reversal of pressure and winds which gives rise to a marked seasonality of rainfall over south and south-east Asia.

This can be seen clearly in India and Bangladesh.

A number of influences have been suggested:

- the effect of the Himalayas on the ITCZ
- reduced CO_2, due to the Tibetan Plateau, leads to cooling in the interior
- differential heating and cooling of land compared to the adjacent seas.

WINTER: NORTH-EAST MONSOON

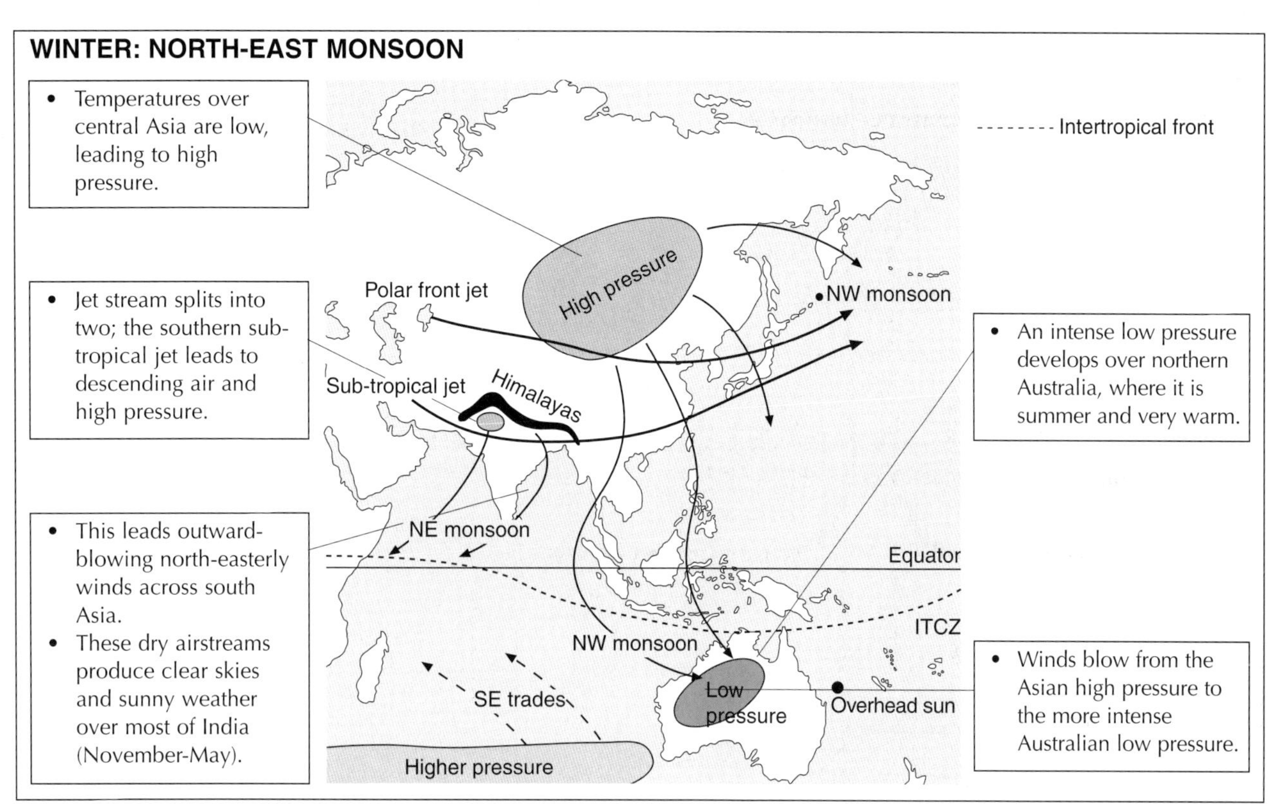

SUMMER: SOUTH-WEST MONSOON

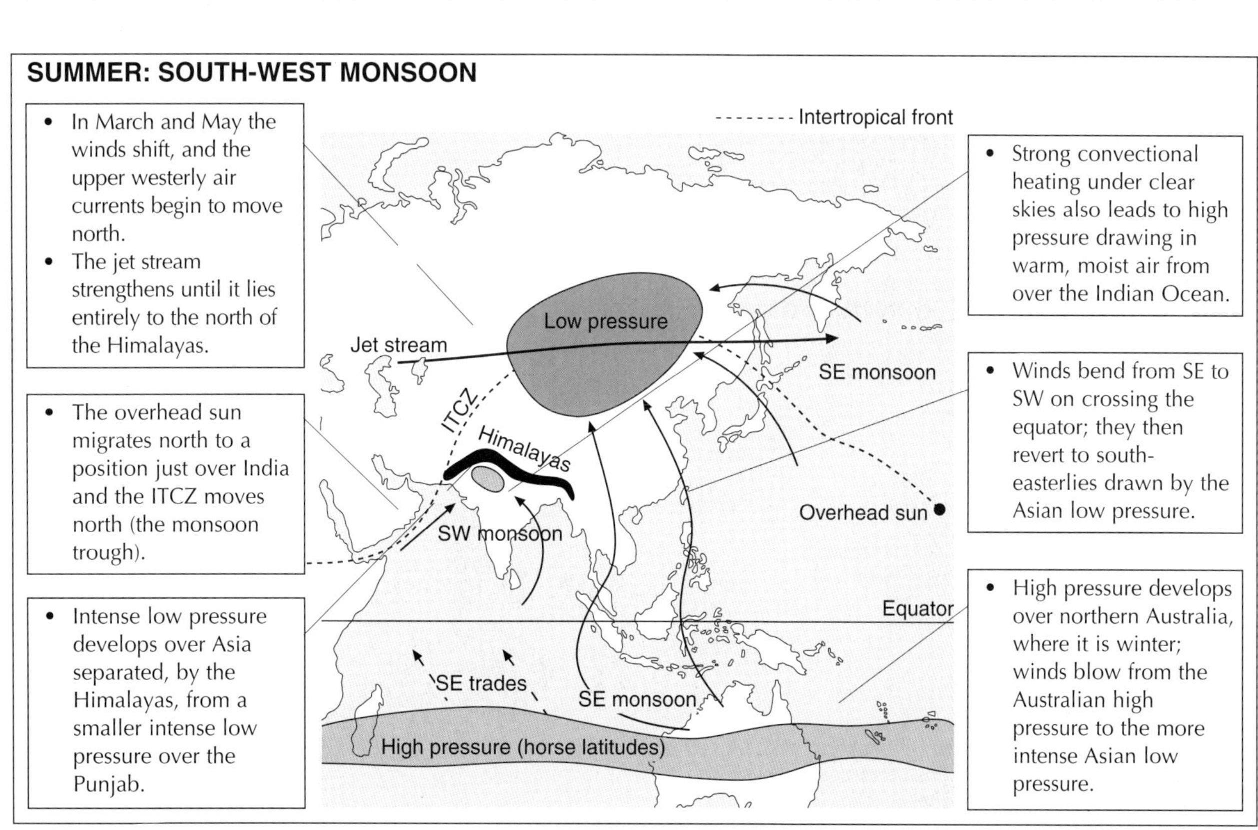

Local winds

MOUNTAIN AND VALLEY WINDS

Anabatic winds (up-valley) form during warm afternoons due to greater heating of valley sides compared with the valley floor.

Katabatic winds (down-valley) reverse the process as the cold, denser air at higher elevations drains into depressions and valleys.

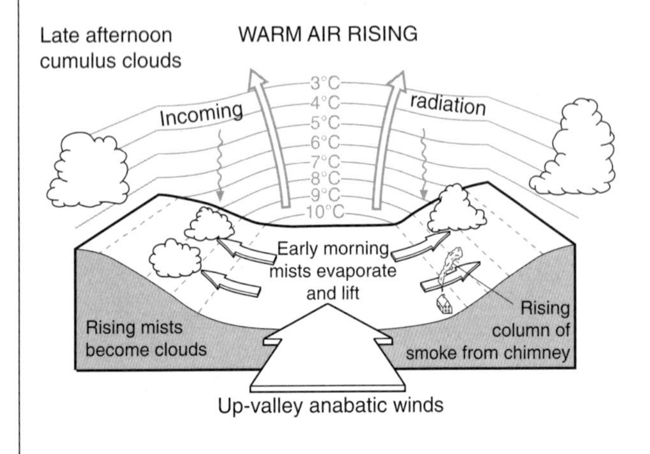

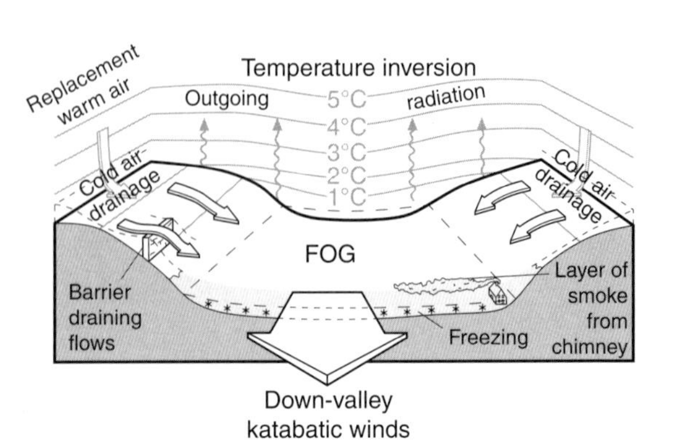

LAND AND SEA BREEZES

A **daytime sea breeze** is an *onshore wind* which occurs because land temperatures rise more rapidly during the day.

A **nocturnal land breeze** is an *offshore wind* generated by the sea cooling less rapidly than the land at night (also downslope winds blowing off the land).

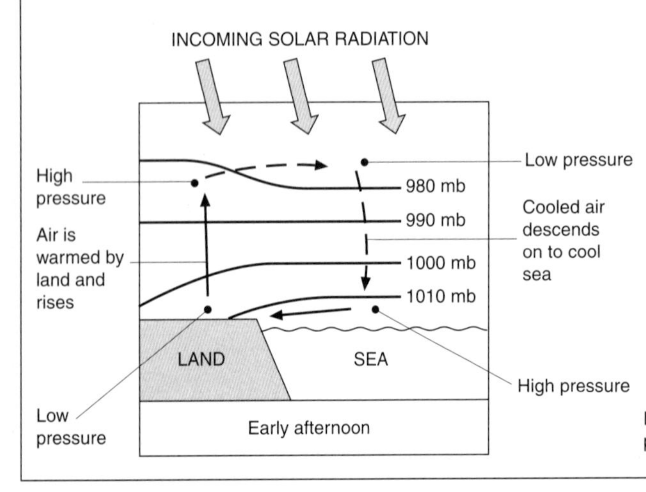

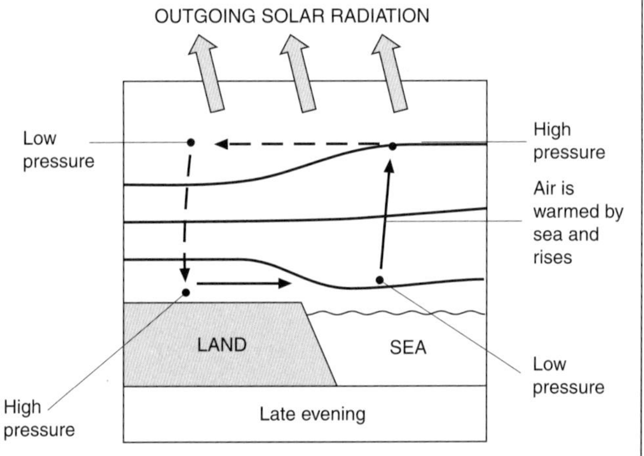

FOHN WINDS

Flows of air over mountains can create a **fohn effect**. This can result in a rapid increase in temperature and falls in relative humidity. The winds melt snow, leading to avalanches and flooding. The low humidity fohn wind can also dry forest areas, causing fires.

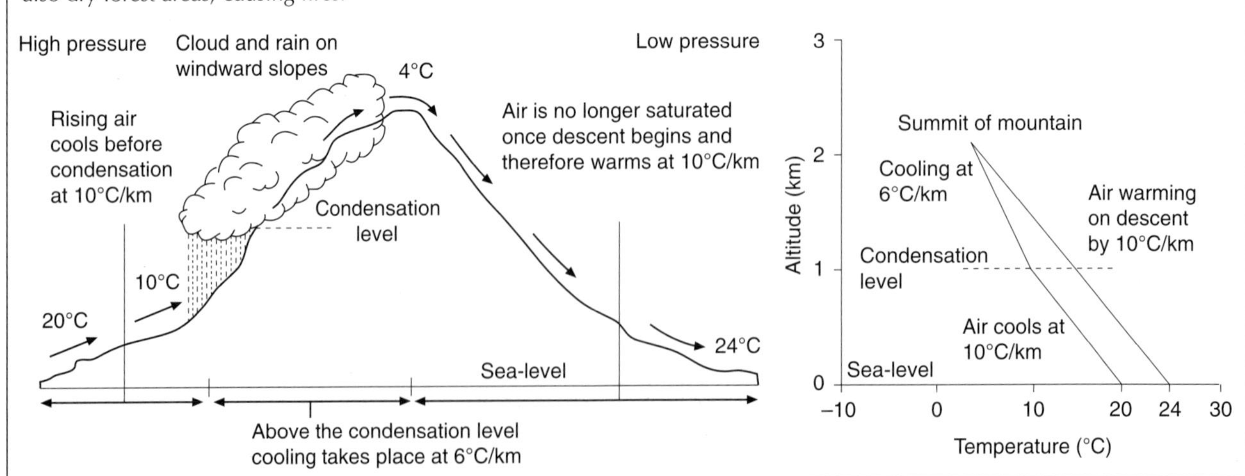

Microclimates

Structure of the air above the urban area

Greater amounts of dust mean increasing concentrations of hygroscopic particles. Less water vapour, but more CO_2 and higher proportions of noxious fumes owing to combustion of imported fuels. Discharge of waste gases by industry.

Structure of the urban surface

More heat-retaining materials with lower albedo and better radiation absorbing properties. Rougher surfaces with a great variety of perpendicular slopes facing different aspects. Tall buildings can be very exposed, and the deep streets are sheltered and shaded.

The effect of city morphology on radiation received at the surface

(a) Isolated buildings

Isolated building

Sunny side heated by insolation, reflected insolation, radiation, and conduction

Heat stored and re-radiated

Shaded side

(b) Low buildings

Street collects reflected radiation

(c) High buildings

Very little radiation reaches street level. Radiation reflected off lower walls after reflection from near tops of buildings

The structure of the urban climatic dome

Prevailing wind

Urban boundary layer

Urban plume develops downwind

Urban canopy layer below roof level

Rural boundary layer

RURAL | SUBURBAN | URBAN | SUBURBAN | RURAL

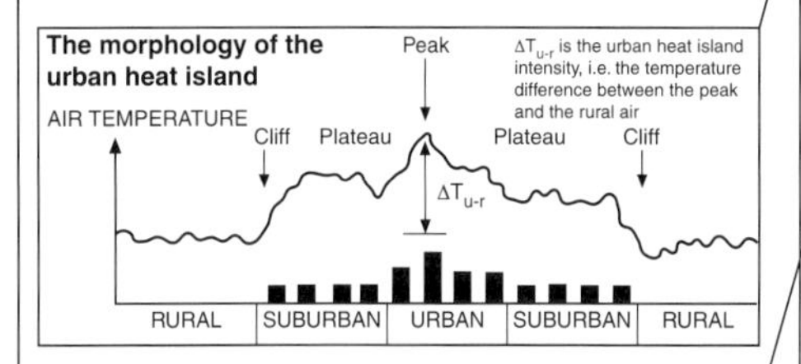

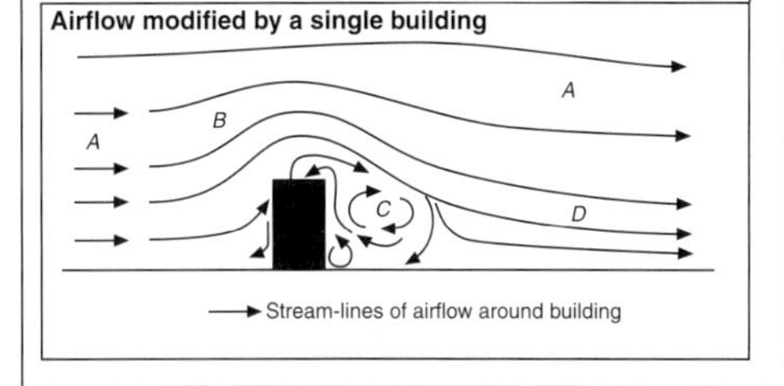

Resultant processes

1. Radiation and sunshine

Greater scattering of shorter-wave radiation by dust, but much higher absorption of longer waves owing to surfaces and CO_2. Hence more diffuse sky radiation with considerable local contrasts owing to variable screening by tall buildings in shaded narrow streets. Reduced visibility arising from industrial haze.

2. Clouds and fogs

Higher incidence of thicker cloud covers in summer and radiation fogs or smogs in winter because of increased convection and air pollution respectively. Concentrations of hygroscopic particles accelerate the onset of condensation (see 5 below). Day temperatures are, on average, 0.6°C warmer.

3. Temperatures

Stronger heat energy retention and release, including fuel combustion, gives significant temperature increases from suburbs into the centre of built-up areas creating heat 'islands'. These can be up to 8°C warmer during winter nights. Heating from below increases air mass instability overhead. Big local contrasts between sunny and shaded surfaces, especially in the spring. Snow in rural areas increases its albedo.

4. Pressure and winds

Severe gusting and turbulence around tall buildings causing strong local pressure gradients from windward to leeward walls. Deep narrow streets much calmer unless aligned with prevailing winds to funnel flows along them - the 'canyon effect'.

5. Humidity

Decreases in relative humidity occur in inner cities owing to lack of available moisture and higher temperatures there. Partly countered in very cold stable conditions by early onset of condensation in low-lying districts and industrial zones (see 2 above).

6. Precipitation

Perceptibly more intense storms, particularly during hot summer evenings and nights owing to greater instability and stronger convection above built-up areas. Probably higher incidence of thunder in appropriate locations. Less snowfall and briefer covers even when uncleared.

THE URBAN HEAT ISLAND

Urban areas are generally warmer than those of the surrounding countryside. Temperatures are on average 2-4°C higher in urban areas. This creates an **urban heat island**. It can be explained by heat and pollution release.

- Lower wind speeds due to the height of buildings and urban surface roughness.
- Urban pollution and photochemical smog can trap outgoing radiant energy.
- Burning of fossil fuels for domestic and commercial use can exceed energy inputs from the sun.
- Buildings have a higher capacity to retain and conduct heat and a lower albedo.
- Reduction in thermal energy required for evaporation and evapotranspiration due to the surface character, rapid drainage, and generally lower wind speeds.
- Reduction of heat diffusion due to changes in airflow patterns as the result of urban surface roughness.

Lapse rates

A lapse rate is the rate of temperature decrease with altitude. The environmental lapse rate (ELR) is, on average, 6.5°C per 1000 m. However, lapse rates vary with local conditions such as:

- height – lapse rates are lower near ground level
- humidity – they are lower on wet (humid) air
- season – lapse rates are lower in winter
- type of air mass
- time – there are differences between night and day, depending upon the balance of radiation and re-radiation.

Adiabatic lapse rates
Adiabatic means internal change. Air loses or gains heat as a result of rising (and expanding) or sinking (and contracting). No heat is gained from external sources.

Dry adiabatic lapse rate (DALR) - this is the rate at which unsaturated (dry) air cools - 9.8°C/1000 m, usually rounded up to 10°C/km.

Saturated adiabatic lapse rate (SALR) - this is the rate at which saturated air cools. Saturated air releases heat through condensation (cloud formation) - this offsets the cooling process. The SALR cools at a rate of between 4°C/km for warm air, and 9°C/km for cold air. (Warm air contains more moisture - therefore more heat is released during condensation - this reduces the impact of cooling.)

DEFINITIONS

Dew point is the temperature at which air is saturated.
Condensation level is the altitude at which dew point is reached.
Saturation means that the air is holding the maximum amount of moisture. At any given temperature, air can hold a certain amount of water, e.g. air at 30°C can hold just over 30 g of moisture; at 10°C it can hold 9.4 g. This is known as saturated vapour pressure.
Absolute humidity is the amount of moisture g/m^3 held in the air.
Relative humidity is the amount contained compared to the saturated vapour pressure

$$\text{Relative humidity} = \frac{\text{Absolute humidity}}{\text{Saturated vapour pressure}} \times 100\%$$

e.g. air of 30°C containing 15.02 g of moisture has a relative humidity of $\frac{15.02}{30.04} \times 100\% = 50\%$

When dew point and condensation level are reached, the relative humidity is 100%, and air cools at the SALR.

STABILITY

Stability (stable conditions) occurs when a rising parcel of air cools more quickly than air surrounding it (the ELR). As it is colder than the ELR it is denser, and therefore sinks back to position. Stable air is associated with calm (high pressure) dry conditions.

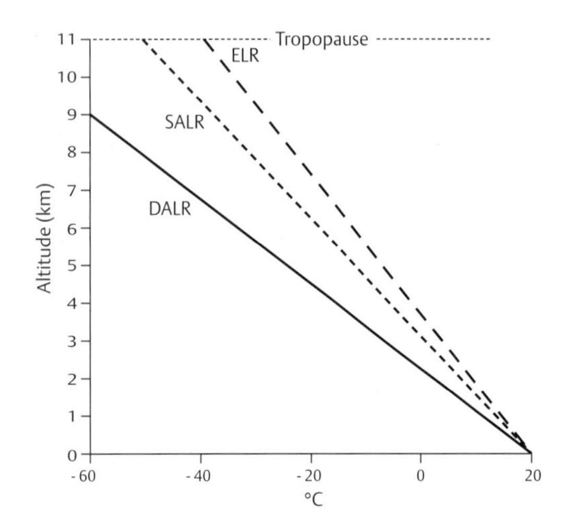

CONDITIONAL INSTABILITY

Conditional instability occurs when the ELR is lower than the DALR but higher than the SALR. Rising pockets of air, cooling at the DALR become cooler than surrounding air, and should sink down to ground. However, they may be forced to rise - for example over a hill. This may cause the air to cool to its dew point. Once saturation occurs, condensation takes place. Thus the air begins to cool at the SALR. If it becomes warmer than the surrounding air, it will continue to rise. The air is unstable on the condition that dew point is reached, and it cools at the SALR.

INSTABILITY

Instability is common, especially on hot days. Localised heating raises the temperature of air above it. The air begins to rise - as it is warmer and less dense than the surrounding air it continues to rise. If it rises to sufficient height, condensation cloud development and rain may occur. A good example of local instability is the heated air above school - or college - kitchens.

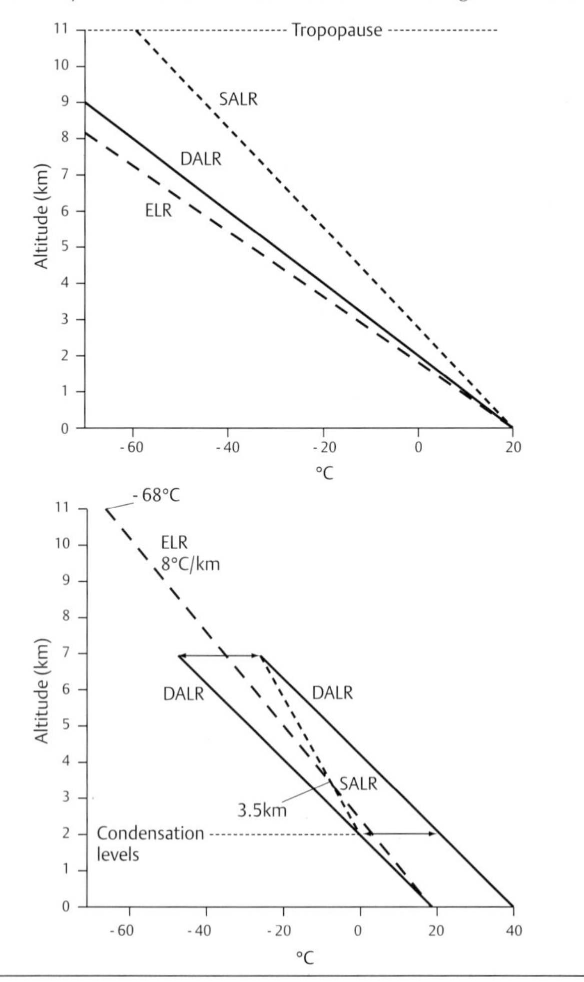

The greenhouse effect

The **greenhouse effect** is the process by which certain gases absorb outgoing long-wave radiation from the earth, and return some of it back to earth. In all, greenhouse gases such as carbon dioxide, methane, CFCs, nitrous oxides, and water vapour, raise the earth's temperatures by about 33°C (compared to the moon which has no atmosphere).

Greenhouse gases vary in their abundance and contribution to global warming.

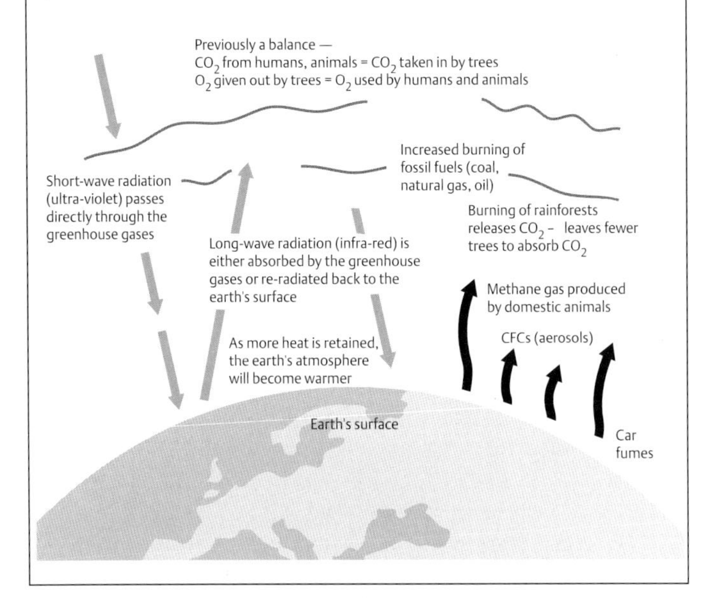

GLOBAL WARMING

The concern about global warming is the build-up of greenhouse gases. CO_2 levels have risen from about 315 ppm in 1950 to 355 ppm and are expected to reach 600 ppm by 2050. The increase is due to human activities - burning fossil fuel and deforestation. Deforestation of the tropical rainforest is a double blow - not only does it increase atmospheric CO_2 levels, it removes the trees which convert CO_2 into oxygen.

Methane is the second largest contributor to global warming, and is increasing at a rate of 1% per annum. It is estimated that cattle convert up to 10% of the food they eat into methane, and emit 100 million tonnes of methane into the atmosphere each. Natural wetland and paddy fields are another important source. Chlorofluorocarbons (CFCs) are synthetic chemicals that destroy ozone, as well as absorb long-wave radiation. CFCs are increasing at a rate of 6% per annum, and up to 10 000 times more efficient at trapping heat than CO_2.

The effects of global warming are mixed:

- sea levels will rise, causing flooding in low lying areas such as the Netherlands, Egypt, and Bangladesh
- an increase in storm activity (owing to more atmospheric energy)
- changes in agricultural patterns - e.g. a decline in USA's grain belt, but an increase in Canada's growing season
- reduced rainfall over the USA and southern Europe
- extinction of wildlife
- a considerable change to Britain's natural, social, and economic environment.

PROPERTIES OF GREENHOUSE GASES

	Average atmospheric concentration (ppmv)	Rate of change % per annum	Direct global warming potential (GWP)	Lifetime (years)	Type of indirect effect
Carbon dioxide	355	0.5	1	120	none
Methane	1.72	0.6—0.75	11	10.5	positive
Nitrous Oxide	0.31	0.2—0.3	270	132	uncertain
CFC 11	0.000 255	4	3 400	55	negative
CFC 12	0.000 453	4	7 100	116	negative
CO				months	positive
NOx					uncertain

EFFECTS OF THE GREENHOUSE EFFECT

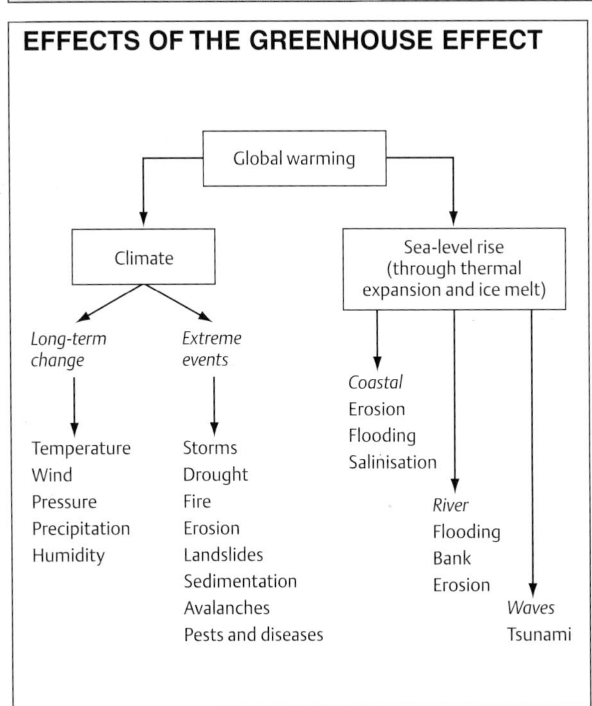

REDUCING THE GREENHOUSE EFFECT

The energy sector

- introduce a carbon tax on electricity generation
- introduce a higher carbon dioxide tax and retain the existing energy tax on non-energy-intensive industry
- introduce an energy tax on combined heat and power
- introduce measures to save electricity (including standards for domestic appliances)

The transport sector

- new rules on company cars and tax
- expand public transport systems
- set carbon divide emission limits on light vehicles
- further develop environmental classification systems for vehicles and fuels
- reduce average speeds on roads
- subject all transport plans and infrastructure investments to environmental impact assessments
- experiment with the introduction of electric vehicles

Other greenhouse gases

- reduce agricultural use of nitrogen fertilisers
- expand methane extraction from waste tips
- reduce emissions of fluorocarbons from aluminium smelters, and ban their use as chemicals; reduce use of hydrofluorocarbons

The ozone hole

Ozone (O_3) is continuously being created and destroyed in the atmosphere. This is a natural process. Oxygen (O_2) is broken down into individual atoms by ultraviolet radiation. Some of these atoms combine with oxygen to form ozone. Ozone is also broken down by ultraviolet radiation - so there is a natural cycle of growth and decay.

OZONE LEVELS AND CFCS

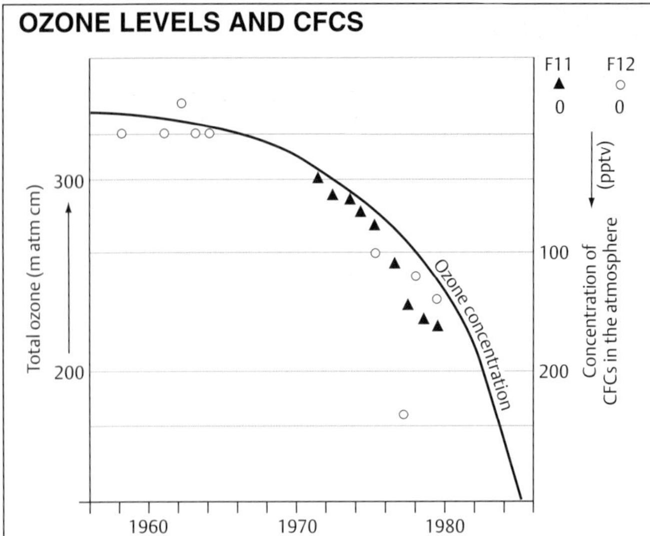

Increasing use of chlorofluorocarbons (CFCs) - synthetic gases that can last for decades - is destroying the ozone layer. The link is very strong. When they are broken down in the atmosphere they release chlorine, which destroys ozone. As this destruction of ozone is faster than its natural regeneration - the amount of ozone is decreasing.

DESTRUCTION OF THE OZONE LAYER

Ozone is important as it filters out harmful ultraviolet radiation. Increased levels of ultraviolet radiation are associated with raised levels of skin cancer and reduced crop yields. For every 1% decrease in ozone, skin cancer will rise by 5%.

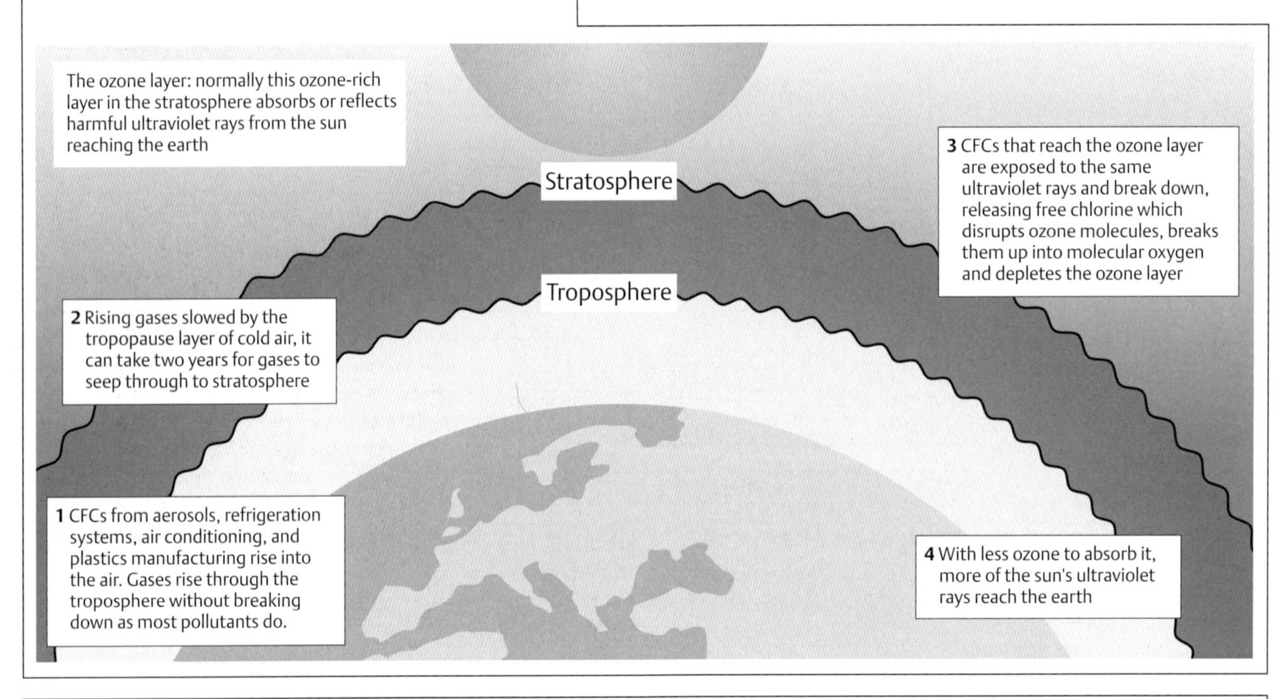

THE OZONE HOLE

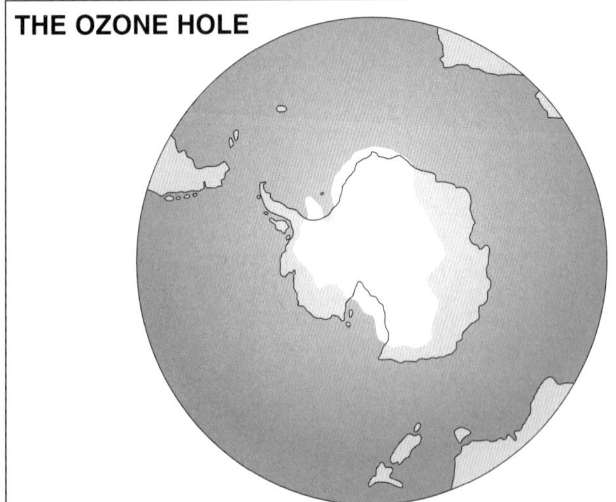

The ozone 'hole' is a large area over Antarctica (and to a lesser extent over the Arctic) where there are very low levels of O_3 - less than half normal levels.

The effects of reduced O_3 include:

- increased risk of skin cancer
- more eye diseases such as cataracts
- crop yields to decline by 25% (if O_3 declines by 25%)
- decline of oceanic plankton - disrupting marine ecosystems.

The Montreal Protocol (1987) requires cuts of CFCs by 1999. However, many developing countries are wary of signing - if it will affect their attempts to develop. It means that alternatives need to be found to CFCs, until then, reduction of leaks, improvements in handling, and elimination of unnecessary uses are needed.

El Niño

El Niño - the 'Christ Child' - is a warming of the eastern Pacific that occurs at intervals between two and ten years, and lasts for up to two years. Originally, El Niño referred to a warm current that appeared off the coast of Peru, but it is now realised that this current is part of a much larger system.

NORMAL CONDITIONS IN THE PACIFIC OCEAN

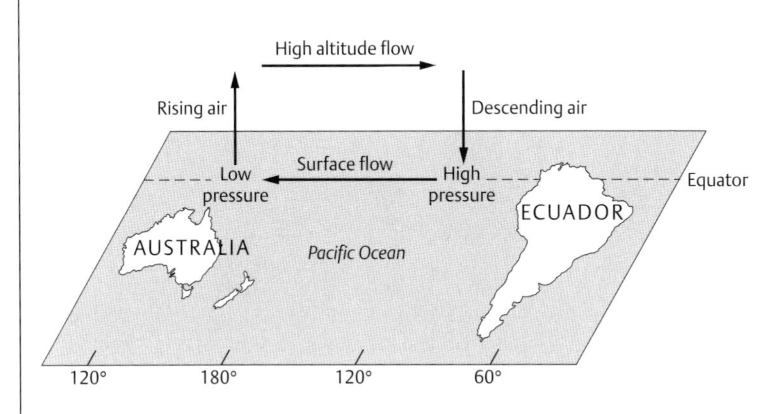

The Walker circulation is the east-west circulation that occurs in low latitudes. Near South America winds blow off-shore, causing upwelling of the cold, rich waters. By contrast, warm surface water is pushed into the western Pacific. Normally sea surface temperatures (SSTs) in the western Pacific are over 28°C, causing an area of low pressure, and producing high rainfall. By contrast, over coastal South America SSTs are lower, high pressure exists, and conditions are dry.

EL NIÑO CONDITIONS IN THE PACIFIC OCEAN

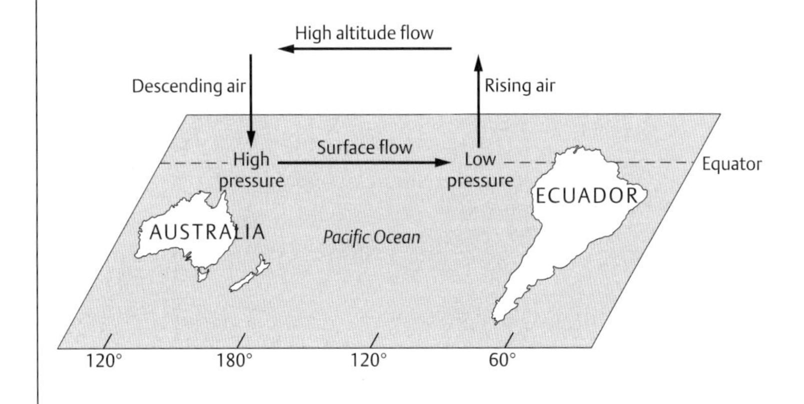

During El Niño episodes, the pattern is reversed. Water temperatures in the eastern Pacific rise as warm water from the western Pacific flows into the east Pacific. During El Niño events, SST of over 28°C extend much further across the Pacific. Low pressure develops over the Eastern Pacific, high pressure over the west. Consequently, heavy rainfall occurs over coastal South America whereas Indonesia and the western Pacific experience warm, dry conditions. Some can be disastrous.

THE 1997-98 EL NIÑO SEASON

The unusual weather events of the late 1990s, such as the flooding in central Europe and the drought in Korea and China have been linked to an early and forceful appearance of the El Niño weather system. This is a large-scale change in ocean temperatures and ocean currents and disrupts the world's climates. The El Niño event that started in the summer of 1997 could be even more catastrophic than the 1982-3 El Niño that claimed nearly 2000 lives and caused over $13 billion damage to property and crops.

In July 1997 the sea surface temperature in the eastern tropical Pacific was 2.0 - 2.5°C above normal, breaking all previous climate records. The El Niño's peak continued into early 1998 before weather conditions returned to normal.

Predicted effects

- a stormy winter in California (the 1982-83 event took 160 lives and caused $2 billion damage in floods and mudslides)
- above average rainfall in the south of the USA
- worsening drought in Australia, Indonesia, the Philippines, southern Africa, and north-east Brazil
- increased risk of malaria in South America
- lower rainfall in northern Europe
- higher rainfall in southern Europe.

Reading climographs

Climatic graphs or climographs describe the seasonal pattern of rainfall and temperature. The diagram shows:

- the mean monthly average
- the mean monthly maximum - the average of all the maximum temperatures for each day of the month
- the mean monthly minimum - an average of all the minimum temperatures recorded for that month
- rainfall - generally shown as a series of bars
- different scales – in this example the temperature scale is shown on the left-hand margin and the rainfall scale is shown on the right.

Reading the climograph

Look out for:

- total rainfall
- seasonality - when most of the rain occurs
- maximum temperature
- minimum temperature
- range of temperature (maximum - minimum temperature)
- length of time (if any) below freezing.

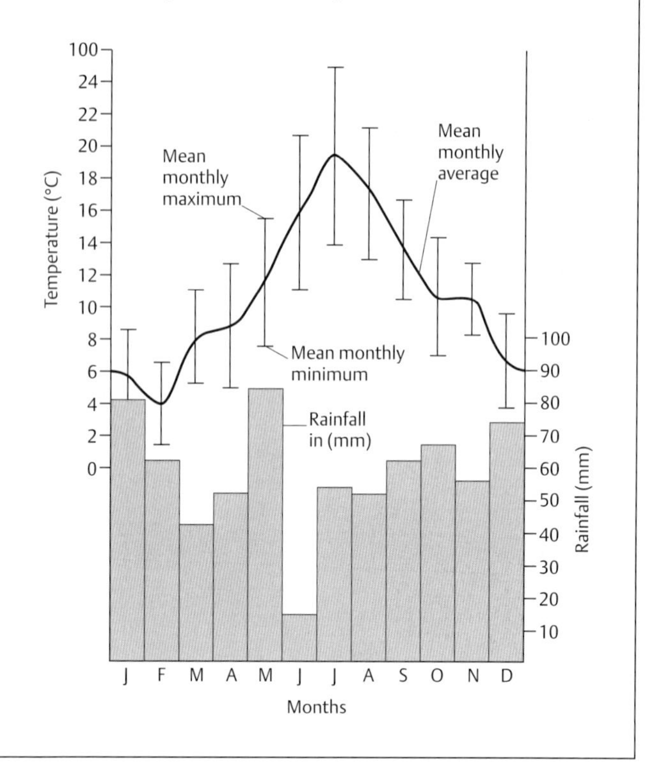

WORLD CLIMATES

Koppen's climatic classification is the most widely used global classification. He classified climate with respect to two main criteria, temperature and seasonality of rainfall. Five of the six main climatic types are based on mean monthly temperature.

A Tropical rainy climate - coldest month > 18°C
B Dry (desert)
C Warm temperate rainy climate - coldest month -3°C - 18°C; warmest month > 10°C
D Cold boreal forest climate - coldest month ≥ -3°C; warmest month ≥ 10°C
E Tundra - warmest month 0-10°C
F Polar - warmest month <0°C.

These temperatures are important for a number of reasons:

- 18°C is the critical winter temperature for tropical forests;
- 10°C is the poleward limit of forest growth;
- -3°C is generally associated with 2-3 weeks of snow annually.

In addition, there are subdivisions due to rainfall characteristics:

f - no dry season
m - monsoonal i.e. short dry season and heavy rains in the rest of the year
s - summer dry season
w - winter dry season

For category B (deserts)
h - mean annual temperatures > 18°C
k - mean annual temperatures < 18°C but warmest month above 18°C
k′ - mean annual temperatures < 18°C and warmest month above 18°C.

CLIMOGRAPHS FROM DIFFERENT CLIMATIC REGIONS

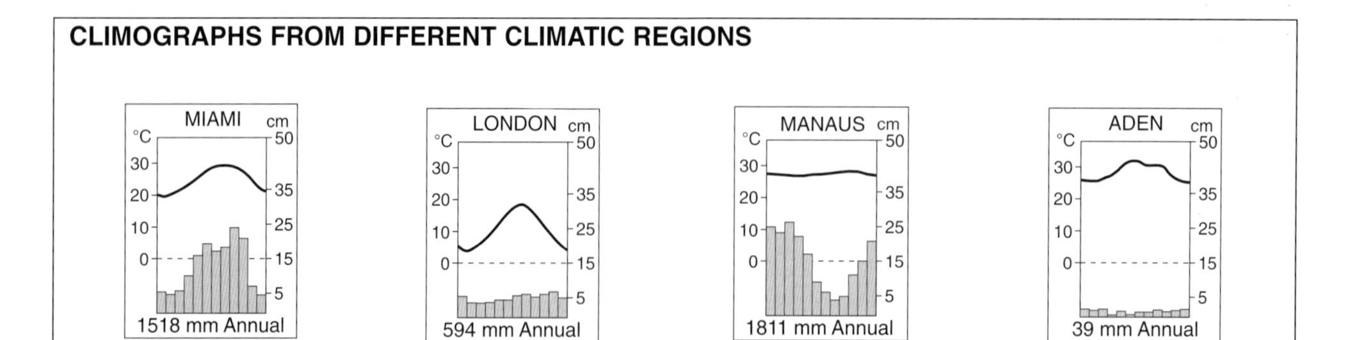

Sub-tropical climate

Temperate climate

Tropical climate

Desert climate

World climatic regions

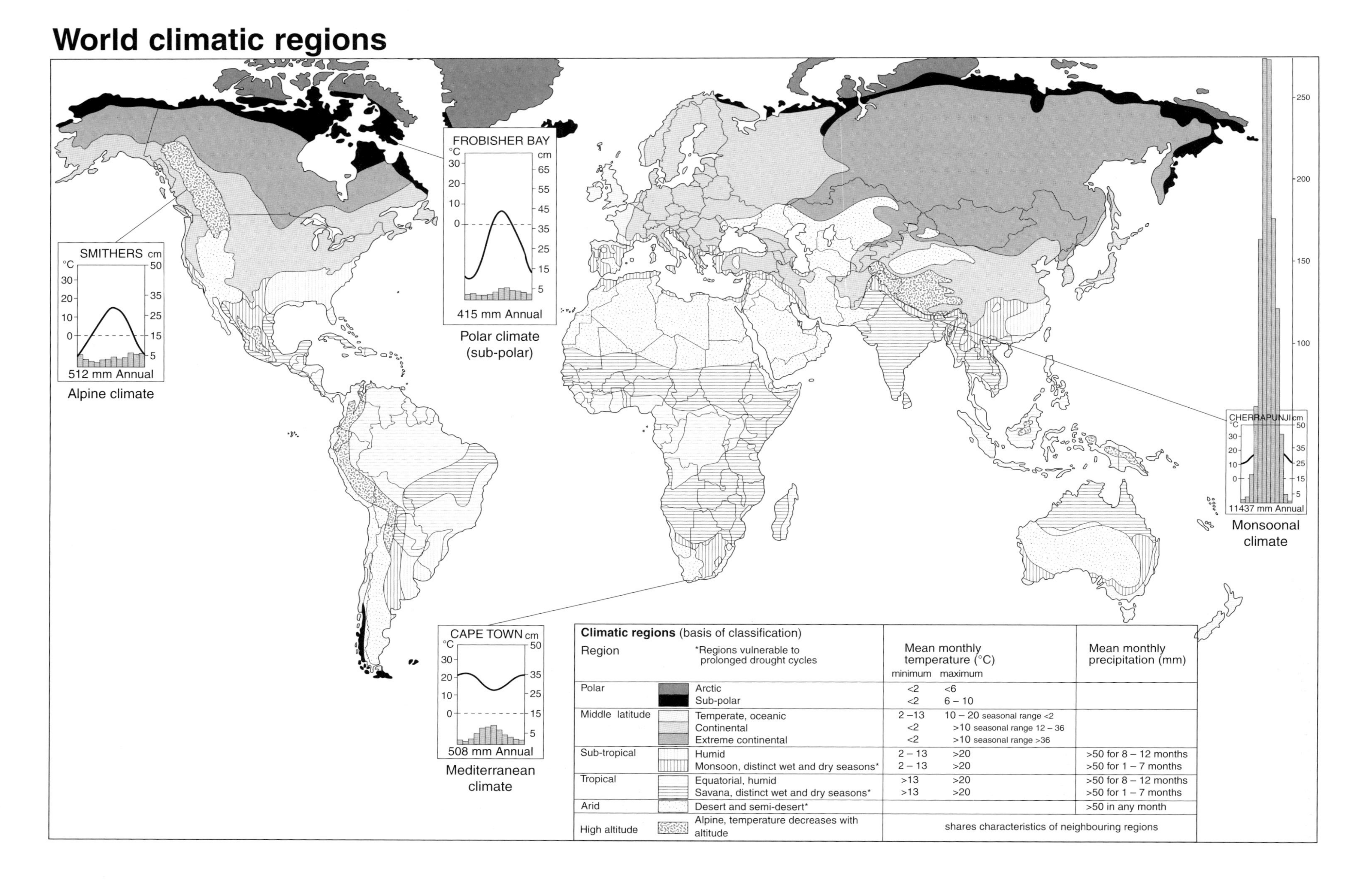

Climatic regions (basis of classification)

*Regions vulnerable to prolonged drought cycles

Region		Mean monthly temperature (°C) minimum	maximum	Mean monthly precipitation (mm)
Polar	Arctic	<2	<6	
	Sub-polar	<2	6 – 10	
Middle latitude	Temperate, oceanic	2 –13	10 – 20 seasonal range <2	
	Continental	<2	>10 seasonal range 12 – 36	
	Extreme continental	<2	>10 seasonal range >36	
Sub-tropical	Humid	2 – 13	>20	>50 for 8 – 12 months
	Monsoon, distinct wet and dry seasons*	2 – 13	>20	>50 for 1 – 7 months
Tropical	Equatorial, humid	>13	>20	>50 for 8 – 12 months
	Savana, distinct wet and dry seasons*	>13	>20	>50 for 1 – 7 months
Arid	Desert and semi-desert*			>50 in any month
High altitude	Alpine, temperature decreases with altitude	shares characteristics of neighbouring regions		

Climate change

Climate and weather vary on a range of scales - daily, seasonally, annually, over decades, centuries, and millennia.

Evidence for long-term climate change comes from a variety of sources:

- pollen analysis
- dendrochronology
- oxygen isotope analysis
- surveys of river activities
- analysis of river sediments
- historical records, including literature and cave art
- fossil remains
- studies of changes in sea-level.

Pollen analysis studies the pollen found especially in peat bogs (due to the waterlogging and lack of oxygen, decomposition is extremely slow). It is an excellent method for showing changes in vegetation (which responds to changes in climate) over the last 10 000 years.

Dendrochronology (the study of tree rings) is used to date climatic change over centuries, although it has been used to monitor climate changes as far back as 4000 years. Wider rings indicate more growth (warmer and/or wetter conditions).

TEMPERATURE CHANGES AT DIFFERENT SCALES

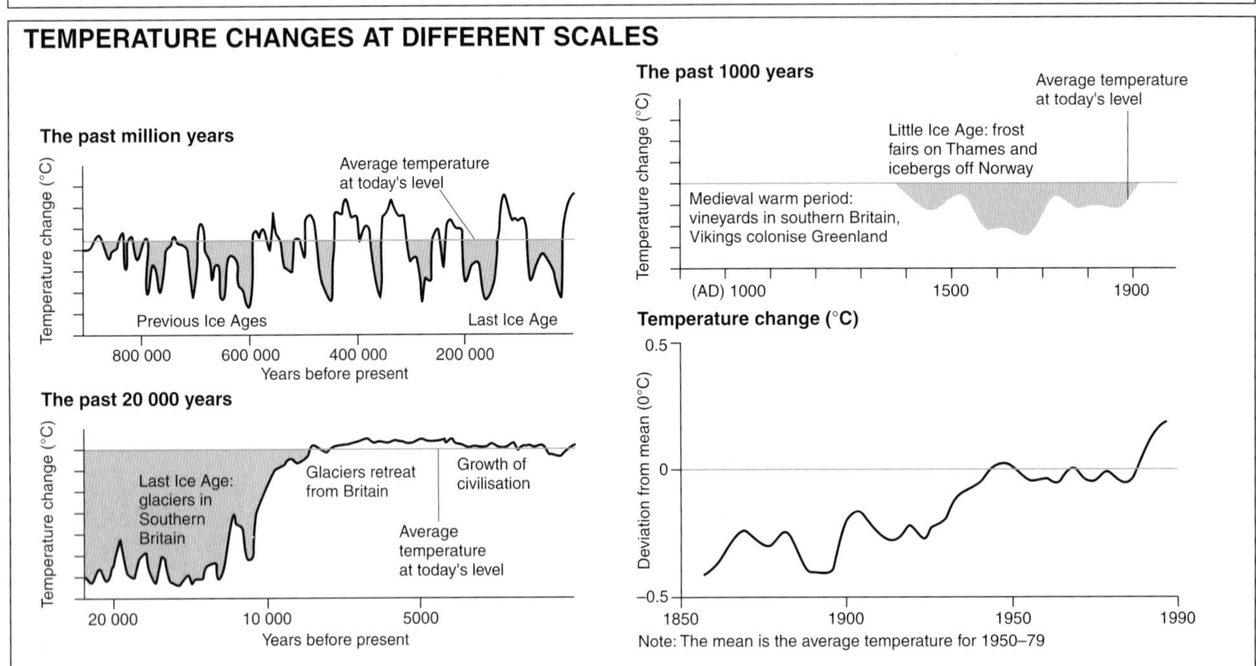

After the last glacial (10 000 years BP) conditions warmed and peaked at about 5000 BP. Around 2000 BP conditions cooled again. Cold periods are associated with medieval villages being abandoned, and with the Little Ice Age (c.1550-1750). Thereafter, climate warmed between 1750 and 1945, and, after a cooler spell until 1970, climate warmed again. The most recent changes are linked to human activity.

THE CAUSES OF CLIMATIC CHANGE

Climate changes for a number of reasons:

- changes in the earth's orbit every 100 000 and 400 000 years
- changes in the tilt of the earth's axis - every 41 000 years
- changes in the orientation of the earth's axis - every 21 000 years
- continental drift (reducing the effect of the north Atlantic drift, for example)
- volcanic emissions (volcanic dust and SO_2 block out insolation and increase the earth's albedo)
- sunspot activity (every 11 years)
- human activity (deforestation, burning of fossil fuels - changing atmospheric composition)
- changes in jet stream activity and Rossby Waves (causing blocking anti-cyclones)
- climatic change occurs on a large-scale - as in the case of fluctuations - as well as small-scale, such as changes over decades.

ASTRONOMICAL CYCLES

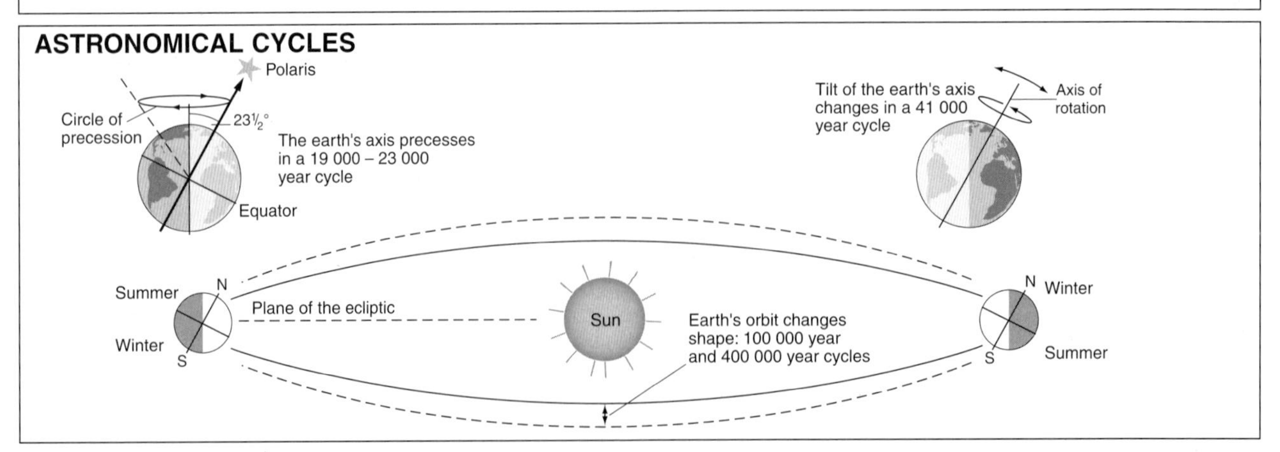

Soils

Soils form the outer-most part of the earth's surface, and are made up of weathered bedrock (regolith), organic matter, air, and water. Soil has matter in all three states - solid, liquid, and gaseous.

SOIL TEXTURE

Soil texture refers to the size of the solid particles in a soil, ranging from gravel to clay. The proportions vary from soil to soil and between horizons within a soil.

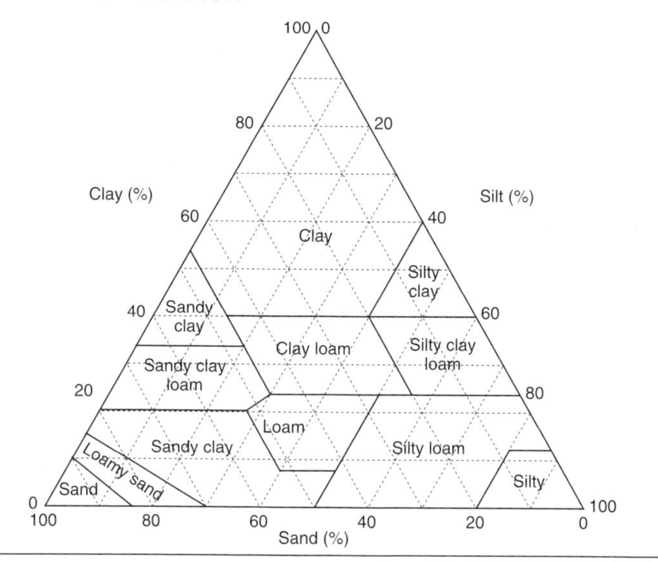

Soil texture is important as it affects:

- moisture content and aeration
- retention of nutrients
- ease of cultivation and root penetration

In general, clay soils become water-logged whereas sandy soils drain rapidly. A loam (mixed soil) is best for plants.

Particle size	diameter (mm)
Clay	<0.002
Silt	<0.02
Sand	<2
Gravel	>2

SOIL MOISTURE

Saturation

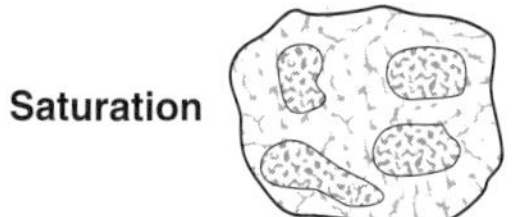

- all pore spaces filled with water
- some water drains as a result of gravity

Field capacity

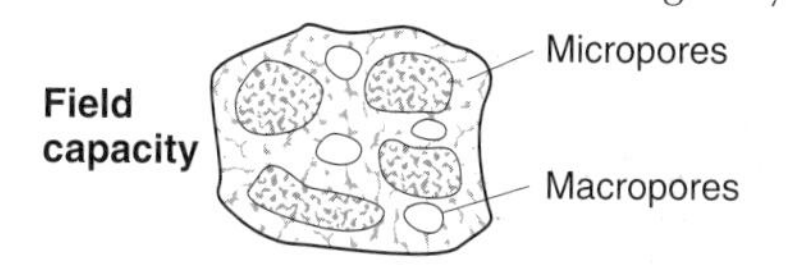

- small micropores filled with water and held by suction
- macropores (large pores) filled with air
- water available to plants

Wilting

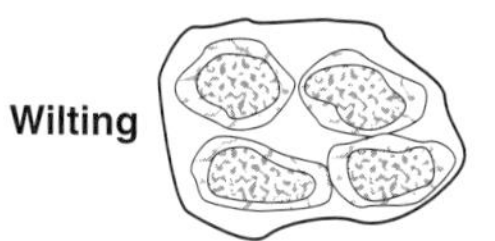

- water present only in small quantities, held by soil hygroscopically

SOIL CHEMISTRY

Nutrients in the soil are called **bases**. Plants use bases for growth and in return provide the soil with hydrogen ions. Consequently, soils become more acidic over time (acidity refers to the proportion of exchangeable hydrogen ions present). However, bases can be returned via leaf fall, application of fertilisers, or the weathering of soft base-rich rocks such as chalk. The cation exchange capacity (CEC) is the ability to retain positively-charged ions. A soil with a high CEC is more fertile than a soil with a low CEC.

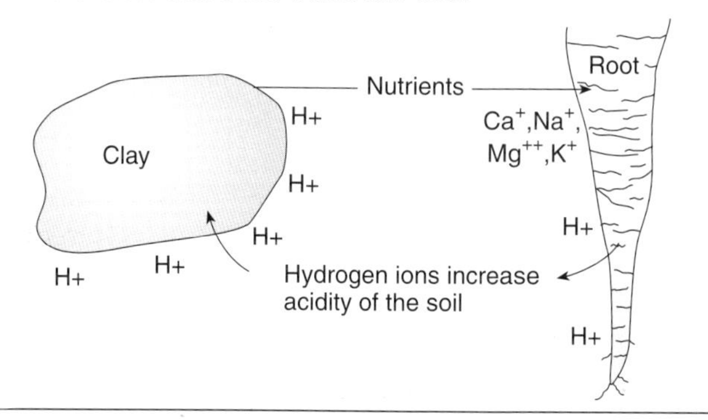

SOIL HORIZONS

O l
f
h
A
E
B
C

O Organic horizon
l Undecomposed litter
f partly decomposed (fermenting) litter
h well decomposed humus
A Mixed mineral-organic horizon
h humus
p ploughed, as in a field or a garden
g gleyed or waterlogged
E Eluvial or leached horizon
a strongly leached, ash coloured horizon, as in a podzol
b weakly leached, light brown horizon, as in a brown earth
B Illuvial or deposited horizon
fe iron deposited
t clay deposited
h humus deposited
C Bedrock or parent material

Soil horizons are the layers within a soil. They vary in terms of texture, structure, colour, pH, and mineral content. The top layer of vegetation is referred to as the Organic (O) horizon. Beneath this is the mixed mineral-organic layer (A horizon). It is generally a dark colour due to the presence of organic matter. In some soils leaching (removal) takes place. This removes material from the E horizon. Consequently, the layer is much lighter in colour.

The B horizon is the deposited or illuvial horizon. At the base of the horizon is the parent material or bedrock.

Soil formation

CLIMATE AND MAJOR SOIL TYPES

Two important mechanisms exist:
- precipitation effectiveness
- temperature

Precipitation effectiveness is a measure of how far precipitation (Ppt) exceeds potential evapotranspiration (P.Evt).
- If Ppt > P.Evt there is a downward movement of materials in the soil, and the soil is leached.
- If Ppt < P.Evt there is an upward movement of material.

Temperature affects the rate of biological and chemical action. In general, as temperatures increase so too does chemical and biological activity.

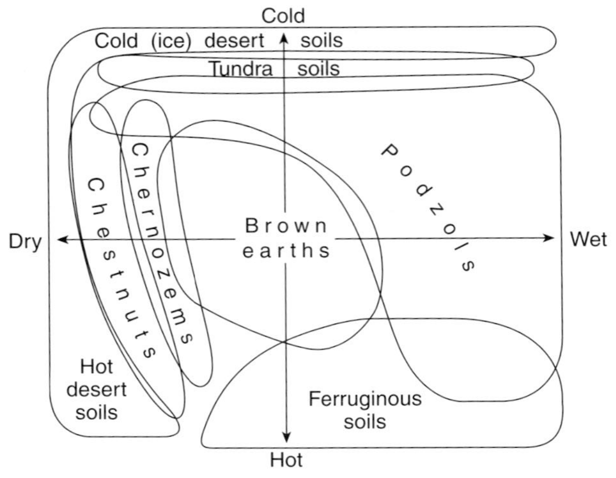

TIME

Time is not a causative factor, but allows processes to operate. The time needed for soil formation varies. Sandstones develop soils more quickly than granite or basalt. Some British soils have evolved since the last glaciation. Soils which have not had enough time to properly mature are termed **azonal** soils.

PLANTS AND ANIMALS

The effects of plants and animals include micro-organisms breaking down leaf litter, worms mixing soils, vegetation returning nutrients, and human activities such as adding fertiliser, irrigating, draining, and compacting soils.

Soil formation is affected by a number of factors, notably climate, geology, biological organisms, and topography. These interact over time to produce distinctive soils and soil profiles.

GEOLOGY

Geology has a lasting effect on soils through texture, structure, and fertility. Sandstones produce free, draining soils, whereas clays give much finer soils. On a regional scale, soils often vary with geology, as in the case of the Isle of Purbeck.

The **intrazonal** classification states that within a climatic zone soils vary with rock types.

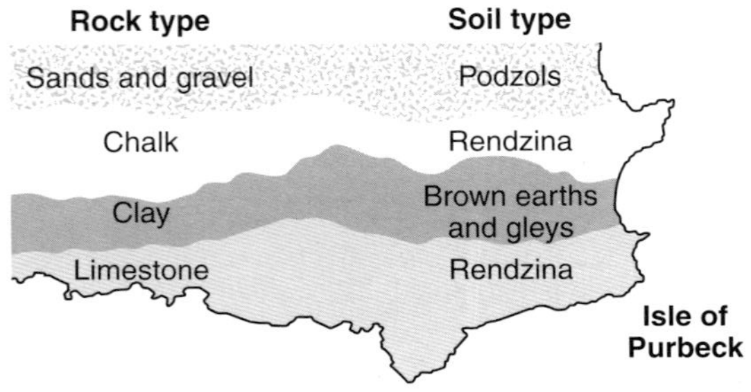

RELIEF

- Steeper slopes have thinner soils.
- Soil erosion increases with slope angle.
- Aspect affects micro-climate (small-scale variations in climate).
- A catena is the variation in soils along a slope owing to changes in slope angle and drainage, climate, and water table. Rock type is constant.

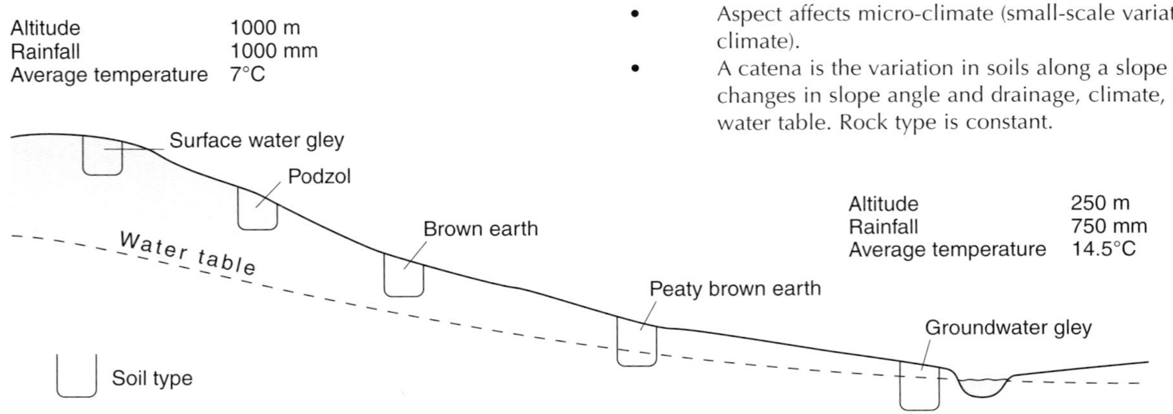

Soil-forming processes

SOIL HORIZONS

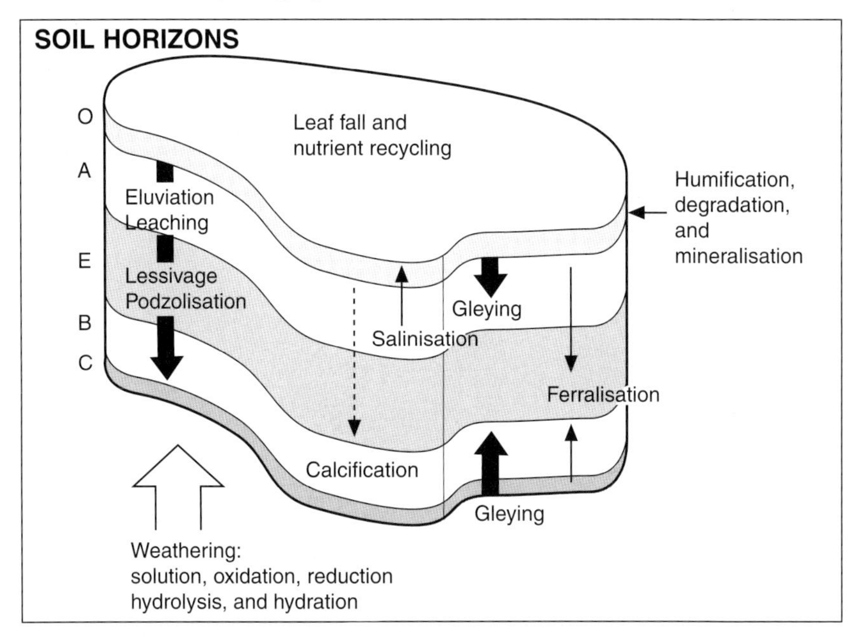

Humification, **degradation**, and **mineralisation** are the processes whereby organic matter is broken down and the nutrients are returned to the soil. The breakdown releases organic acids, **chelating agents**, which break down clay to silica and soluble iron and aluminium.

Cheluviation is the removal of the iron and aluminium sesquioxides under the influence of chelating agents.

Illuviation is the redeposition of material in the lower horizons.

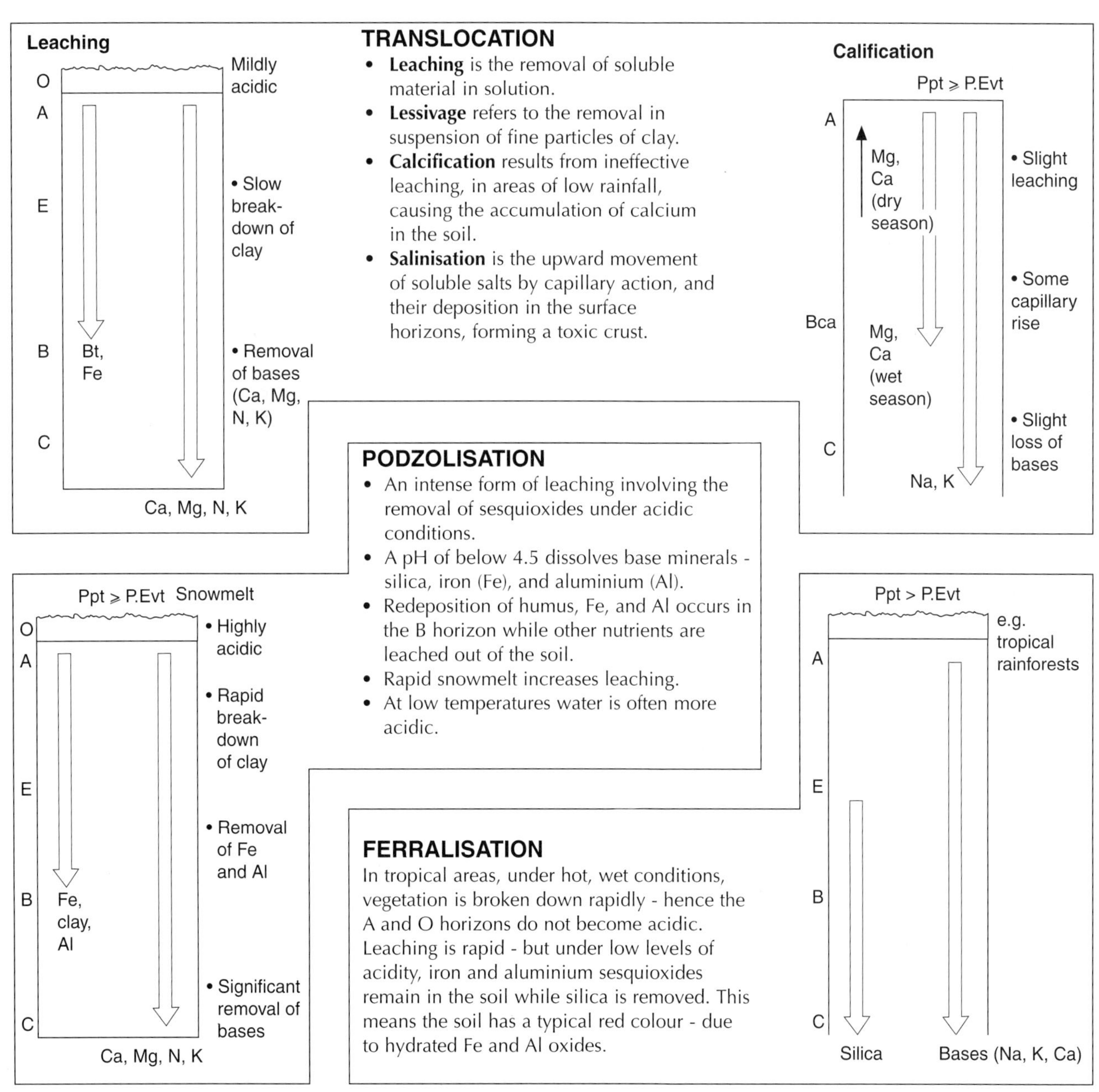

TRANSLOCATION

- **Leaching** is the removal of soluble material in solution.
- **Lessivage** refers to the removal in suspension of fine particles of clay.
- **Calcification** results from ineffective leaching, in areas of low rainfall, causing the accumulation of calcium in the soil.
- **Salinisation** is the upward movement of soluble salts by capillary action, and their deposition in the surface horizons, forming a toxic crust.

PODZOLISATION

- An intense form of leaching involving the removal of sesquioxides under acidic conditions.
- A pH of below 4.5 dissolves base minerals - silica, iron (Fe), and aluminium (Al).
- Redeposition of humus, Fe, and Al occurs in the B horizon while other nutrients are leached out of the soil.
- Rapid snowmelt increases leaching.
- At low temperatures water is often more acidic.

FERRALISATION

In tropical areas, under hot, wet conditions, vegetation is broken down rapidly - hence the A and O horizons do not become acidic. Leaching is rapid - but under low levels of acidity, iron and aluminium sesquioxides remain in the soil while silica is removed. This means the soil has a typical red colour - due to hydrated Fe and Al oxides.

Soil types

SPODOSOLS	Podzols
ALFISOLS	Brown earths (brunizems)
MOLLISOLS	Chernozems
ARIDISOLS	Desert and semi-desert
ULTISOLS	Ferruginous soils of the savanna (pronounced wet and dry seasons)
OXISOLS	Ferralitic soils of the tropical rainforest (latosols)

Podzolisation
Spodosols
Alfisols
Mollisols
Aridisols
Ultisols
Alfisols
Oxisols
Calcification
Laterisation

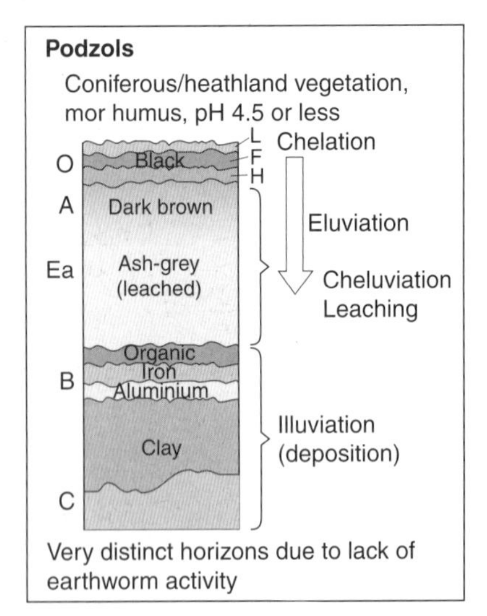

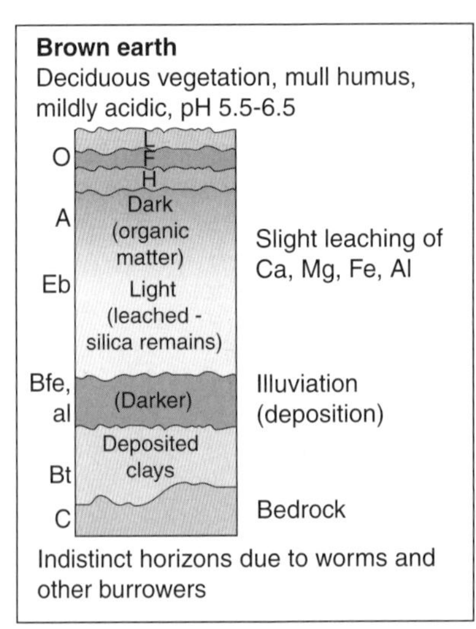

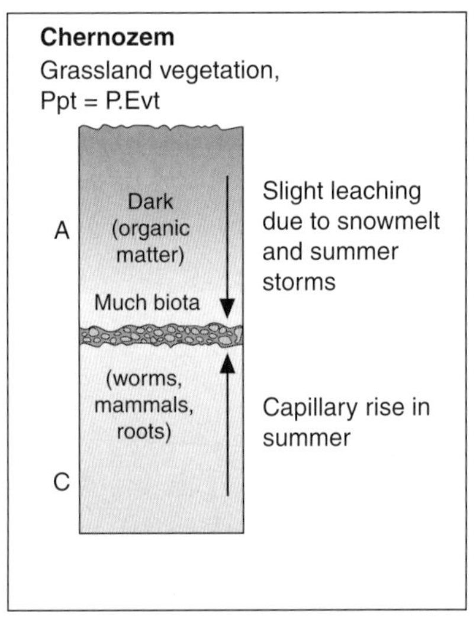

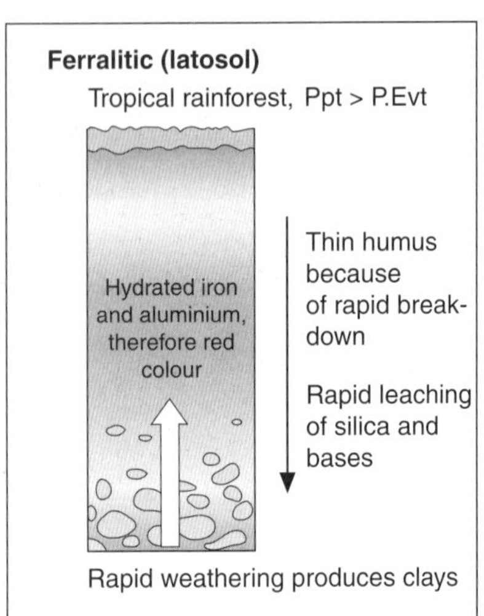

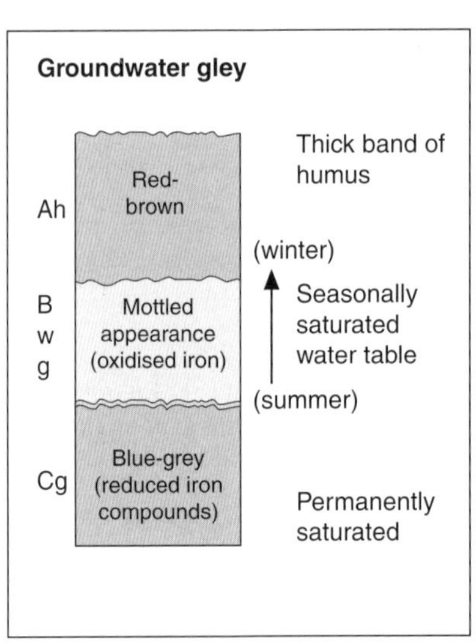

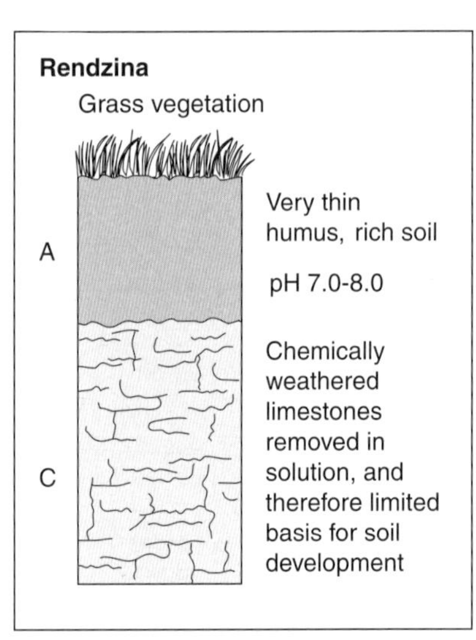

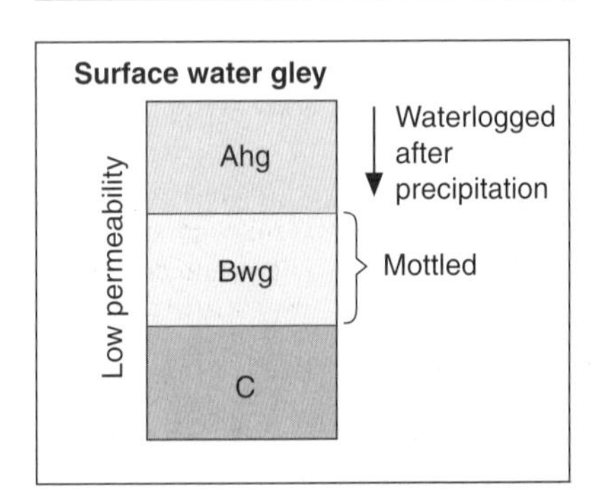

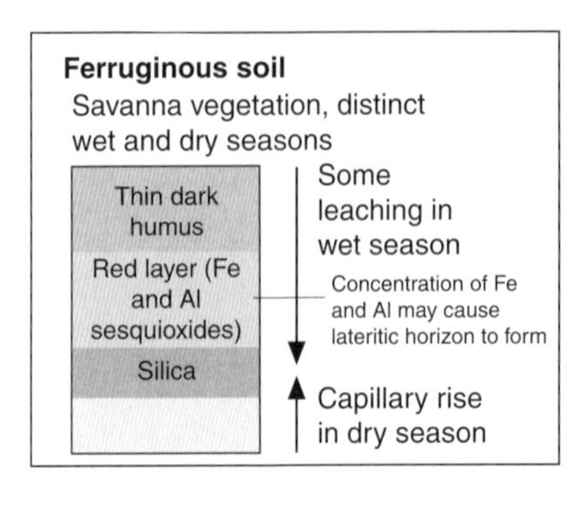

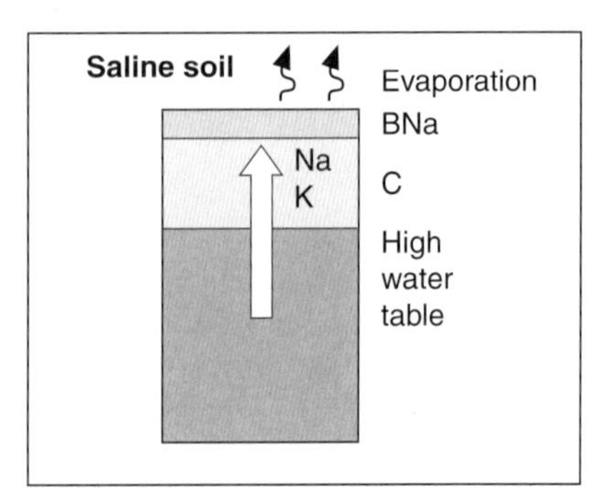

Human impact on soils

SOIL EROSION

Soil erosion in North America: the effect of land use

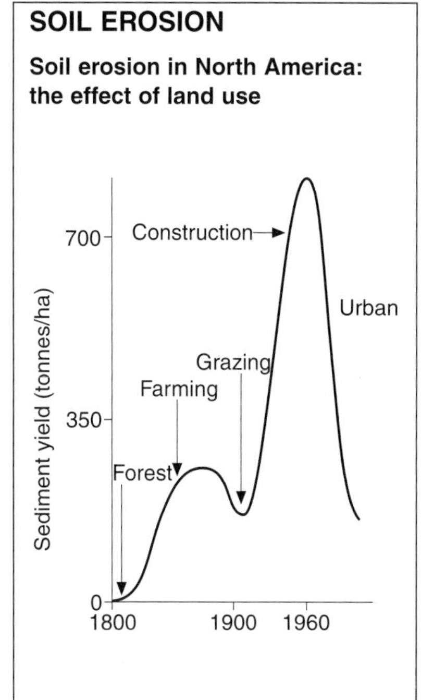

Soil erosion and vegetation cover

	Erosion (mm/yr)
Forest	0.08
Pasture	0.03
Scrub forest	0.1
Barren abandoned land	24.4
Crops (contour ploughing)	10.6
Crops (downslope ploughing)	29.8

Soil erosion rates (tonnes/ha/yr)

UK	
South Downs	250
Norfolk	160
West Sussex	150
Shropshire	120
World	
China (loess)	250
Nepal	70
Ethiopia	42
Burkina Faso	35

INTENSIVE AGRICULTURE AND ITS EFFECT ON SOIL PROFILES

Soil characteristic	Natural state	Intensive agriculture
Organic content	A horizon high (7%) B horizon 0%	Uniform (3-5%) in ploughed horizon
Carbonates	A horizon low/zero B/C horizon maximum	Uniform if limed and tilled
Nitrogen	Medium/low	High (nitrate fertiliser)
Biological activity	High	Medium
Exchangeable cation balance	Ca 80% K 5% P 3% H 7%	Ca 85% K 10% P 5% H 0%

SOIL COMPACTION

A change in structure can occur, turning a free-draining soil or horizon into a compact, impermeable soil or horizon. A **plough pan** forms, caused when damp soil is ploughed and moulded. Compaction by the weight of modern machinery increases the problem.

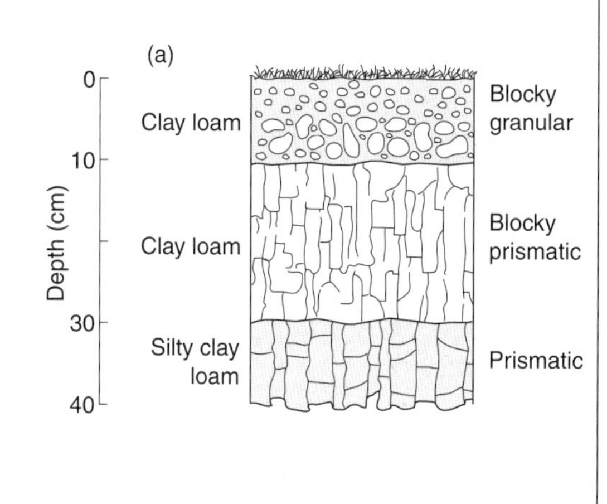

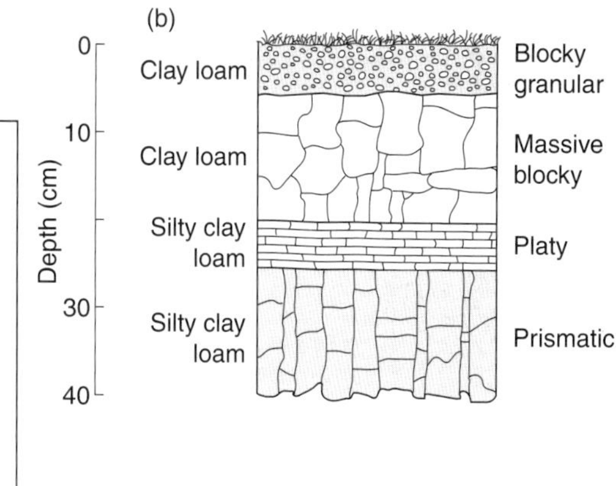

SALINISATION

This occurs when excessive irrigation water causes the water table to rise to the surface. As the water evaporates, soluble salts are left forming a toxic saline crust.

PRINCIPLES OF GOOD SOIL MANAGEMENT

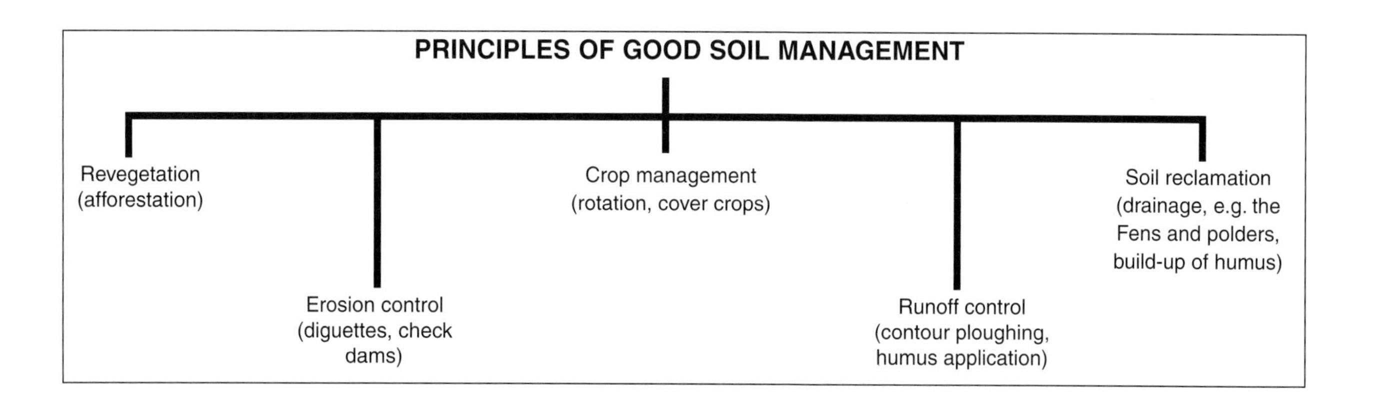

Ecosystems

An **ecosystem** is the interrelationship between plants, animals, and their living and non-living environments. **Biogeography** is the geographic distribution of soils, vegetation, and ecosystems - where they are and why they are there.

Ecosystems can be divided into two main components:

- **Abiotic** elements (non-living), e.g. air, water, heat, nutrients, rock, and sediments.
- **Biotic** elements (living), e.g. plants and animals. These can be divided into:
  - **Autotrophs** (or producers) - organisms capable of converting sunlit energy into food energy by photosynthesis.
  - **Heterotrophs** (or consumers) - organisms that must feed on other organisms, e.g.
    herbivores - plant eaters
    carnivores - meat eaters
    omnivores - plant and meat eaters
    detritivores - decomposers

THE TROPHIC PYRAMID

The **trophic classification** or system is based on feeding patterns. Typically there is a **trophic pyramid**, showing a larger plant biomass and a smaller consumer biomass. This occurs because:

- no energy transfer is 100% efficient - the transfer of light energy to food energy is only 1% efficient;
- there are large losses of energy at each trophic level due to respiration, growth, reproduction, mobility, and so on.

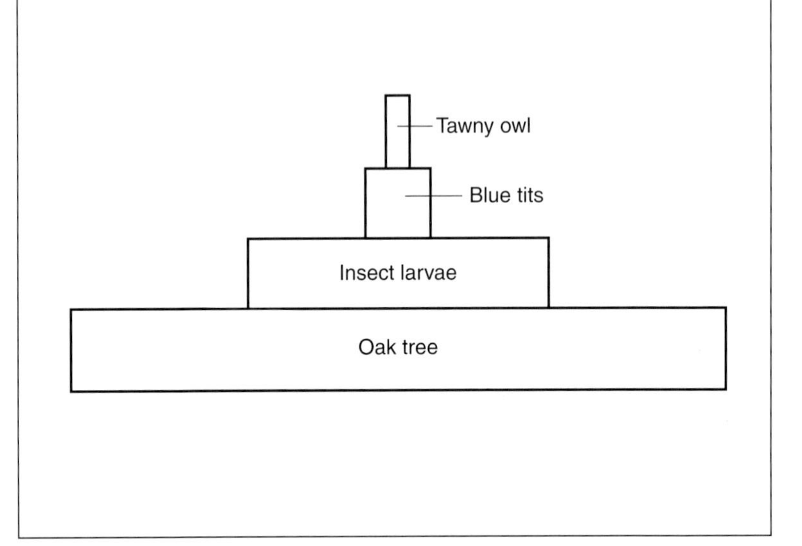

THE FOOD WEB IN A DECIDUOUS FOREST

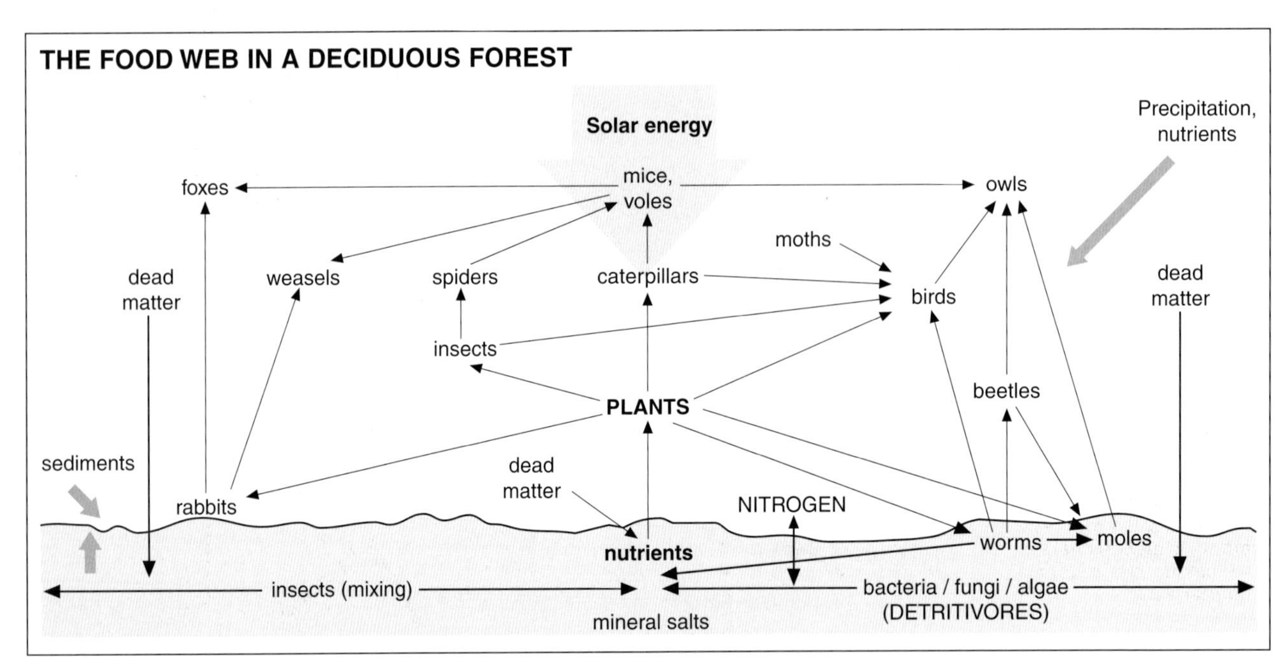

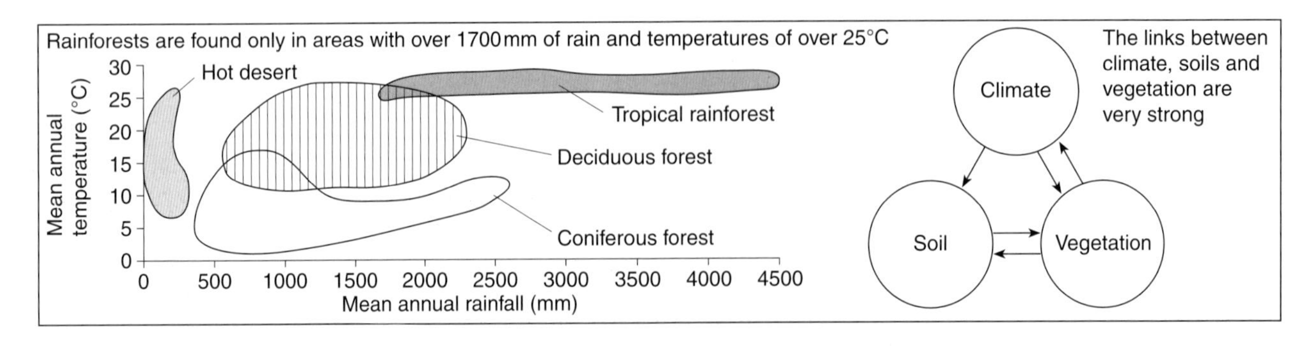

Ecosystems and nutrient cycles

Productivity refers to the rate of energy production, normally on an annual basis.

Primary productivity refers to plant productivity.

Secondary productivity refers to that produced by animals.

Gross productivity is the total amount of energy fixed.

Net productivity is the amount of energy left after losses to respiration, growth, and so on, are taken into account.

Net primary productivity (NPP) is the amount of energy made available by plants to animals at the herbivore level. It is normally expressed as $g/m^2/yr$. NPP depends upon the amount of **heat**, **moisture**, **nutrient availability**, and **competition**, the number of **sunlight hours**, the **age of plants**, and the **health of plants**. In geographic terms NPP increases towards the equator, water permitting, and declines towards the poles.

Ecosystem	Mean NPP ($kg/m^2/yr$)	Mean biomass (kg/m^2)
Tropical rainforest	2.2	45
Tropical deciduous forest	1.6	35
Tropical scrub	0.37	3
Savanna	0.9	4
Mediterranean sclerophyll	0.5	6
Desert	0.003	0.002
Temperate grassland	0.6	1.6
Temperate forest	1.2	32.5
Boreal forest	1.2	32.5
Tundra and mountain	0.14	0.6
Open ocean	0.12	0.003
Continental shelf	0.36	0.001
Estuaries	1.5	1

GERSMEHL'S NUTRIENT CYCLES

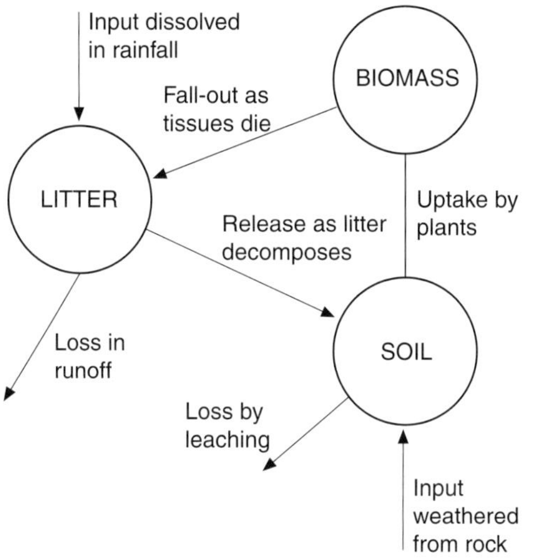

Input dissolved in rainfall

BIOMASS

Fall-out as tissues die

LITTER

Release as litter decomposes

Uptake by plants

SOIL

Loss in runoff

Loss by leaching

Input weathered from rock

B Biomass
L Litter
S Soil

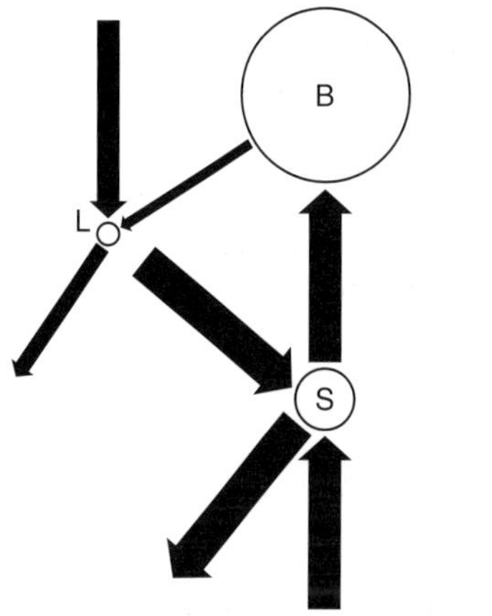

Selvas (equatorial rainforest) environment

The size of the nutrient stores (B, L, S) are proportional to the quantity of nutrients stored. The thickness of the transfer arrows is proportional to the amount of nutrients transferred.

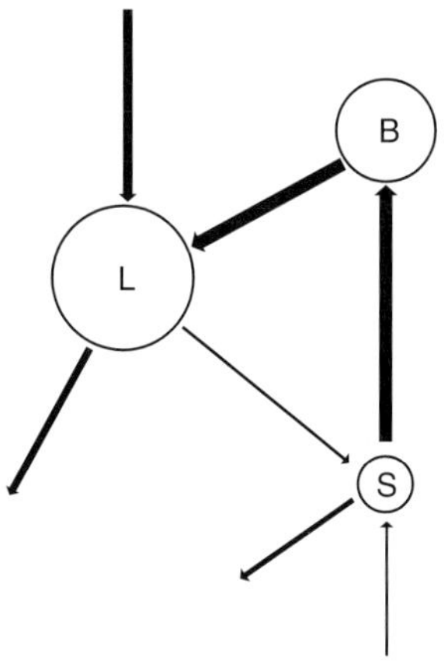

Taiga (northern coniferous forest) environment

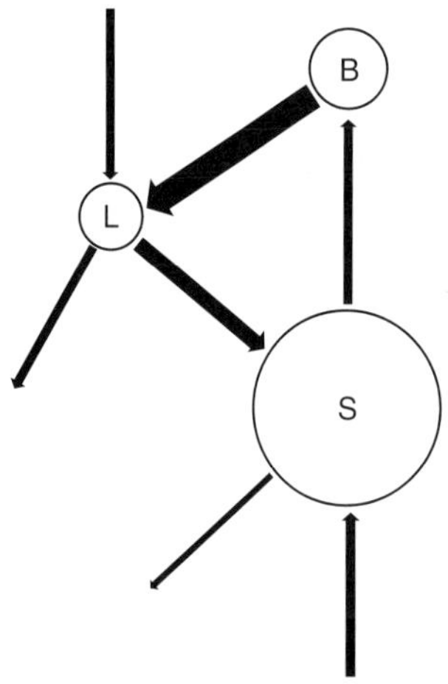

Steppe (mid-latitude continental grassland) environment

Nutrient cycles can be **sedimentary based**, i.e. the source of the nutrient is from rocks, or it can be **atmospheric based**, as in the case of the nitrogen cycle. Nutrient cycles can be shown by means of simplified diagrams - indicating the stores of nutrients as well as the transfers. The most important factors which determine these are availability of moisture, heat, fire (in grasslands), density of vegetation, competition, and length of growing season.

Succession

Succession, or **prisere**, is the sequential change in species in a plant community as it moves towards a **seral climax**. Each **sere** is an association or group of species, which alters the micro-environment and allows another group of species to dominate. The **climax community** is the group of species that are at a **dynamic equilibrium** with the prevailing environmental conditions - in the UK, under natural conditions, this would be oak woodland. On a global scale, climate is the most important factor in determining large scale vegetation groupings or **biomes**, e.g. rainforest, temperate grassland, and so on. However, in some areas, vegetation distribution may be influenced more by soils than climate. This is known as **edaphic** control. In savanna areas forests dominate clay soils, grassland sandy soils. Soils may also affect plant groupings on a local scale, within a climatic region, e.g. on the Isle of Purbeck grassland is found on limestone and chalk rendzina soils, forest on the brown earths, and heathland on the podzols associated with sands and gravels.

SUCCESSION AND SPECIES SELECTION

r-species are the initial colonisers - large numbers of a few species. Highly adaptable, rapid development, early reproduction, small in size, short life, highly productive.

k-species are diverse, and are specialists - a few of many species. Slower development, delayed reproduction, larger size, longer living, less productive.

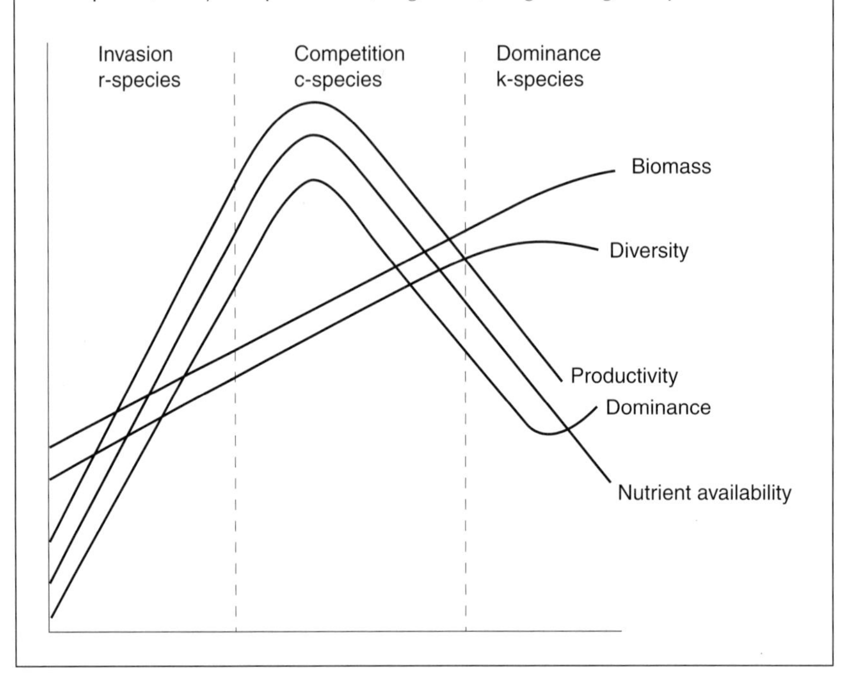

A **plagioclimax** refers to a plant community permanently influenced by people, i.e. it is prevented from reaching climatic climax by burning, grazing, and so on. Britain's **moorlands** are a good example: deforestation, burning, and grazing have replaced the original oak woodland.

A MODEL OF SUCCESSION

Rooted plants increase sedimentation and nutrient availability. They alter the micro-environment so that reeds, fen, carr, and oak are increasingly able to tolerate the thickening and drying soil.

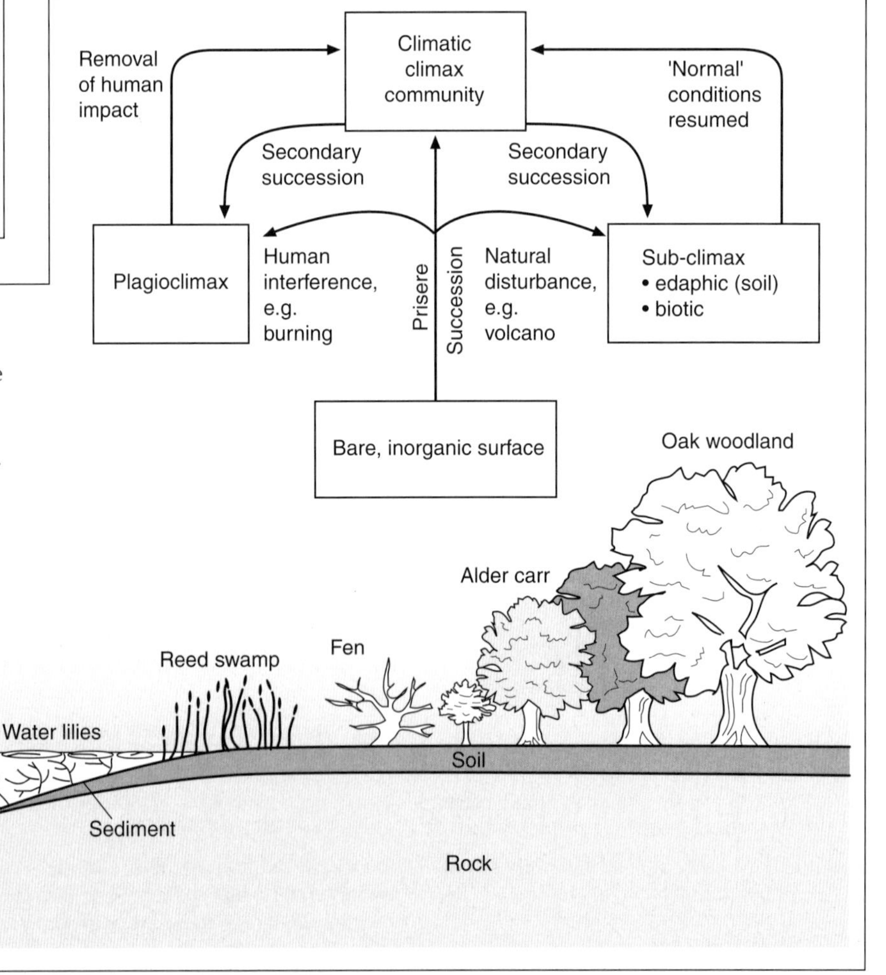

Tropical rainforest

CLIMATE

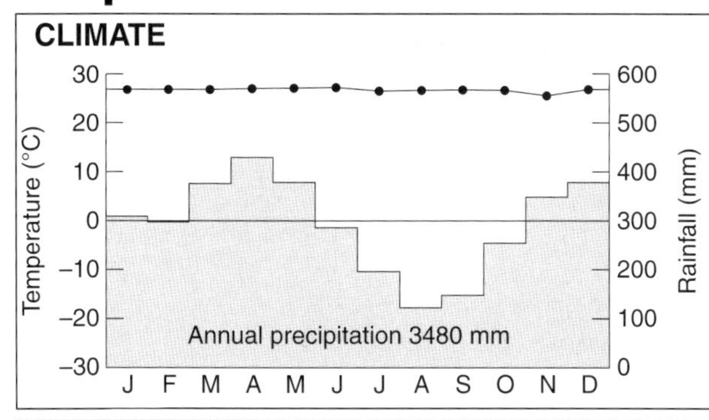

Some features of climates in rainforest areas:

- Annual temperatures are high (26-27°C), owing to the equatorial location of rainforest areas.
- Seasonal temperature ranges are low, 1-2°C, and diurnal (daily) ranges are greater,10-15°C.
- Rainfall is high (>2000 mm per year), intense, convectional, and occurs on about 250 days each year.
- Humidity levels on the ground are high, often 100%.
- The growing season is year-round.

VEGETATION

- The vegetation is evergreen, enabling photosynthesis to take place year-round.
- It is layered, and the shape of the crowns vary with the layer, in order to receive light.
- Rainforests are a very productive ecosystem: NPP is about 2200 g/m^2/yr, and there is a large amount of stored energy (biomass 45 kg/m).
- The ecosystem is diverse - there can be as many as 200 species of tree per hectare, including figs, teak, mahogany, and yellow-woods.

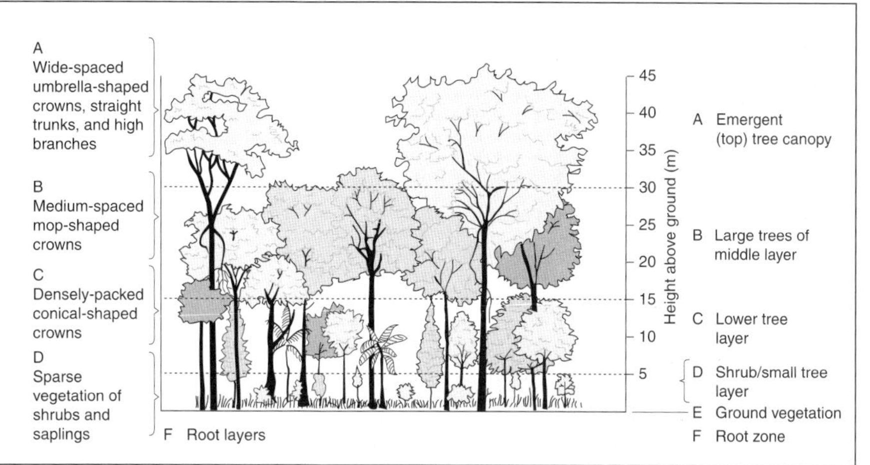

SOILS: TROPICAL LATOSOLS

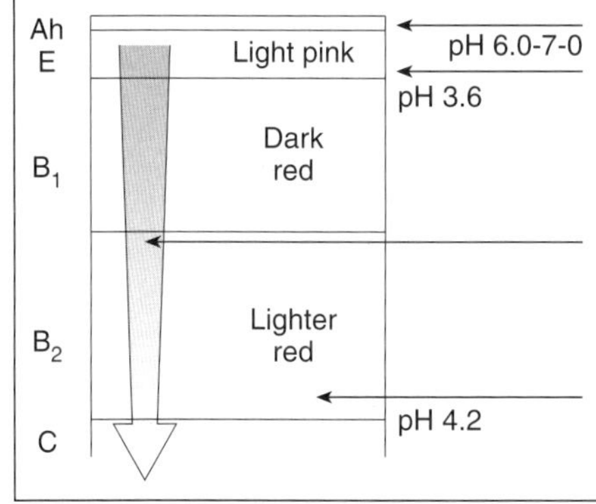

Thin layer of humus. Continuous supply of leaves: rate of humus turnover 1%/day. Very active soil fauna.

Accumulation of iron and aluminium gives the soil a red colour. Concentration of aluminium may form bauxite nodules.

Hot, wet conditions cause rapid removal of fine clay and silicate particles.

Loss of N, Ca, Mg, and K from the soil by leaching.

Bedrock intensely weathered - due to hot, wet conditions, and lack of disturbance in glacial times.

THE EFFECT OF SHIFTING CULTIVATION ON SOIL FERTILITY

In many tropical rainforests shifting cultivation is the main type of farming. This involves clearing a plot of land, and cultivating it for a few years. When the soil fertility has dropped and farming can no longer continue, they move to another plot.

Until recently, many geographers believed that rainforest soils regained their fertility quickly - about 20 years or so after being used for farming. Now it is realised that fertility does not recover and continued use of the rainforest leads to a long term decline in soil fertility. Other effects of deforestation include: high rates of soil erosion, leaching of nutrients from the soil, increased number of landslides, poor quality water - full of sediment and unfit to use, increased sediment in river, more water in river, increased overland run-off, and greater chance of flooding.

a
SC SC
Soil fertility
PA PA
Traditional view
Time ⟶

SC Shifting cultivation
PA Plot abandoned

b
Long term declining fertility
SC SC SC
Soil fertility
PA PA PA
Modern view
Time ⟶

Savanna

CLIMATE

- There is no such thing as a 'typical' savanna climate.
- Rainfall in savanna areas ranges from 500-2000 mm per year with a drought lasting between one and eight months.
- Annual temperatures are high (>25°C).
- Summers are hot and wet, winters hot and dry.
- Convectional rain occurs in summer.
- High temperatures year-round lead to high evapotranspiration losses.

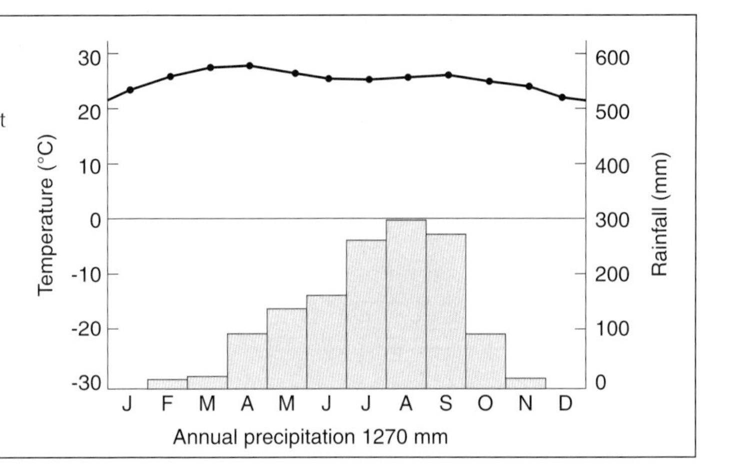

SOILS: FERRUGINOUS

Savanna soils are influenced by distinct seasonal changes in processes. Moreover, they vary with topography. Frequently, sandy and/or leached soils predominate on the upper slopes, clay-based soils on lower slopes. This **catena** is reflected by changes in vegetation. In places **laterite**, a hardened layer of iron/aluminium, may limit further soil development and agricultural practices.

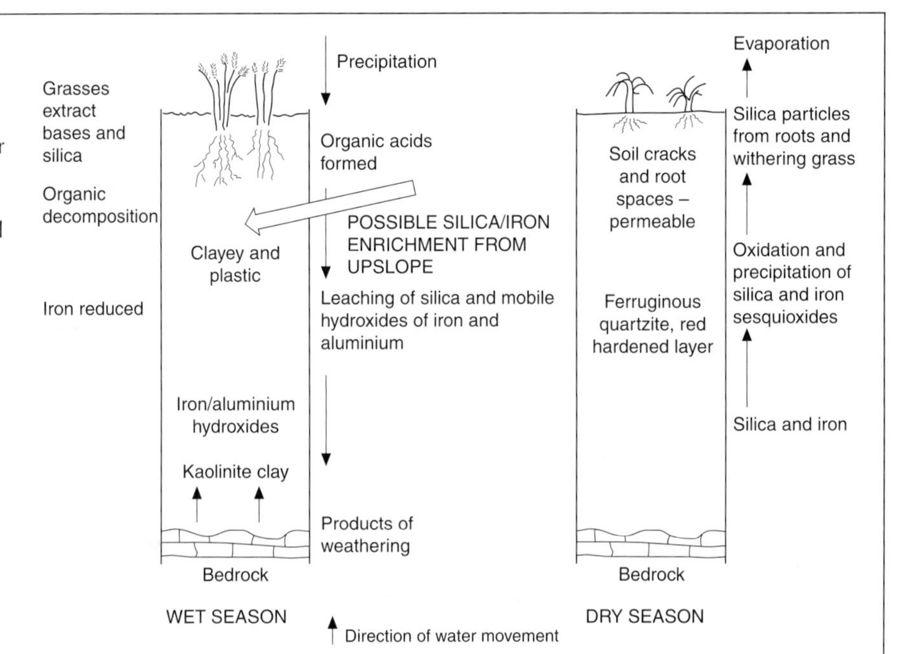

VEGETATION

Savanna vegetation is **xerophytic** (adapted to drought) and **pyrophytic** (adapted to fire). Grasses predominate on sandy, leached soils, while trees may be found in moister areas, such as valleys. Growth is rapid in the summer (NPP 900 g/m^2/yr).

Fire (natural and as a result of human activity) reduces the biomass store (4 kg/m). Grasses are well adapted because the bulk of their biomass is beneath ground level and they regenerate quite quickly after burning.

NUTRIENT CYCLE

- The biomass store is less than that of the tropical rainforest due to a shorter growing season.
- The litter store is small due to fire. This means that the soil store is relatively large.

The savanna nutrient cycle differs from the tropical rainforest nutrient cycle due to the combined effects of the seasonal drought and the occurrence of fire. Consequently there is:

- a lower nutrient availability
- a reduced biomass store
- a small litter store
- a relatively large soil store.

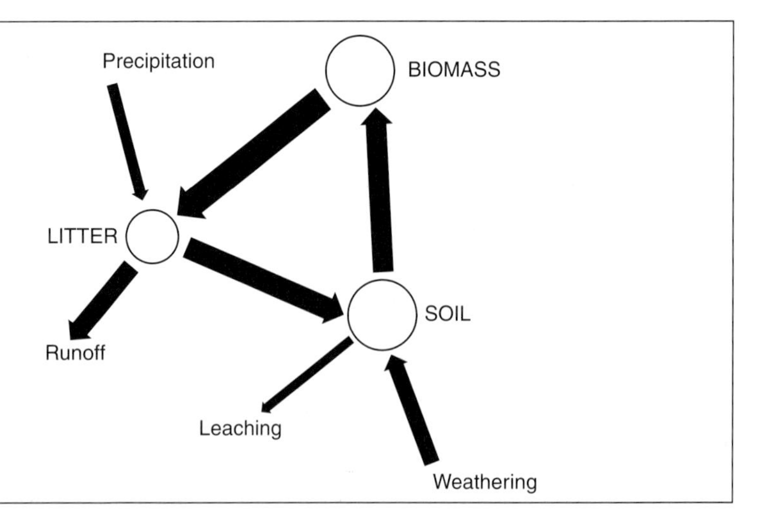

Temperate grassland

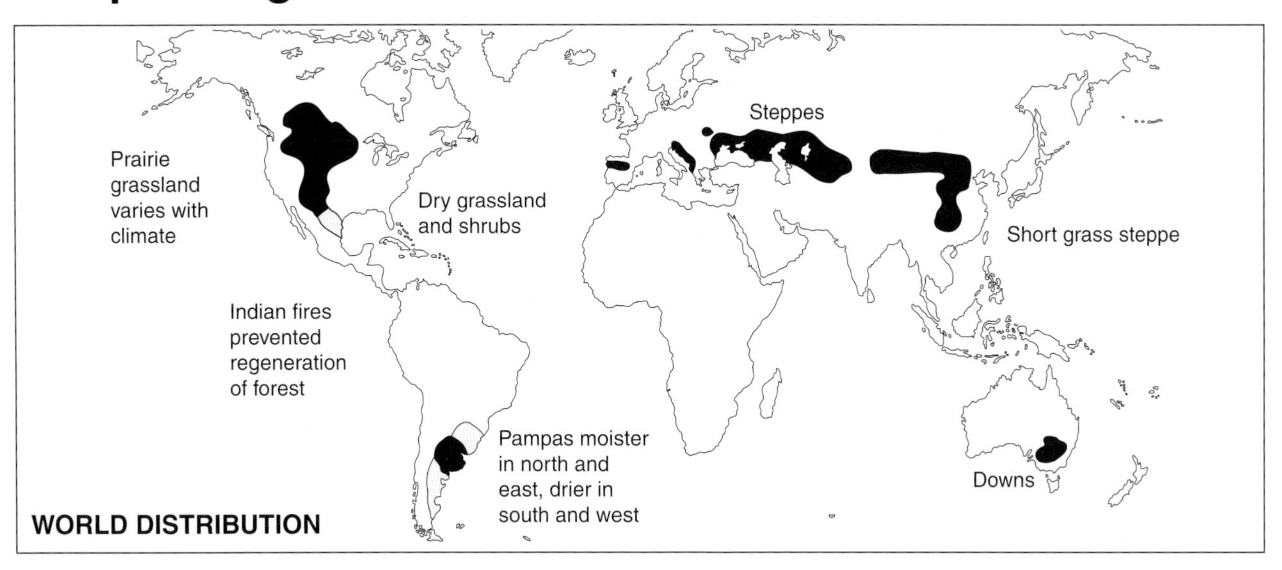

CLIMATE

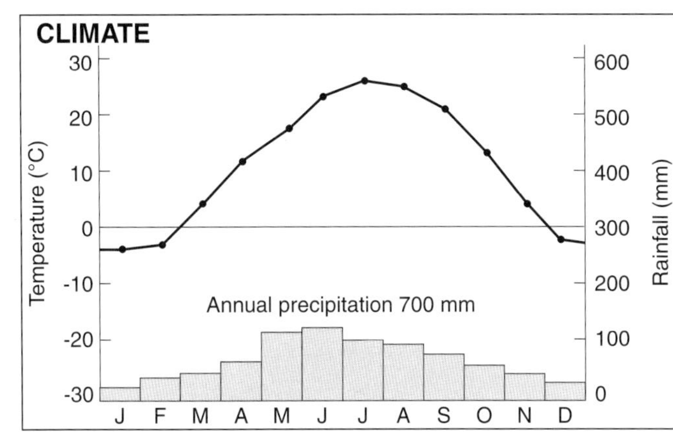

- Climate is continental in character, i.e. hot summers, up to 30°C, and cold winters, with temperatures below freezing for up to 6 months a year.
- Precipitation is low (250-750 mm per year), with snow in winter and convectional rain in summer.
- Evaporation rates are high, especially in summer.
- In the USA, precipitation decreases from east to west, e.g. Virginia and Tennessee 750 mm, Nevada 250 mm.

SOIL: CHERNOZEM

Grasslands are dominated by black earth or **chernozem** soils. These are frequently developed on wind-blown loess deposits, rich in **calcium carbonate** ($CaC0_3$). Burrowing animals and soil fauna mix the soil. There may be some concentration of $CaCO_3$ due to spring leaching (snowmelt) and summer capillary rise. In the USA, soils vary from east to west. In the eastern areas of higher rainfall, brown earths are formed, in the drier regions of the west, chestnut soils are found.

A

Dark colour – much organic matter

Nodules of $CaCO_3$

C

1.5 m

Slight leaching due to snowmelt and thunderstorms; capillary rise in summer

NUTRIENT CYCLE

The natural nutrient cycle is affected by:

- climate (drought)
- fire
- the highly matted root system
- the grassland vegetation.

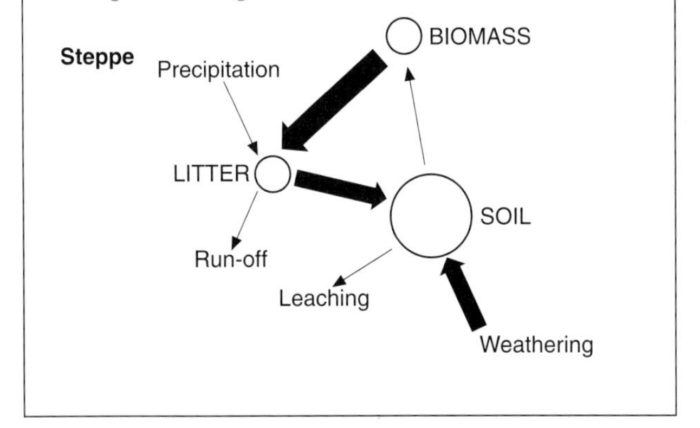

VEGETATION

Vegetation is closely related to climate:

- trees in eastern (wetter) areas and along water courses
- height of grass relates to amount of precipitation.

Tall grass, e.g. Bluestem (1.5-2.5 m), is found in the east where precipitation is about 750 mm per year.

Mixed grass, e.g. Little Bluestem (0.6-1.5 m), is found in central areas where precipitation is about 400-600 mm per year.

Short bunch, e.g. Buffalo Grass (<0.5 m), is found in the west where precipitation is about 250 mm per year.

Vegetation is both **xerophytic** and **pyrophytic**.

- NPP is 600 $g/m^2/yr$
- Biomass 1.6 kg/m (due to the lack of woody species).

Temperate deciduous woodland

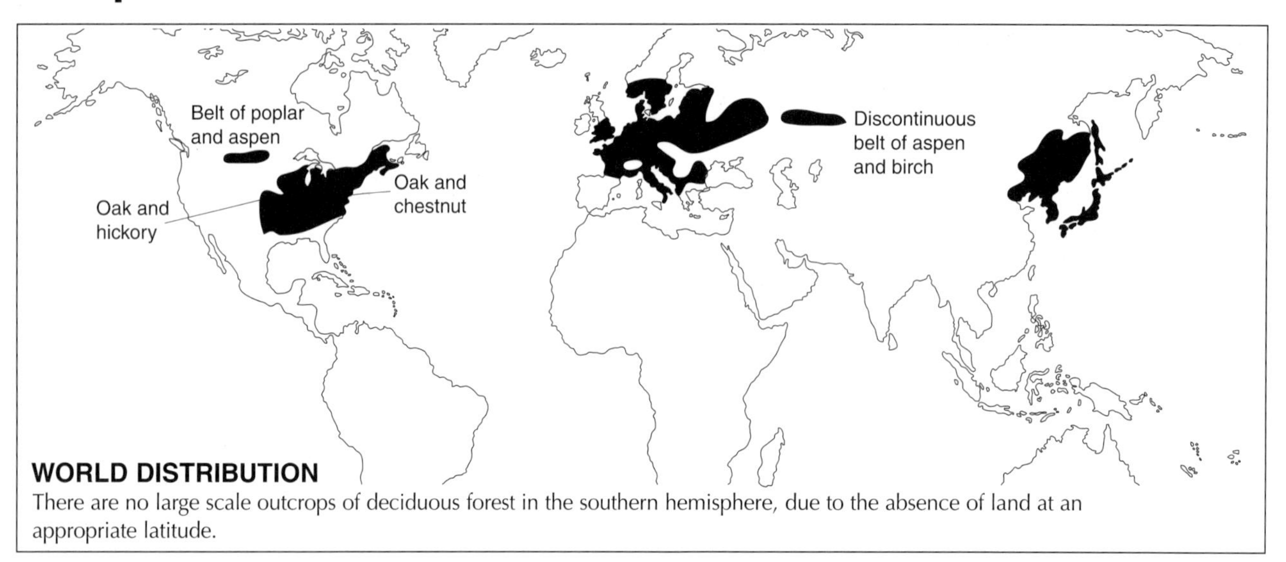

WORLD DISTRIBUTION

There are no large scale outcrops of deciduous forest in the southern hemisphere, due to the absence of land at an appropriate latitude.

CLIMATE

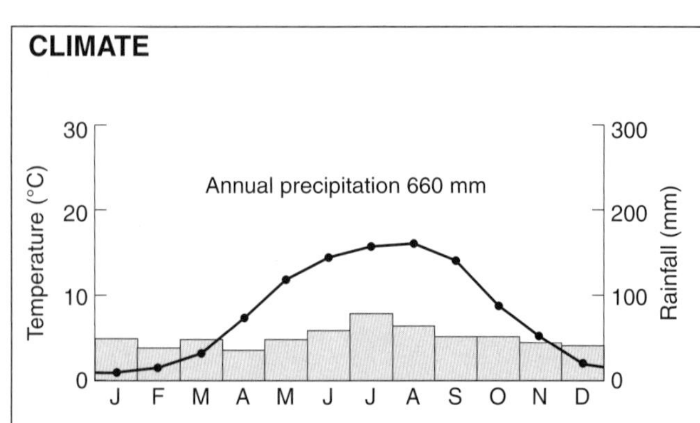

- There are wide variations in climate, e.g. between North-East USA and South-West Ireland.
- Rainfall is 500-1500 mm per year, with a winter maximum. It is mostly frontal (cyclonic) rainfall.
- Precipitation is greater than evapotranspiration, e.g. in Ireland, precipitation is 1000 mm per year and evapotranspiration is 450 mm per year.
- Winters below freezing (for 2-3 months in eastern China and North-East USA), although they are milder in western Europe due to the Gulf Stream.
- Cool summers, 15-20°C.

SOIL: BROWN EARTH

- Soils are generally quite fertile.
- The mull humus is mildly acidic (pH 5.5-6.5).
- Soil fauna such as earthworms flourish, mixing the layers and nutrients.
- Decomposition of leaves takes up to 9 months.
- Blurred horizons due to earthworm activity.

Horizon		Description
O (L, F, H)		Mull humus, mildly acidic
A	Brown	
E	Light colour (leached of minerals)	Slight leaching of Ca, Mg, Fe, and Al
B	Deposition of clay, humus, and iron	Illuviation
C		Bedrock

NUTRIENT CYCLE

There is a large store of nutrients in the soil due to:

- slow growth in winter
- a low density of vegetation compared with the tropical rainforest.

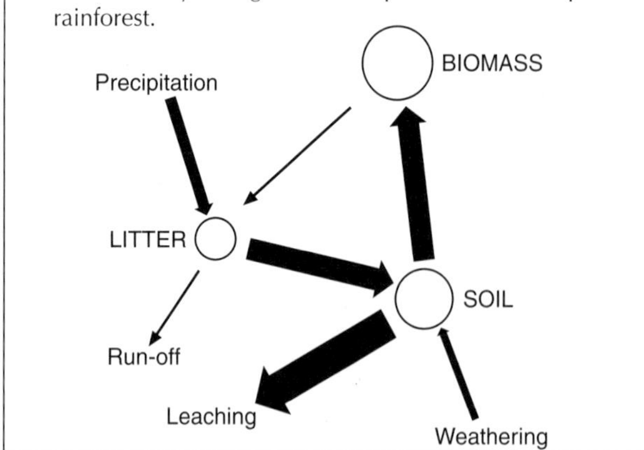

VEGETATION

- **Deciduous** trees shed their leaves in winter to retain moisture, conserve nutrients, and avoid damage by snow/ice.
- **NPP** is high - 1200 g/m^2/year due to high summer temperatures and the long hours of daylight.
- **Biomass** is high, 35 kg/m^2, due to large amounts of woody material.
- Vegetation varies with **soil** type, e.g. on more acidic soils birch and rowan are found, on more alkaline soils box and maple (oak is a generalist).
- Shrub vegetation varies with light, e.g. **heliophytes** need light, such as the wood anenome which flowers early, while **sciophytes** tolerate darkness, e.g. dog's mercury and ivy.

Temperate coniferous (boreal) forest

WORLD DISTRIBUTION

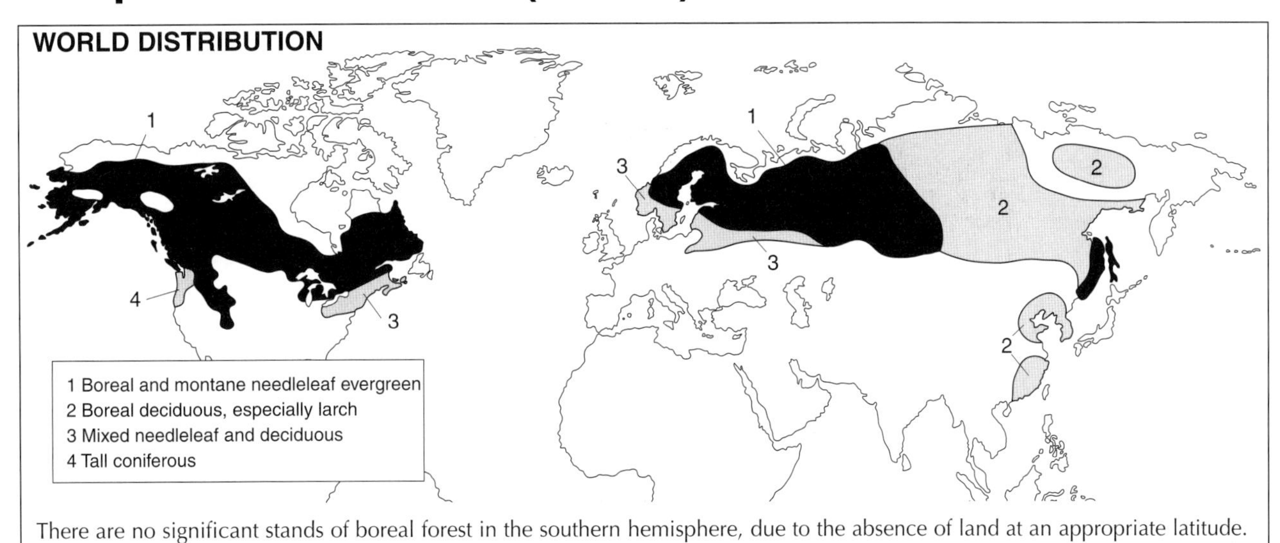

There are no significant stands of boreal forest in the southern hemisphere, due to the absence of land at an appropriate latitude.

CLIMATE

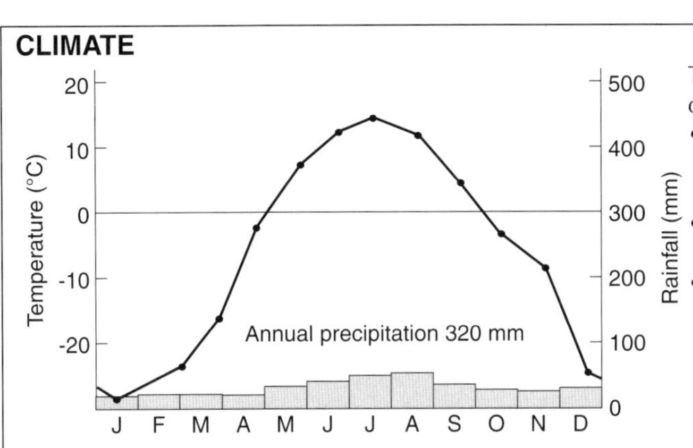

The climate is a cool temperate or cold continental climate, depending on the proximity of the ocean.

- Rainfall is low (<500 mm per year), with no real seasonal pattern. Snowfall is frequent in winter, and frosts occur in summer.
- Summer temperatures 10-15°C, with winter temperatures below freezing for up to 6 months.
- Growing season is limited to a maximum of 6-8 months, but the long hours of daylight in summer (16-20 hours) allow photosynthesis.

SOIL: PODZOL

- Precipitation greater than potential evapotranspiration.
- Leaching by snowmelt.
- Raw acid mor humus (pH 4.5-5.5).
- Acid water releases iron and aluminium oxides, transfers them down the soil, and may form an impermeable iron pan.
- Very few earthworms, due to acidity, and therefore little mixing of horizons.
- Thick litter layer due to the low temperatures, and resistant acidic nature of needles.

Thin organic layer
Organic matter
Silica
Ash-grey alleviated horizon — Ea
Dark brown depositer layer containing humus, clay, Fe, and Al — Bfe
Bedrock — C

NUTRIENT CYCLE

Low temperatures and slow rates of weathering produce large stores of nutrients in the litter layer.

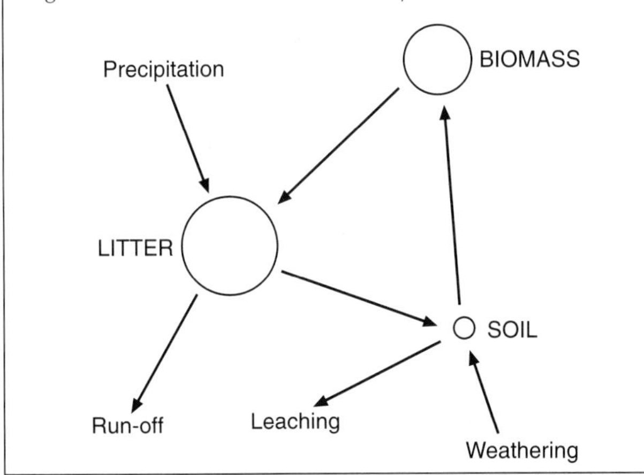

VEGETATION

- **Evergreen:** green throughout the year, and therefore able to photosynthesise when temperatures rise above 3°C.
- Conicle **shape** enables trees to shed snow and reduce rocking by wind.
- **Needle leaves** - small surface area, and therefore water loss is reduced.
- Generally occur in **stands** of one species.
- Pine favours sandy **soils**, spruce damper soils.
- Ground vegetation limited - it is too dark.
- NPP 800 g/m^2/yr.
- Biomass 20 kg/m (much woody matter).

Hot deserts and mediterranean woodland

HOT DESERTS

CLIMATE

- hot throughout the year (30°C–40°C)
- large diurnal ranges (50°C–0°C)
- low and unreliable rainfall (≤ 250 mm per year).

VEGETATION

Vegetation from desert margins is often referred to as scrub. There are two main types.

- **Tropical scrub** on the margins of hot deserts include acacia, cactus, succulents, tuberous rooted plants, and herbaceous plants that only grow with rain. Special types are mulga in Australia (dense acacia thickets), spinifax in Australia ('porcupine grass'), and chaparral in Chile (spiny shrubs).
- **Temperate scrub** on the margins of temperate deserts includes maquis, chaparral, and garrigue, malee in Australia (dense dwarf eucalyptus), brigalow (acacias), and mulga (acacias). Sage brush is a mass of heath-like shrubs about 2 m high.

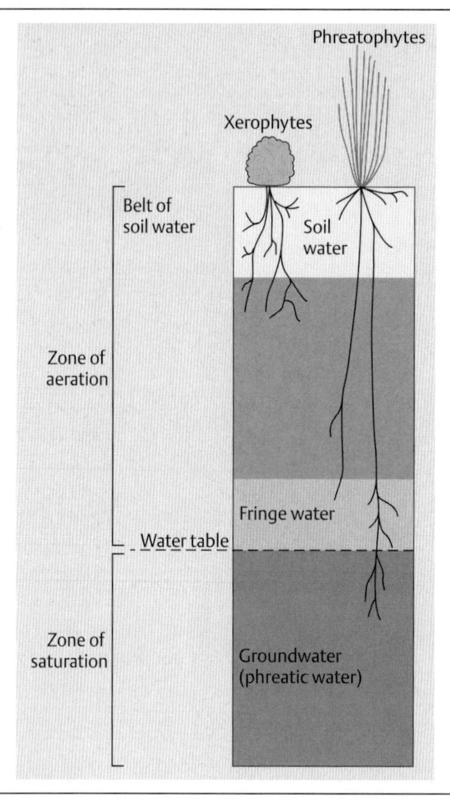

PLANT ADAPTIONS

Plants here are adapted to drought by:

- growing long, water-seeking roots
- having few leaves to reduce transpiration
- storing water (succulents).

SOILS

- little organic or moisture content
- few nutrients.

ANIMALS

Animals are adapted to living in the desert in a number of ways. These include:

- nocturnal (night time) activity to avoid the heat of the day
- panting and /or large ears to reduce body heat
- burrowing by day
- secreting highly concentrated uric acid thereby reducing water loss
- seasonal migration
- long-term aestivation (dormancy) that ends only when triggered by moisture and temperature conditions.

MEDITERRANEAN WOODLAND

The proper term for the vegetation is **sclerophyll**, that is, evergreen and drought-resistant.

CLIMATE

- Typically warm, western margin or Mediterranean.
- > 26°C in summer, and over 6°C in winter.
- Rainfall occurs mostly in winter; there are long, dry, hot, sunny summers (20–30°C) and drought is common.
- Temperatures are never too cold for plant growth, but growth is checked by drought in summer, and winters are too cool for vigorous growth.

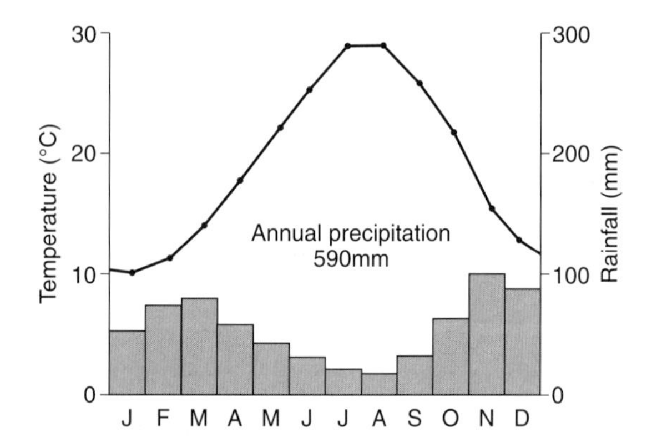

SOIL

- The soil is frequently a red terra rosa soil.
- It often has with a thin litter layer and is generally lacking in phosphorus.

VEGETATION

Typical vegetation is adapted to drought by thick leathery leaves, sunken stomata, and waxy cuticles. Many trees possess needle-like leaves which reduce evapotranspiration losses. In addition to the trees, are dense, impenetrable scrub. Herb species are also common. Key aspects of the vegetation include:

- NPP is about 700 $g/m^2/year$
- biomass is about 6 kg/m
- evergreen trees such as eucalyptus, pine, and cedar
- low-grazing evergreen shrubs include laurel and myrtle
- annual plants have a quick cycle through growth, flowering and seeding in spring; plants with tuberous roots, such as the lily and tulip, flower in spring and then die down.

PLANT ADAPTATIONS

Trees and plants withstand the summer drought by:

- having long roots to bring water from a depth
- storing water by means of bulbous roots
- reducing transpiration through small, shiny leaves that reflect sunlight, woody stems, and thick rough bark.

HUMAN IMPACT

There has been a long history of human impact in Mediterranean areas. These impacts include: deforestation, cultivation, terracing, irrigation, urbanisation, and fire.

Tundra ecosystems

CLIMATE

- Long winters. Temperatures below freezing point for up to 11 months of the year.
- Short summers. Temperatures between 6°C and 10°C for up to 3 months of the year.
- Little precipitation – about 250 mm per year. Occasional frosts in summer, snow in winter.
- Strong winds, causing severe windchill.

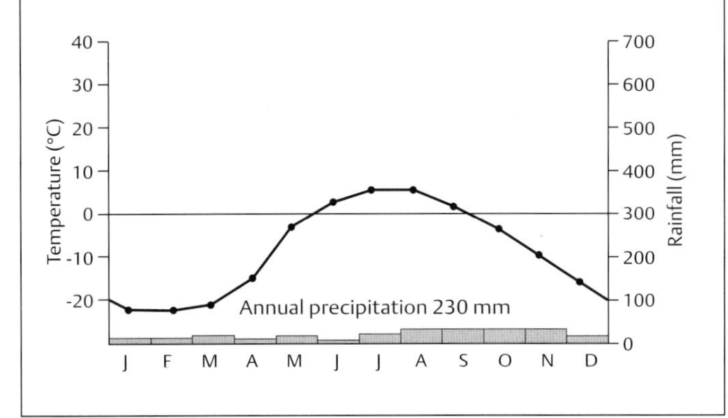

VEGETATION

Low-lying plants with wide roots have evolved in response to the strong winds and impermeable permafrost. Coarse (cotton) grass, herbs, lichen, and moss are the main plant-types. In the southern parts of the tundra and in the sheltered areas, on warmer south-facing slopes some trees may be found, such as dwarf willow and dwarf birch.

- NPP is low, 140 g/m^2/year, due to the short growing season.
- Biomass is low, 0.6 kg/m, also due to the short growing season.
- Plants have to tolerate acidic soils.
- Many plants are able to photosynthesise at low temperatures (2–3°C) to make use of the long daylight hours in summer.
- Most plants adopt a cushion-shaped form as a protection against wind and excessive transpiration, for example, mosses, lichens, saxifrages, and bilberry.
- Plants are adapted to the short growing seasons by pollinating in one year, and germinating and flowering in the next year.

HUMAN IMPACT

Human activities in tundra areas are increasing rapidly. The impact is especially important as it takes a long time for the ecosystem to recover. This is largely due to the low temperatures, short growing season, and low rates of biological activity. The main activities include:

- oil drilling, such as in Alaska
- iron ore mining, as at Kiruna in Sweden
- gold mining in Alaska
- tourism in Alaska and in Lapland
- transport, such as the Trans Alaska Highway and the Trans Siberian express
- military exercises, especially in Norway.

Precipitation

Biomass

Litter

Run-off

Soil

Weathering

NUTRIENT CYCLE IN THE TUNDRA

ANIMALS

There are a number of adaptations that animals make in order to survive tundra environments. They include:

- pigmentation (colour changes) in Arctic hares, Arctic foxes, weasels, and ptarmagins; these turn white in winter and brown in summer
- insulation with lots of fur
- small size and shape to reduce heat and moisture loss
- migration to find food supplies, such as by caribou and by wild fowl.

SOILS

Soils in tundra areas are developed upon permafrost. They thaw out in summer, but only to a depth of a few centimetres. In addition, soils are often waterlogged because of the low temperatures and hence, low evaporation rates. The most common soils are gleys on the flatter slopes and podzols or gleyed podzols on steeper slopes. Soils are:

- shallow
- infertile
- acidic
- waterlogged
- peaty.

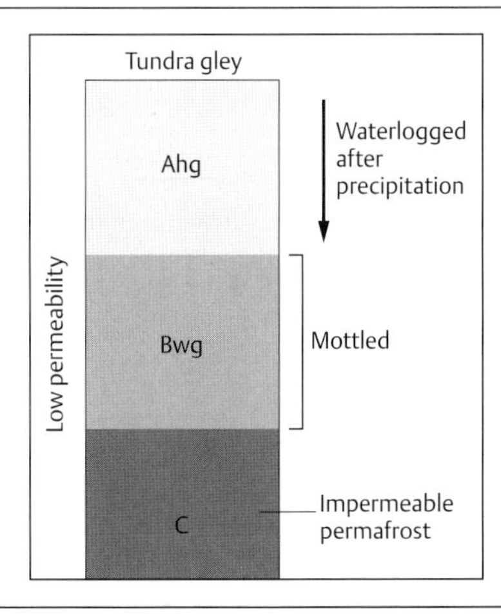

There is an accumulation of organic matter due to the low temperatures and the acidic nature of the soil. The acidity hinders the development of soil fauna which help to break down organic matter. Few varieties of plant can survive in the sour acid soil, full of decaying vegetation which is slow to form humus because of the cold climate. There are few nutrients in tundra soils.

Heathland and moorland

A COMPARISON OF HEATHLAND AND MOORLAND		
	Heathland	**Moorland**
Vegetation	heather (erica; ericoid; calluna)	sphagnum (moss)
Soil	podzol	gleyed soil; peaty soils
Geology	outwash sands and gravels, coastal sands	uplands – hard igneous rock; metamorphic, granite, basalts; lowland – clay
Rainfall	800 mm	highland areas >2000 mm; lowlands 800 – 1000 mm
Altitude	low, below 250 m	high ≥300m except clay e.g. Otmoor, Oxfordshire
Human impact	urbanisation; mineral extraction; recreation	not very good for agriculture nor urbanisation, but increasingly used for afforestation, recreation (grouse moors)
Adaptions of plants	fire-regeneration (burning heather increases productivity)	sphagnum extracts nutrients directly from the atmosphere

THE EVOLUTION OF BOGS AND HEATHS

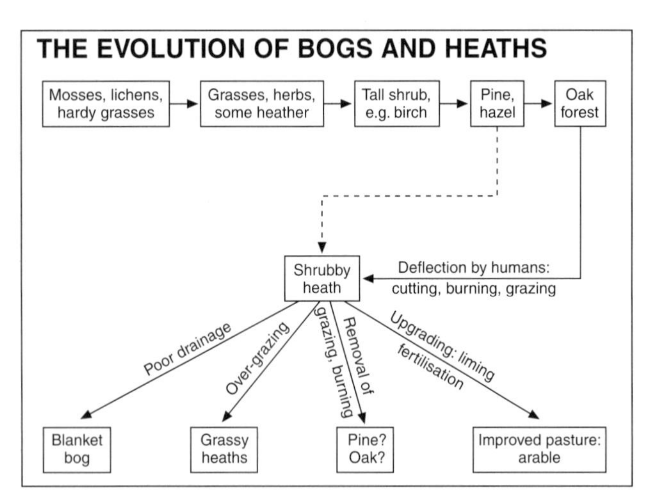

THE DECLINE OF LOWLAND HEATH

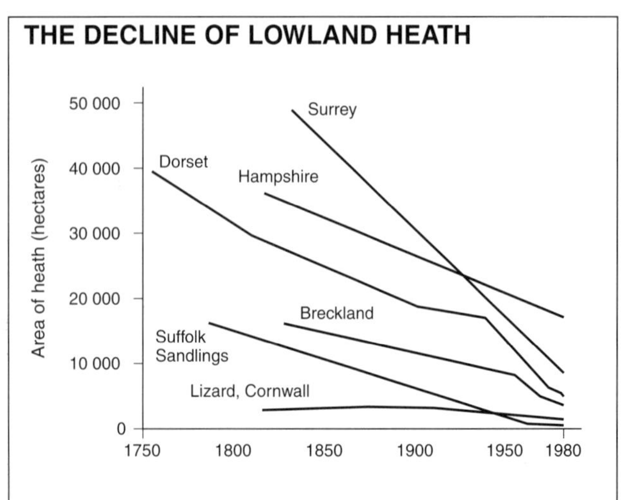

DISTRIBUTION OF HEATHLAND AND MOORLAND

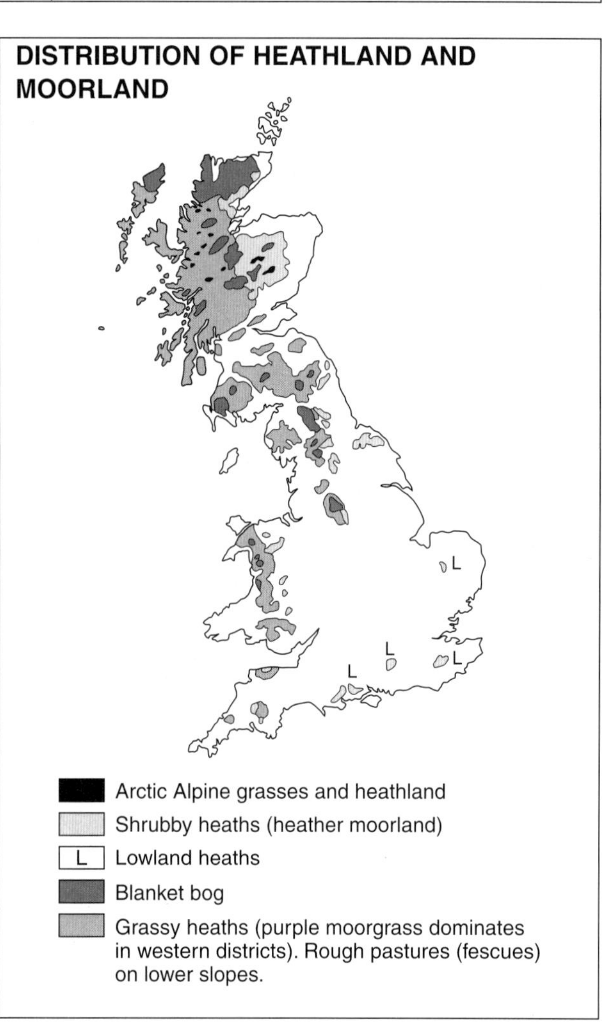

PLANT COMMUNITIES ON MOORLAND AND BOG

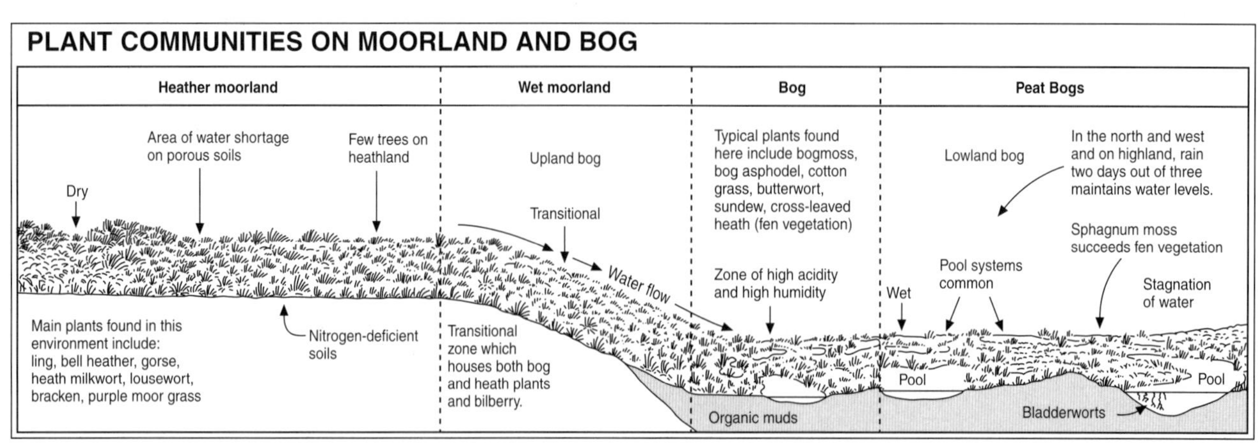

Distribution and change

POPULATION DISTRIBUTION

Population distribution refers to where people live. On a global scale

- 75% of the population live within 1000 km of the sea
- 85% live in areas less than 500 m high
- 85% live between latitudes 68°N and 20°N
- less than 10% live in the southern hemisphere.

The most favoured conditions include:

- fertile valleys
- a regular supply of water
- a climate which is not too extreme
- good communications.

Disadvantaged areas include deserts (too dry), mountains (too steep), high latitudes (too cold), and rainforests (too infertile).

There is no such thing as a 'best' climate - many people live in south east Asia, and this has a monsoonal climate, with hot, wet seasons and hot, dry seasons.

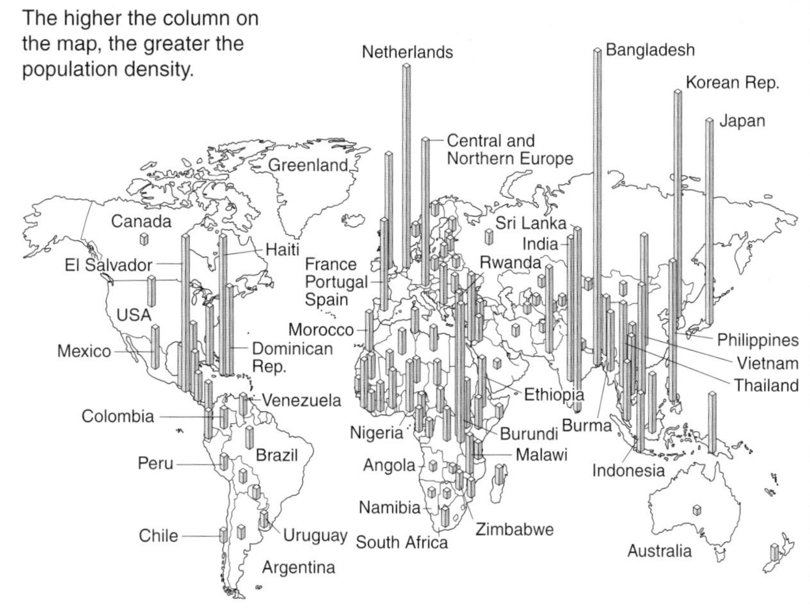

High-rises – relative population densities around the world (people per km^2)

CASE STUDY: NATIONAL DISTRIBUTION IN THE UK, 1991

National level

- The South-East has one-third of the population living on a tenth of the land area.
- Scotland has one-tenth of the population living on one-third of the land area.
- The most congested region is the North-West with an average of 868 persons per square kilometre.

Low density

- Remote upland regions.
- Harsh physical conditions and a lack of economic opportunity.
- In parts of the Scottish Highlands there are large areas with no population.

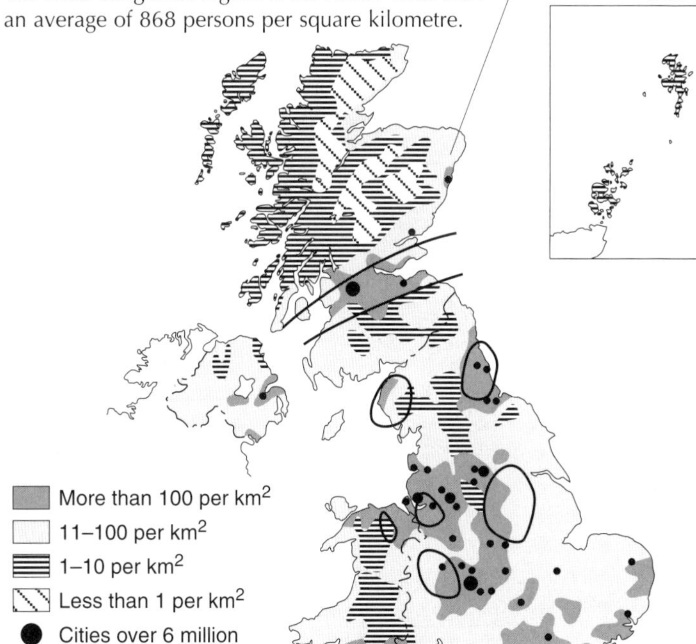

Urban-rural shift

- Counter-urbanisation - the shift of people down the settlement hierarchy.
- Growth of 'rural' counties - Cornwall through to Norfolk and Lincolnshire, with Cambridge growing by 13.7% between 1981 and 1991.
- The reasons are numerous:
  - regional policy - green belt forces people to leapfrog
  - rise of commuting
  - decentralisation of industry
  - social and environmental problems in the cities.

Explanation

- Coalfields were initially important to the rapid growth of industrial towns and subsequent agglomeration - they provided pools of skilled labour, large markets, industrial linkages, and inertia.

High density

- Conurbations have the highest concentration.
- Especially the inner suburbs of the large industrial cities.
- Greater London contains 4308 persons per square kilometre.

Fertility and mortality

FERTILITY

Fertility is the measured capacity of a population to generate births. Two measures are used:

- **crude birth rate** - the number of births in a given year divided by the population and multiplied by 1000 (reliant on accurate data and does not account for age-sex structure)
- **total fertility rate** - the number of children born to 1000 women passing through the child-bearing ages (assuming none of the women die); a fertility rate of 2.1 births per woman indicates replacement level (stable population)

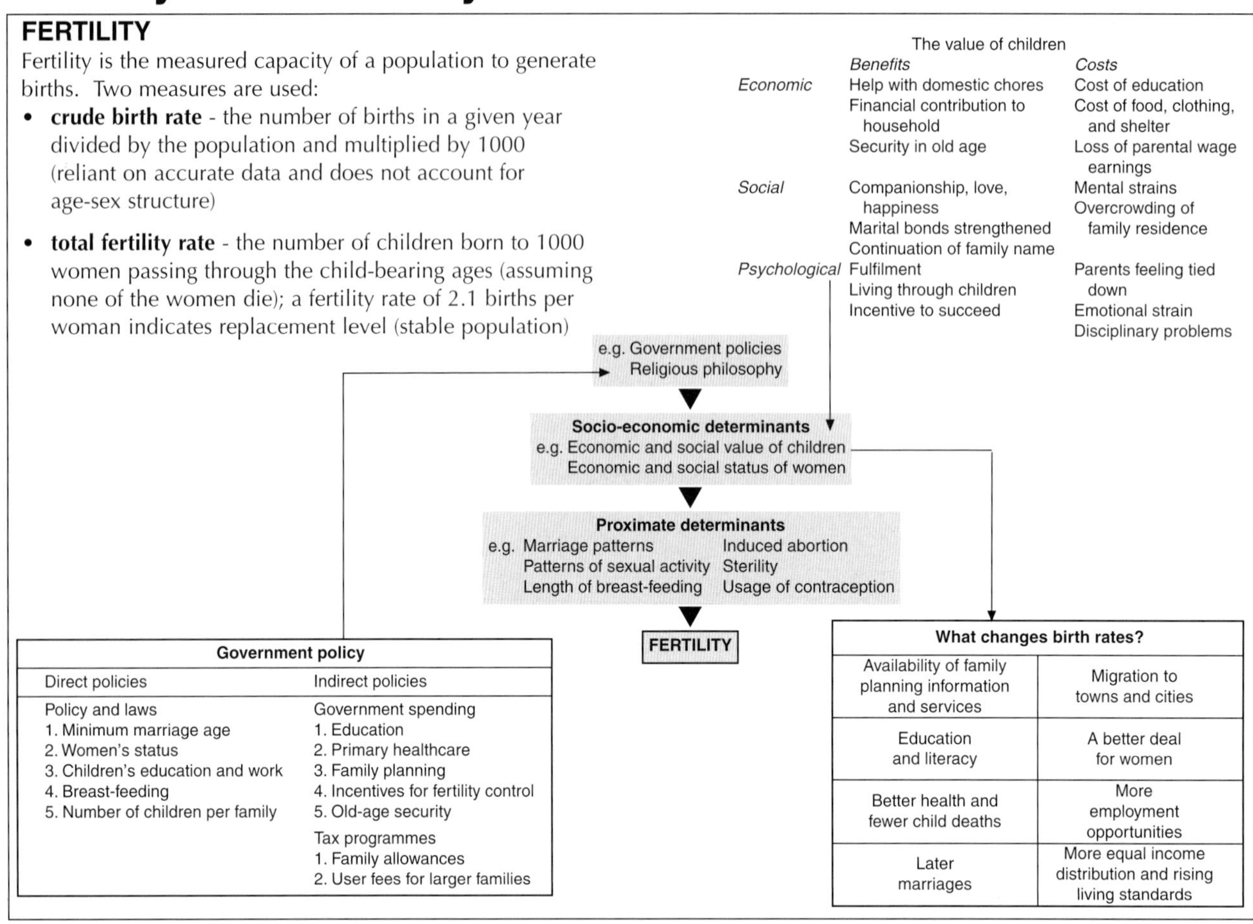

MORTALITY

Mortality is different from fertility as it tends to be more stable and predictable. There are a number of measures:

- **crude death rate** - the number of deaths in a specific period per 1000 of population (distorted by age structure)
- **age-specific mortality rates** - the number of deaths of people of a certain age per 1000, e.g. infant mortality rate
- **life expectancy at birth** - the average number of years a person can expect to live.

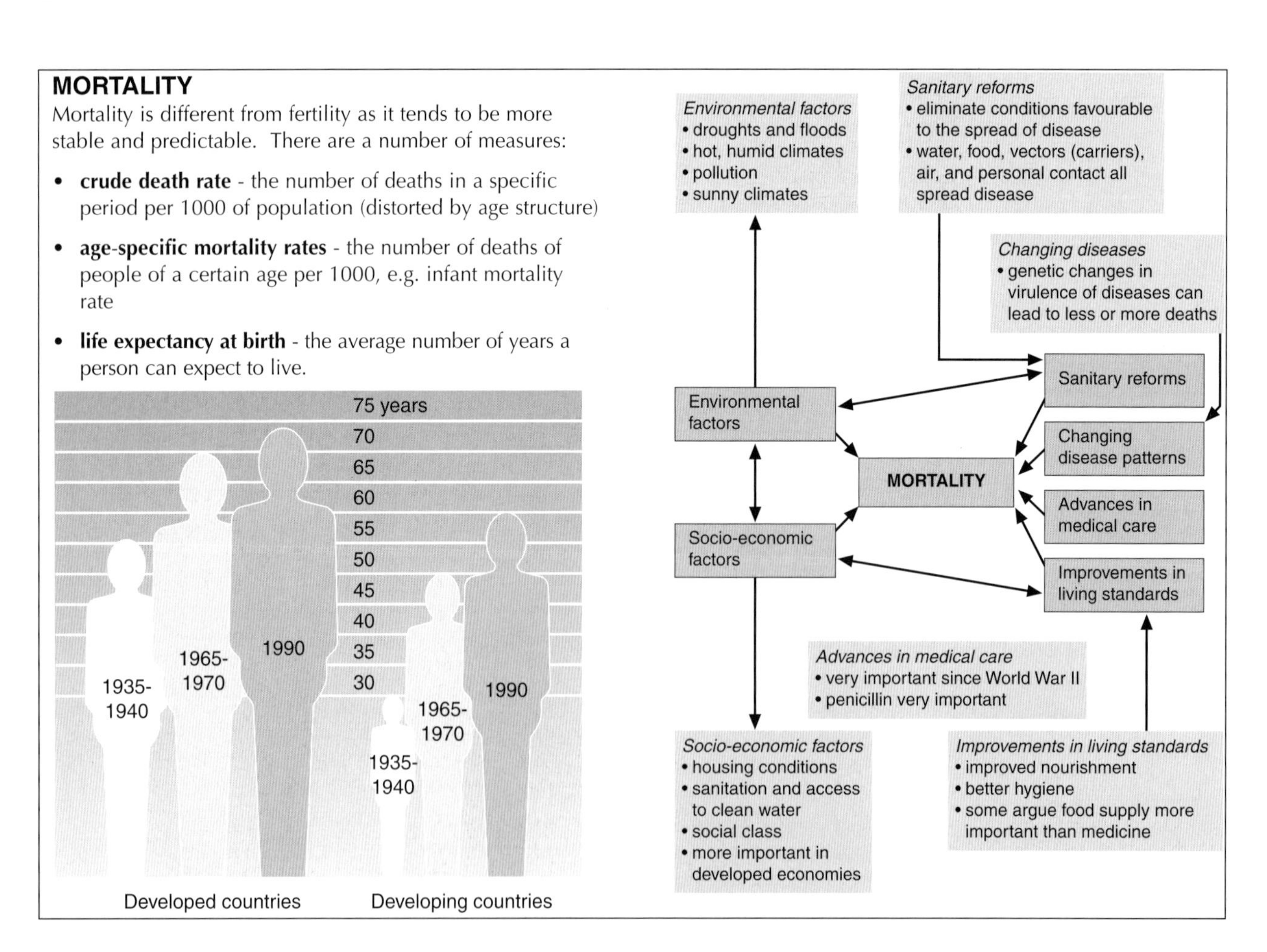

The demographic transition model (DTM)

Stage 1: High stationary
- both birth rate and death rate high
- population fluctuates
- England and Wales were at this stage until 1750
- There are no countries at this stage, although some indigenous (native) tribes still are.

Stage 2: Early expanding
- birth rate stays high
- death rate falls rapidly
- life expectancy increases
- population grows very quickly
- England and Wales passed through this stage 1750-1850
- countries such as Afghanistan, Sudan, and Libya are at this stage.

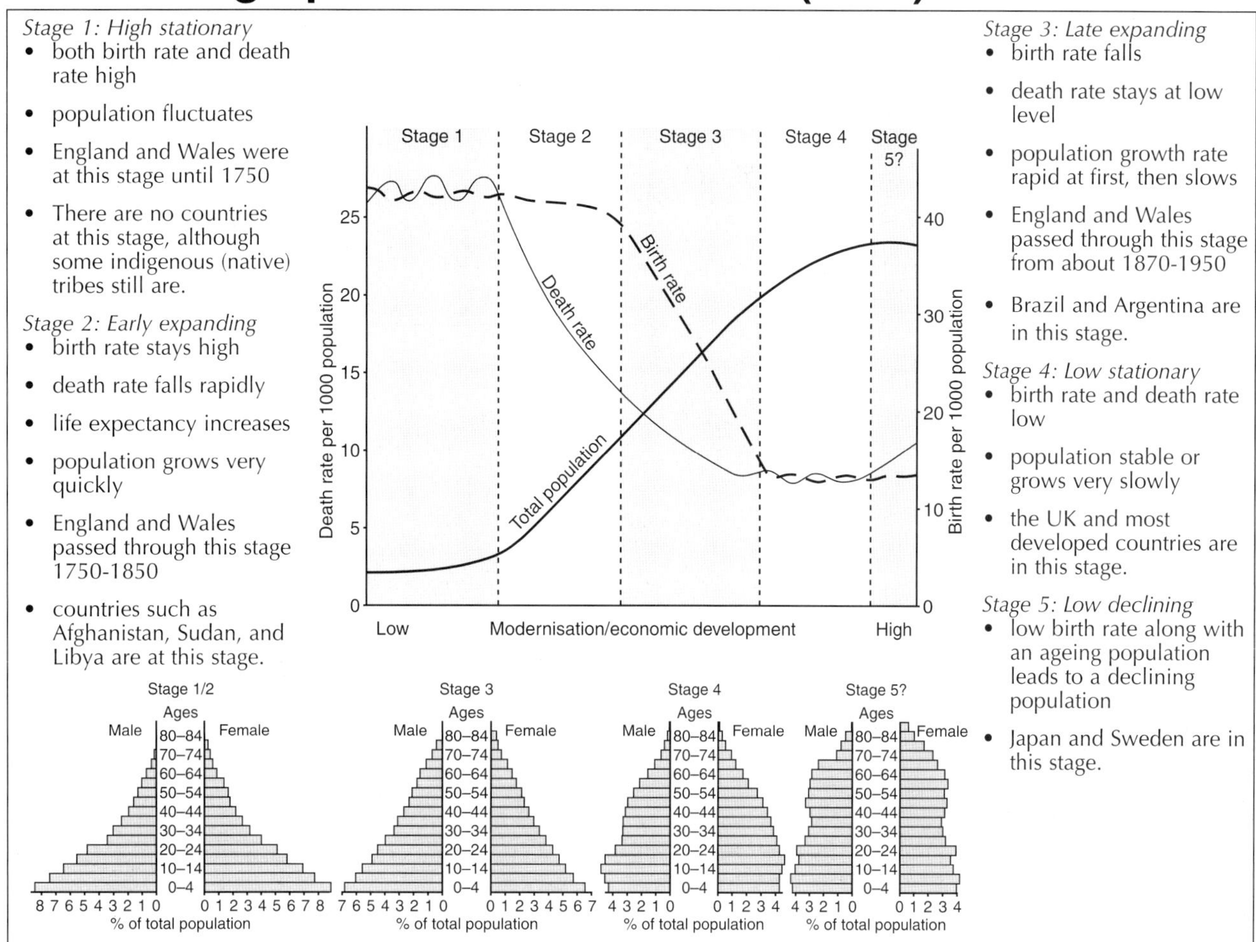

Stage 3: Late expanding
- birth rate falls
- death rate stays at low level
- population growth rate rapid at first, then slows
- England and Wales passed through this stage from about 1870-1950
- Brazil and Argentina are in this stage.

Stage 4: Low stationary
- birth rate and death rate low
- population stable or grows very slowly
- the UK and most developed countries are in this stage.

Stage 5: Low declining
- low birth rate along with an ageing population leads to a declining population
- Japan and Sweden are in this stage.

The demographic transition model states that both a population's mortality and fertility will decline from high to low levels as a result of social and economic development. It is based on the European experience, but has also been used as a predictive tool to explain change in the developing world.

DEMOGRAPHIC TRANSITION AND GLOBAL TRENDS

Group 1 - Africa:
High birth rate, increasing growth rate.

Low income; low status of women; large rural communities; lack of infrastructure; child workers

Group 2 - Asia/Latin America:
Declining birth rates and death rates, stabilising growth rates.

Increased urbanisation and industrialisation; emancipation of women, at least in the workforce; effective family planning measures; compulsory education

Group 3 - the West:
Declining birth rates, declining/fluctuating death rates, falling growth rates.

Ageing population; women marrying later; consumer society; readily available contraception; 'selfish' society

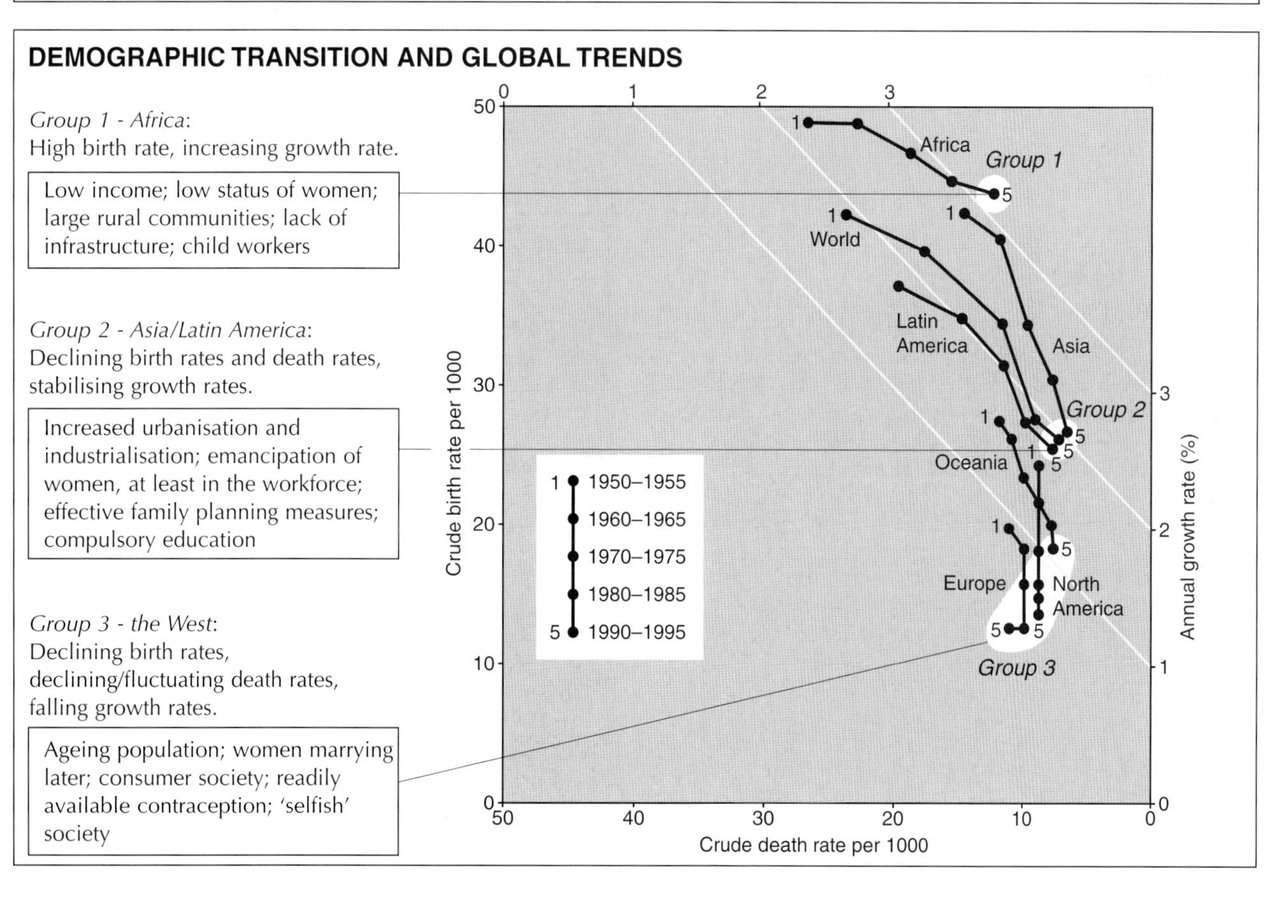

The DTM in Britain

Stage 1
Britain was in stage 1 of the demographic transition model (DTM) until about 1740. Birth rates were high and stable - between 30 and 40 per thousand, while death rates were generally slightly lower. Death rates rose somewhat between 1720 and a 1750, a period associated with cheap gin drinking.

Stage 2
Britain passed through stage 2 between 1740 and about 1880. The death rate fell from 35 per thousand to about 20 per thousand. This was associated with general increases in food availability, and improvements in sanitation and water supply. Death rates increased for a while in the 1820s and 1830s, associated with rapid urbanisation. The 1848 and 1875 Public Health Acts helped improve the supply of clean water and sanitation. At the same time birth rates remained high, 30 to 35 per thousand.

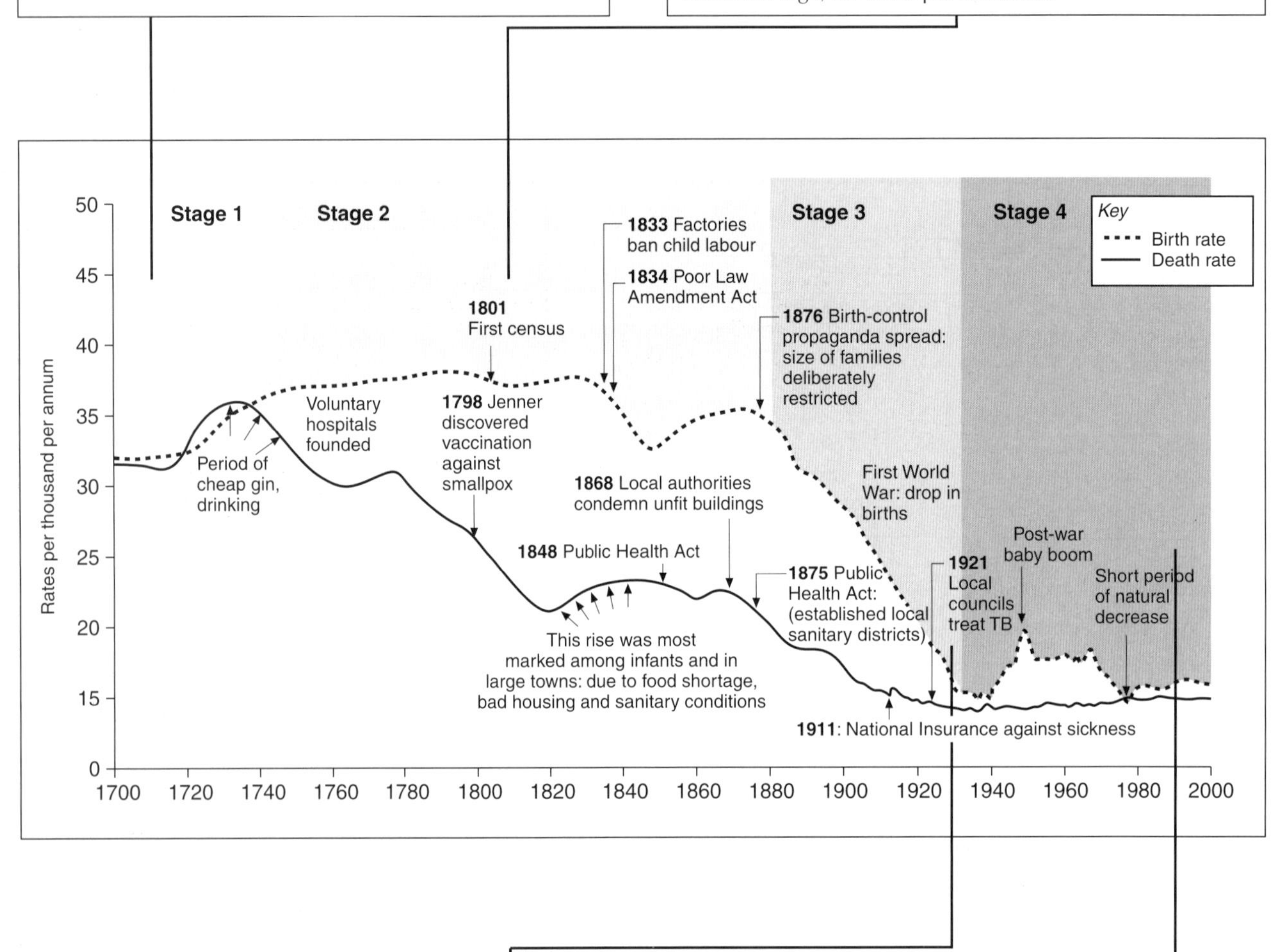

Stage 3
Stage 3 was marked by a fall in birth rates to around 15 per thousand. Birth control was promoted and important changes in British society also helped lower the birth rate. The change from an agricultural to urban society is often associated with a decrease in the birth rate. The death rate continued to decline - national insurance, pensions, and local council treatment of TB - helped raise the health standards of a large proportion of the population.

Stage 4
Stage 4 has been marked by low and variable birth and death rates. There was a marked post-war baby boom in the late 1940s, and there have been short-term increases in the death rate linked with periods of smog and outbreaks of disease. Britain's death rate is likely to increase in forthcoming years as the population ages.

UK: live births 1951–74

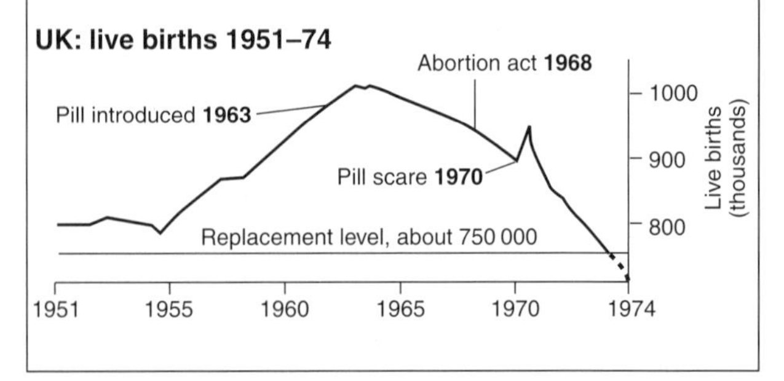

Optimum population and population policy

OPTIMUM POPULATION

Optimum population
- the size of population which permits the full utilisation of the natural resources of an area giving maximum per capita output and standard of living

Underpopulation
- population is too small to develop its resources effectively

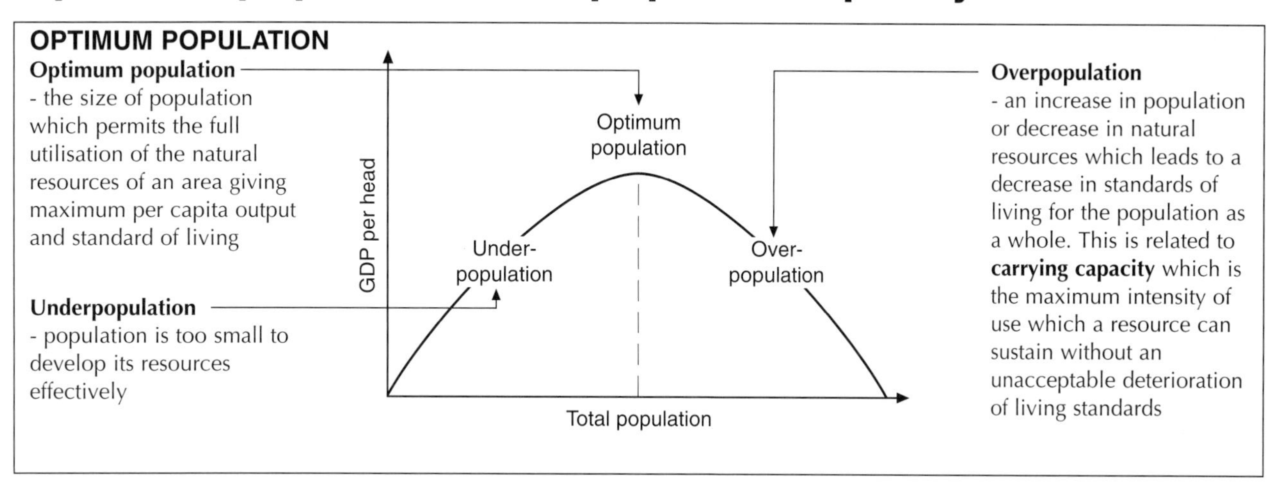

Overpopulation
- an increase in population or decrease in natural resources which leads to a decrease in standards of living for the population as a whole. This is related to **carrying capacity** which is the maximum intensity of use which a resource can sustain without an unacceptable deterioration of living standards

POPULATION-RESOURCE REGIONS (ACKERMAN)

Optimum population is based on the combination of three factors: **population density, resources,** and **technology**. Regions of the world can be classified according to population-resource ratios.

Ackerman's classification shows many of the problems associated with the concept of optimum population:

- only measured in economic terms
- little time given to social or regional inequalities within a country
- population structure (active population, dependency ratio) is not considered.

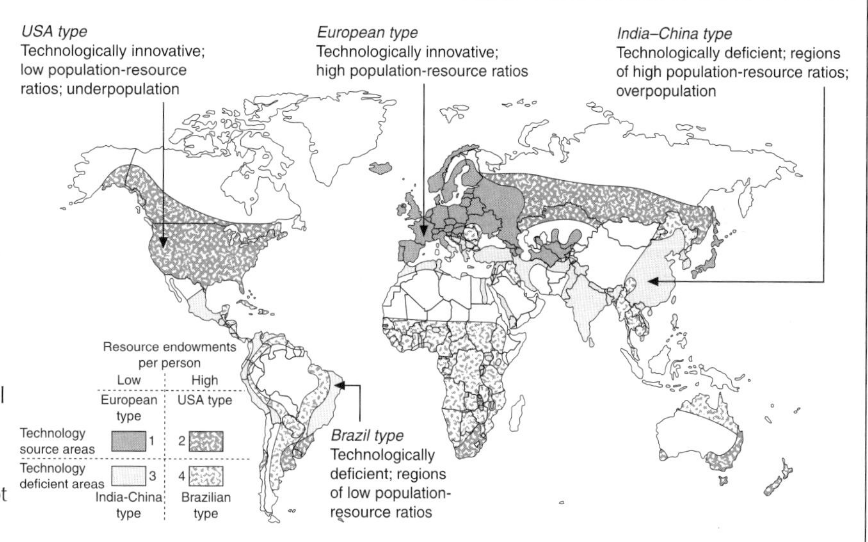

CASE STUDY: POPULATION POLICY IN CHINA

China has used the concept of optimum population to stabilise its population at 1.2 billion by the year 2000 and to reduce the population to a government-set optimum of 700 million within a century. A number of different policies have been initiated to achieve this.

Great Leap Forward (1959-62, not really a 'policy')
- Chinese leadership wanted industrialisation at any cost
- food production ceased to be a top priority
- this coupled with drought led to famine
- population dropped by 14 million and an estimated 25 million babies were not conceived or did not survive
- infant mortality rate rose to 284/1000 (today the figure is 30/1000).

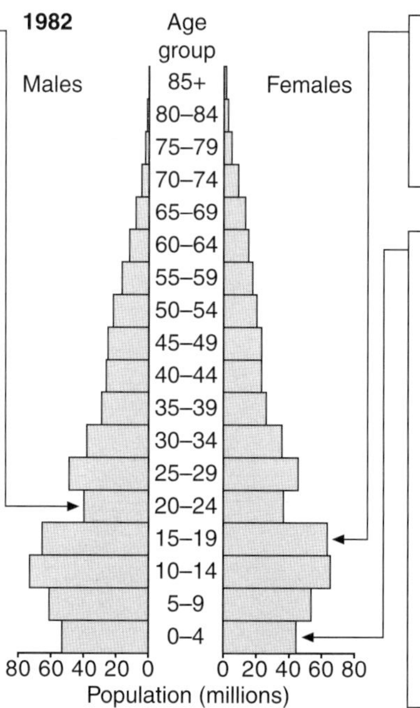

Retirement bulge
- in 50 years time people over 65 will make up about 25% of the population.

One-child policy (late 1970s)
- 'only children' afforded preferential treatment in education, housing, employment
- one-child families rewarded with salary bonuses
- families with two or more children lost 10% of their salaries
- compulsory sterilisation and forced abortions.

Evaluation of policy
- fertility rates have fallen from 2.25 to 1.9
- effectiveness has been concentrated in the large cities of the north-eastern seaboard
- much foreign criticism of the government's aggressively enforced policy
- the one-child policy has also given rise to the 'Little Emperor' syndrome - spoilt and overweight only children
- evidence of resistance, especially in rural areas.

Migration

DEFINITION

There are various forms of spatial mobility, but not all are considered migration. Population movements take four forms:

	Recurrent	Non-recurrent
Local	Commuters	Intra-urban residential relocation
Extra-local	Seasonal or temporary workers, students	MIGRATION

Migration is defined as a movement involving a change in permanent residence with a complete readjustment of the community affiliations of the migrant.

CLASSIFICATIONS

Migration can be classified in terms of distance (internal or international), time (temporary or permanent), and origin (rural or urban).

Conditions which cause migration can involve both 'push' and 'pull' factors.

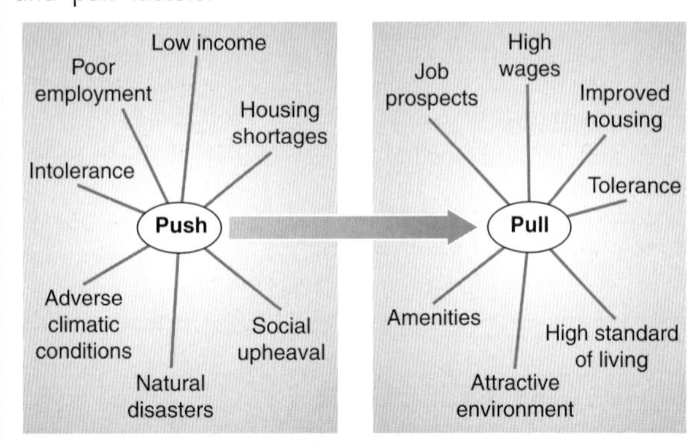

TYPES OF MIGRATION

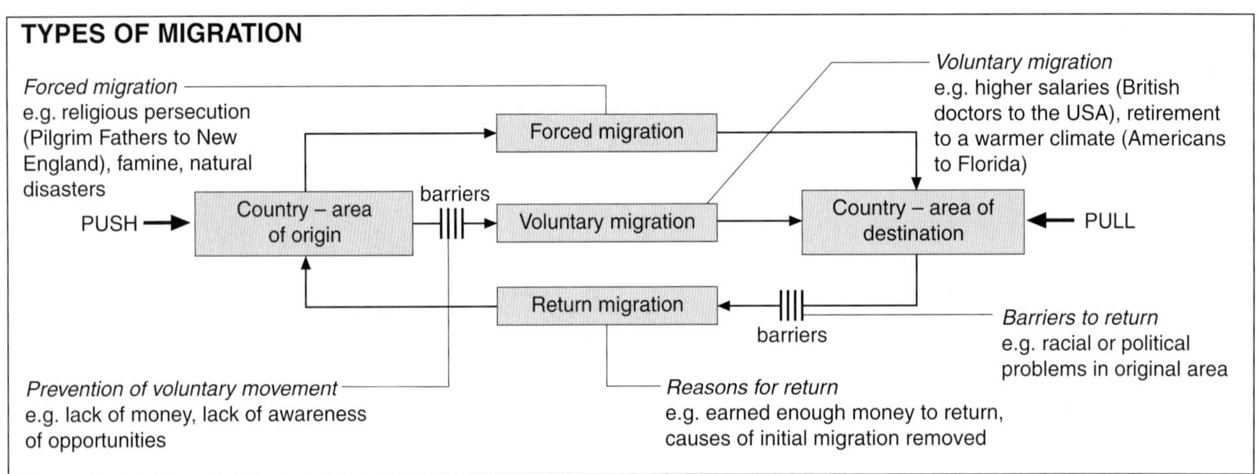

THE IMPACT OF MIGRATION

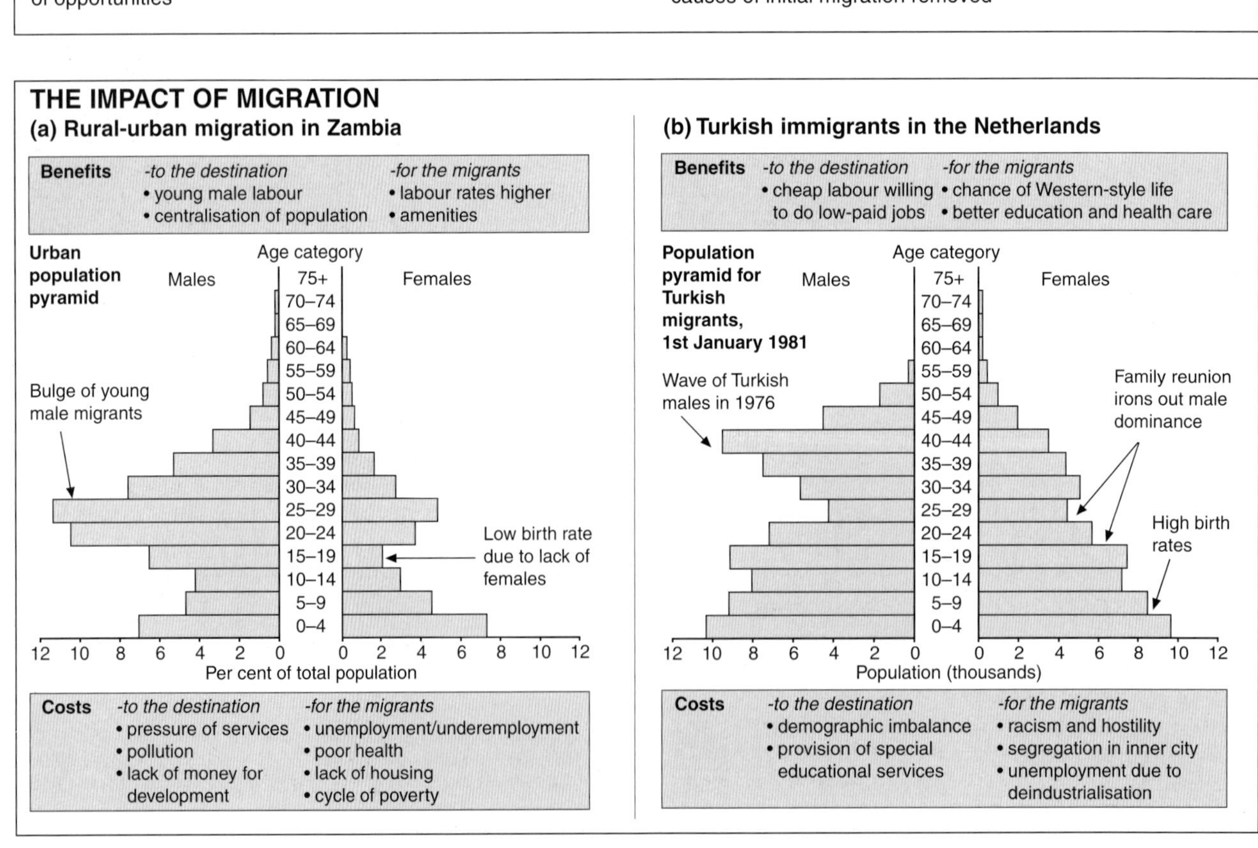

Population and gender

In many countries high rates of population growth are associated with a low status of women in society. This is for a variety of reasons:

- a wife continues to bear children to prove her fertility and to prevent the husband from marrying another wife
- wives in polygamous families compete with each other to produce the largest number of children
- children provide labour for fetching firewood and water and for digging in the field
- children are an investment as they provide security for the parents in their old age
- women have no effect in determining the size of the family.

The UN Decade for Women from 1975 to 1985 recommended three important points for action:

- there should be legal equality for women
- further development needs to improve on the substandard role that women play
- women should receive an equal share of power.

INEQUALITY OF GENDERS

In 1970 Esther Boserup identified women as having been left behind in the development process. The social roles that women play vary from place to place but in most countries women have three important parts:

- biological reproduction
- social reproduction
- economic reproduction.

These three roles create a great deal of physical and psychological stress. It is believed that in Africa:

- up to one third of women are pregnant or breast-feeding at any one time
- women comprise over half the workforce, sometimes over 70%
- women grow over 80% of the food eaten and contribute half of the region's cash crops.

WOMEN AND DEVELOPMENT

A number of approaches to the study of women and development have emphasised welfare, equality, anti-poverty, efficiency, and empowerment. Strategic or political change is needed to attain equality and empowerment. In many countries this is highly unlikely.

Progress for sexual equality has been painfully slow. For example the illiteracy rate is much higher for girls than for boys, over 70% of African countries have no female cabinet minister, and generally, women are becoming poorer.

Number of girls who miss out on primary school

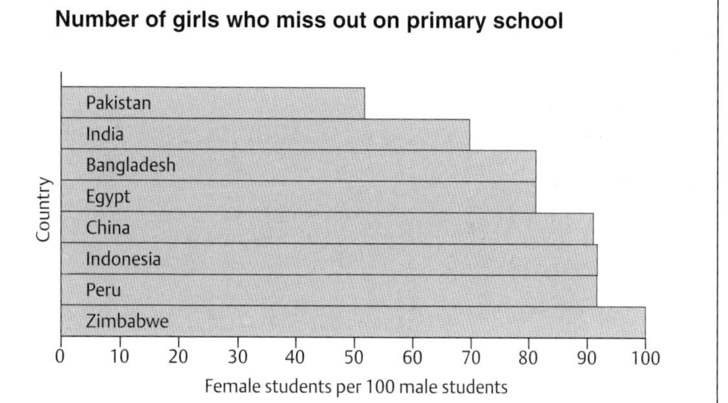

DIFFERENT POLICY APPROACHES TO WOMEN IN AFRICA

Issues	Welfare	Equality	Anti-poverty	Efficiency	Empowerment
Period most popular	1950—70	1975—85	1970s onward	post-1980s	1975 onward
Purpose	Women are given the resources to become better mothers	Women are seen as active participants in the development process	Women's poverty seen as a problem of underdevelopment not of subordination	To ensure development is more efficient and more effective; women's economic participation is linked to equality	To encourage women to be more self-reliant; women's subordination seen as part of colonial oppression
Needs of women met and roles recognised	Food aid, malnutrition addressed and family planning	Reducing inequality with men by allowing political and economic autonomy	Allows women to earn an income in small-scale income-generating projects	Relies on the three roles of women to replace declining social services	Bottom-up role is recognised as women are empowered
Comment	Women are seen in a traditional reproductive role; little change of status	Criticised as western feminism and not popular with governments	Poor women isolated as a separate category; popular with small-scale non-government organisations	Women seen entirely in terms of delivery capacity and ability to extend the working day	Emphasis on self-reliance; largely unsupported at present

THE REASONS FOR SLOW PROGRESS

- Conditions are deteriorating in a large part of Africa. For countries that have managed to borrow money, there are structural adjustment programmes available. These countries have to agree to spend less money thus cuts are made. These cuts are disproportionately borne by women.
- There is a lack of commitment to women by many countries and by donors.
- Age and gender roles are changing. Diseases such as AIDS have had a dramatic impact on populations. AIDS removes a high proportion of young people from the workforce. Women have to work as well as be the head of the household but they have little legal status.

Population composition and population pyramids

POPULATION COMPOSITION

Population composition refers to any characteristic of the population. This includes the age, sex, ethnicity (race), language, occupational structure, and religious make up.

In the UK 19% of the population are aged under the age of 15 years and 16% are aged over the age of 65 years. This is different for ethnic groups. Among Whites 19% are in the under-15 group and 21% over-60 group. By contrast, among Pakistanis and Bangladeshis the proportions are 45% and 3%.

Population composition is important because it tells us about population growth. It helps planners to find out how many services and facilities, such as schools and hospitals, they will need to provide in the future.

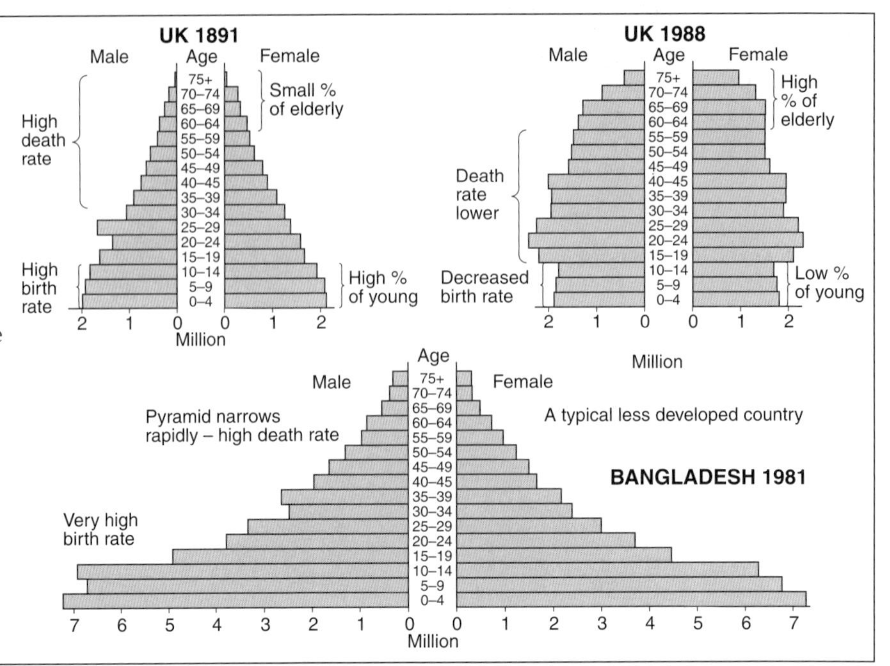

COMBINED POPULATION PYRAMIDS IN SOUTH AFRICA

Overall, South Africa has a population typical of a less developed country.

- low proportion of elderly
- high proportion of young
- high birth rates
- high death rates.

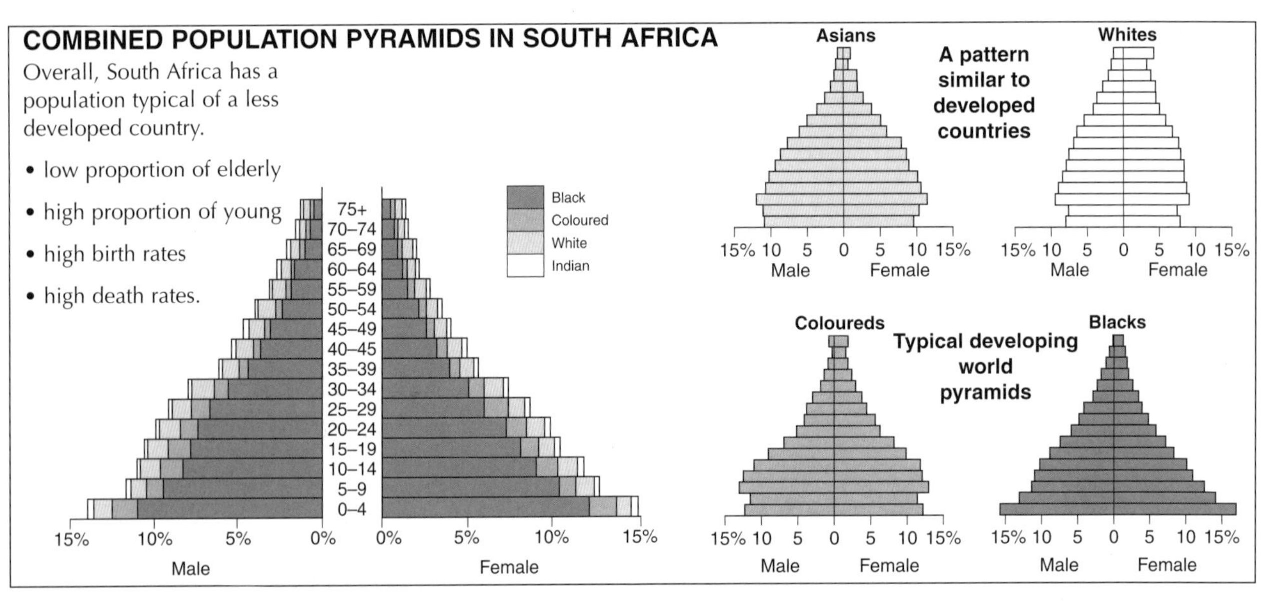

THE DEPENDENCY RATIO

The dependency ratio measures the working population and the dependent population. It is worked out by a formula:

$$\frac{\text{Population aged} \leq 15 + \text{population aged} \geq 60 \text{ (the dependents)}}{\text{Population aged 16-59 (the workers)}}$$

It is very crude. For example, many people stay on at school after the age of 15 and many people work after the age of 60. But it is a useful measure to compare countries.

- In the developed world there is a high proportion of elderly.
- In the developing world there is a high proportion of youth.

These can be shown on a triangular graph.

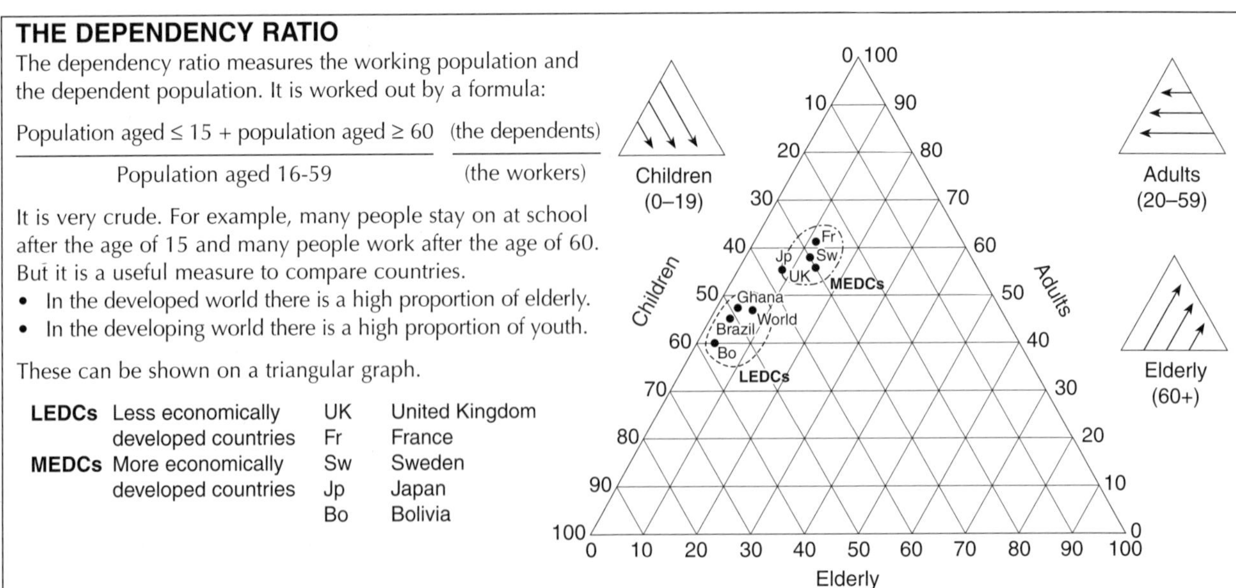

LEDCs	Less economically developed countries	UK	United Kingdom
		Fr	France
MEDCs	More economically developed countries	Sw	Sweden
		Jp	Japan
		Bo	Bolivia

Population and resources

The world's population is growing very rapidly. 95% of population growth is taking place in LEDCs. Population growth creates great pressures on governments to provide for their people; increased pressure on the environment; increased risk of famine and malnutrition; and greater differences between the richer countries and the poorer countries.

THE PRINCIPLES OF THOMAS MALTHUS

In his *Essays on the Principle of Population Growth,* 1798, the Reverend Thomas Malthus predicted that there was a finite population size in relation to food supply, and that any increase in population beyond this point would lead to a decline in the standard of living and to 'war, famine and disease'. His theory was based on two principles:

- food supply at best only increases at an arithmetic rate, i.e. 1, 2, 3, 4, 5 etc.
- population grows at a geometric or exponential rate, in the absence of checks, i.e. 1, 2, 4, 8 etc.

Malthus suggested two ways in which population growth could be reduced:

- delayed age of marriage
- abstinence from sex.

Relationship between population and food supply, after Malthus

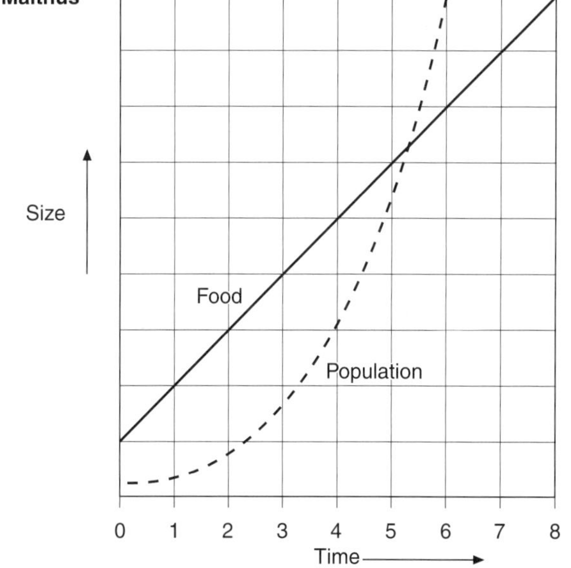

THE THEORIES OF ESTHER BOSERUP

Esther Boserup, 1965, believed that people have the resources of knowledge and technology to increase food production. She suggested that in a pre-industrial society, an increase in population brought about a change in agricultural techniques to allow for an increase in food production. Thus population growth enabled agricultural development to occur.

Boserup assumed that people knew of the techniques required by more intensive systems and used them when the population grew. If knowledge was not available then the agricultural system would regulate the population size in a given area. People would find new ways of increasing food production.

THE RISE OF WORLD POPULATION

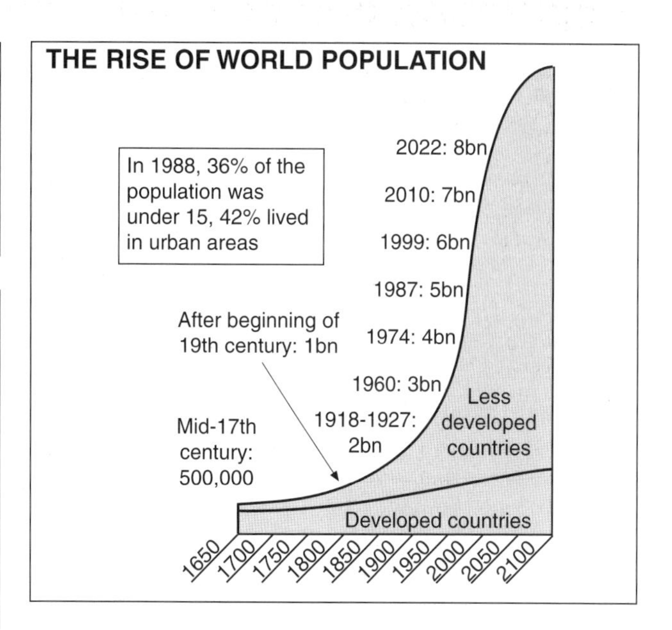

INCREASED FOOD PRODUCTION

There have been many ways since Malthus's time that people have increased food production. These include:

- draining marshlands
- reclaiming land from the sea
- cross breeding of cattle
- developing high yielding varieties of plants
- terracing on steep slopes
- growing crops in greenhouses
- using more sophisticated irrigation techniques
- making new foods such as soya
- making artificial fertilisers
- farming native species of crops and animals
- fish farming.

Views of population growth

Neo-Malthusian

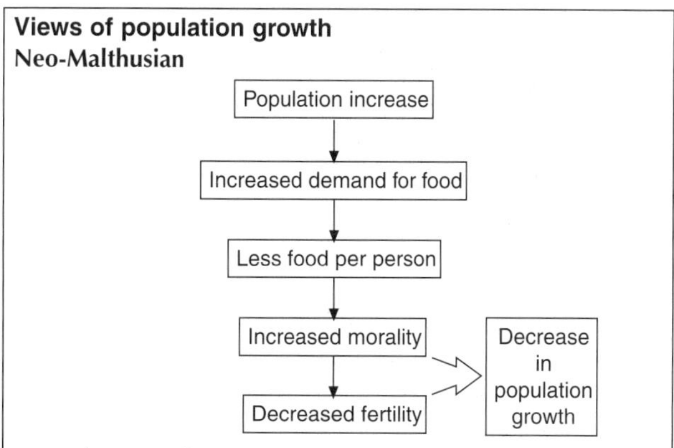

Expanding population means increasing food production causing environmental and financial problems.

Resource optimists (Boserup)

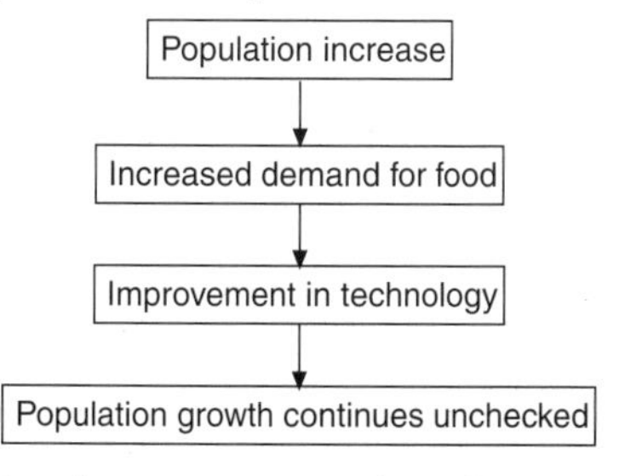

People are the ultimate resource - through innovation or intensification humans can respond to increased numbers.

Limits to growth

The Limits to Growth was a report of the world's resources and the likelihood of those resources running out. The five basic factors that determine and ultimately limit growth on the planet are:

- population
- agricultural production
- natural resources
- industrial production
- pollution.

Many of these factors were observed to grow at an **exponential rate**. Exponential growth is seen when a factor increases by a contrast percentage. The authors of the model illustrated exponential growth by considering the growth of lilies on a pond. The lily patch doubles in area every day. Once half the pond is covered by lilies, it will only be another day for it to be covered totally. This emphasised the apparent suddenness with which the exponential growth of a phenomenon approaches a fixed limit.

ORIGINAL LIMITS TO GROWTH

Food-induced output and population grows exponentially until a rapidly diminishing resource base forces a decline in industrial growth. Natural delays in the system, mean population and pollution continue to increase for some time after peak industrialisation. Population growth is finally halted by a rise in the death rate due to decreased food and medical services.

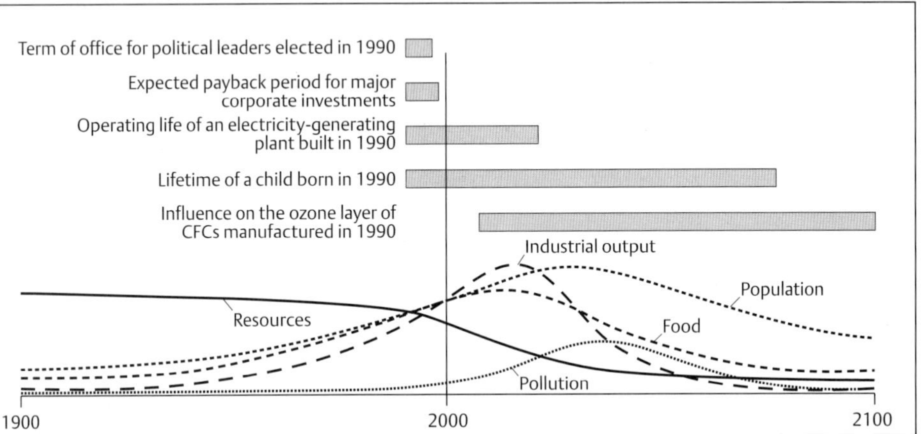

SUSTAINABLE LIMITS TO GROWTH

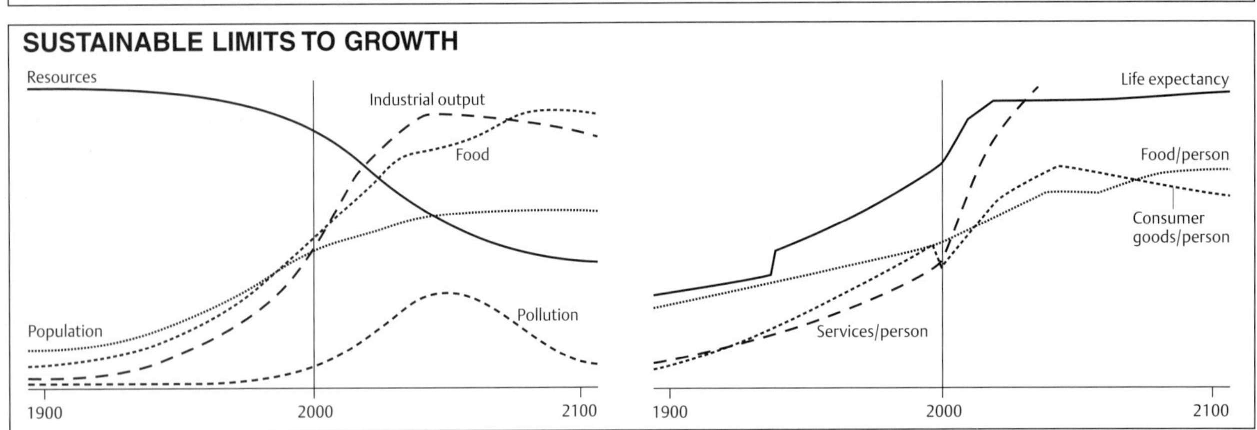

POPULATION GROWTH AND CARRYING CAPACITY

The **carrying capacity** is the number of people that can live at a high standard of living in any given environment. There are three models which show what might happen as a population growing exponentially approaches carrying capacity.

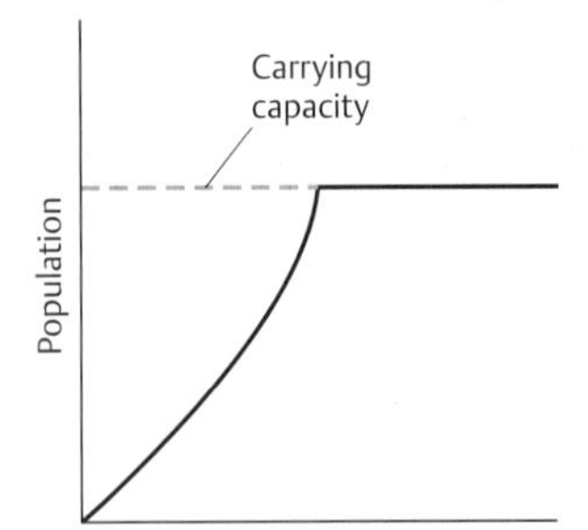

Model 1: The rate of increase may be unchanged until the ceiling is reached, at which point the increase drops to zero. This highly unlikely situation is not supported by evidence from either human or animal populations.

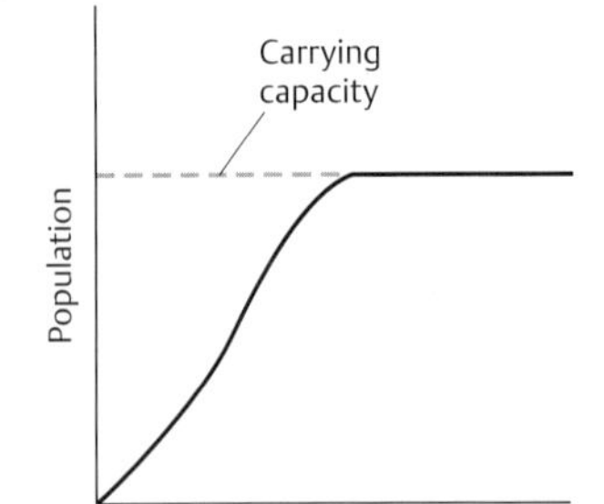

Model 2: The population increase begins to taper off as the carrying capacity is approached, and then levels off when the ceiling is reached. Populations which are large in size, have long lives and low fertility rates, conform to this S-curve pattern.

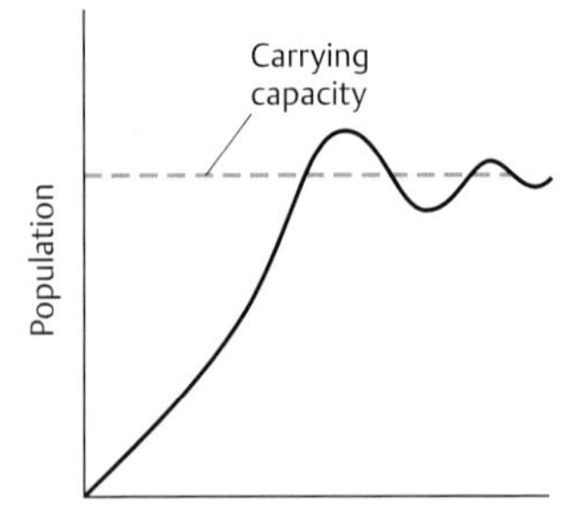

Model 3: The rapid rise in population exceeds the carrying capacity, resulting in a sudden check e.g. famine, birth control. The population recovers and fluctuates, eventually settling down at the carrying capacity. This J-shaped curve is more applicable to populations which are small in number, have short lives and high fertility rates.

Site and situation

SITE

The site is the actual location on which settlements are built. Good sites include:

- well drained, and free from flooding (dry point site)
- close to a reliable source of water (wet point site)
- defendable sites
- south-facing slopes (warm sites)
- sheltered sites
- fertile land

SITUATION

The situation is the location of a settlement relative to other settlements and large physical features. For example, Oxford is situated in the Thames Valley between the Cotswolds and Chiltern Hills, about 60 miles north-west of London.

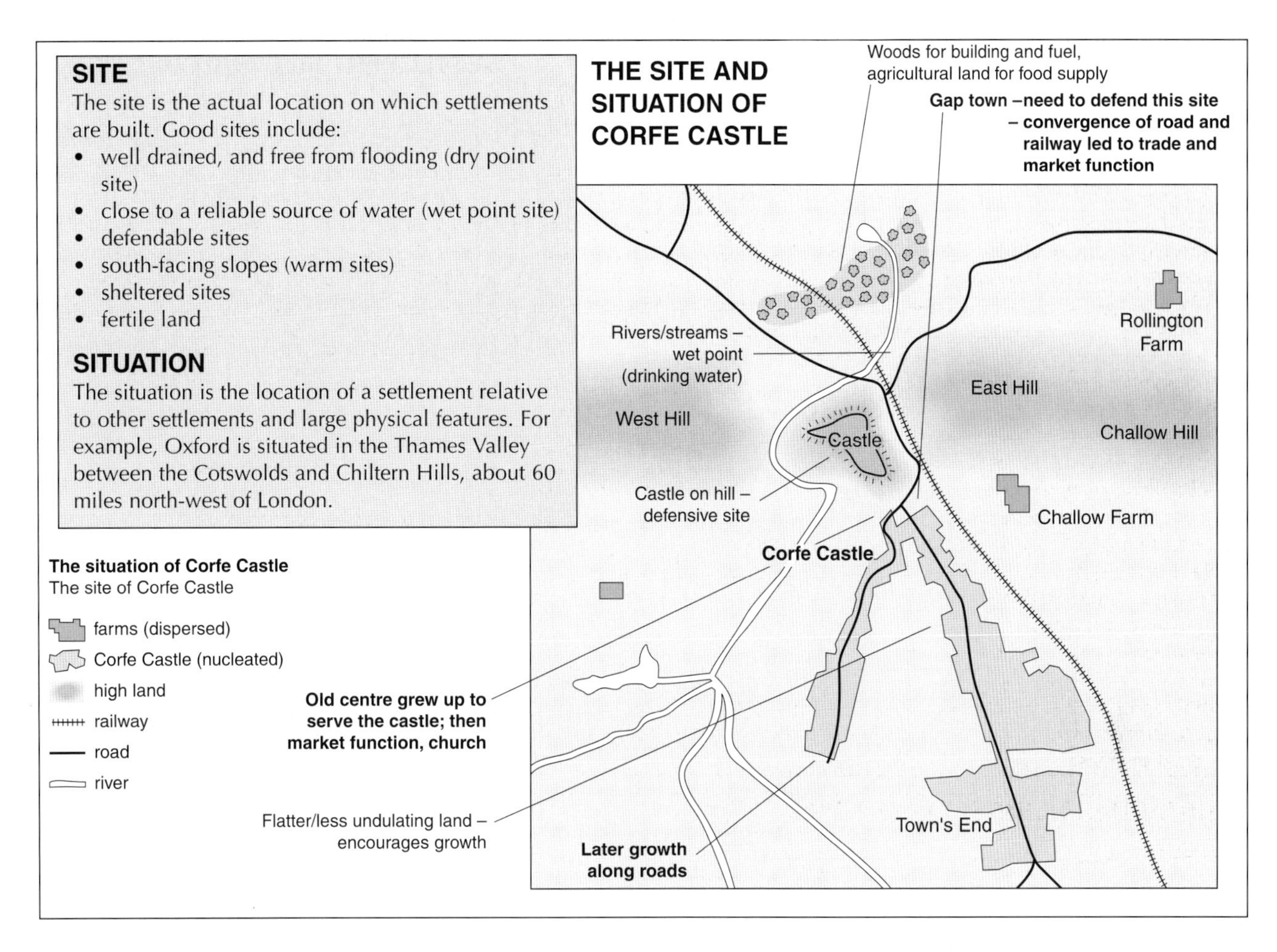

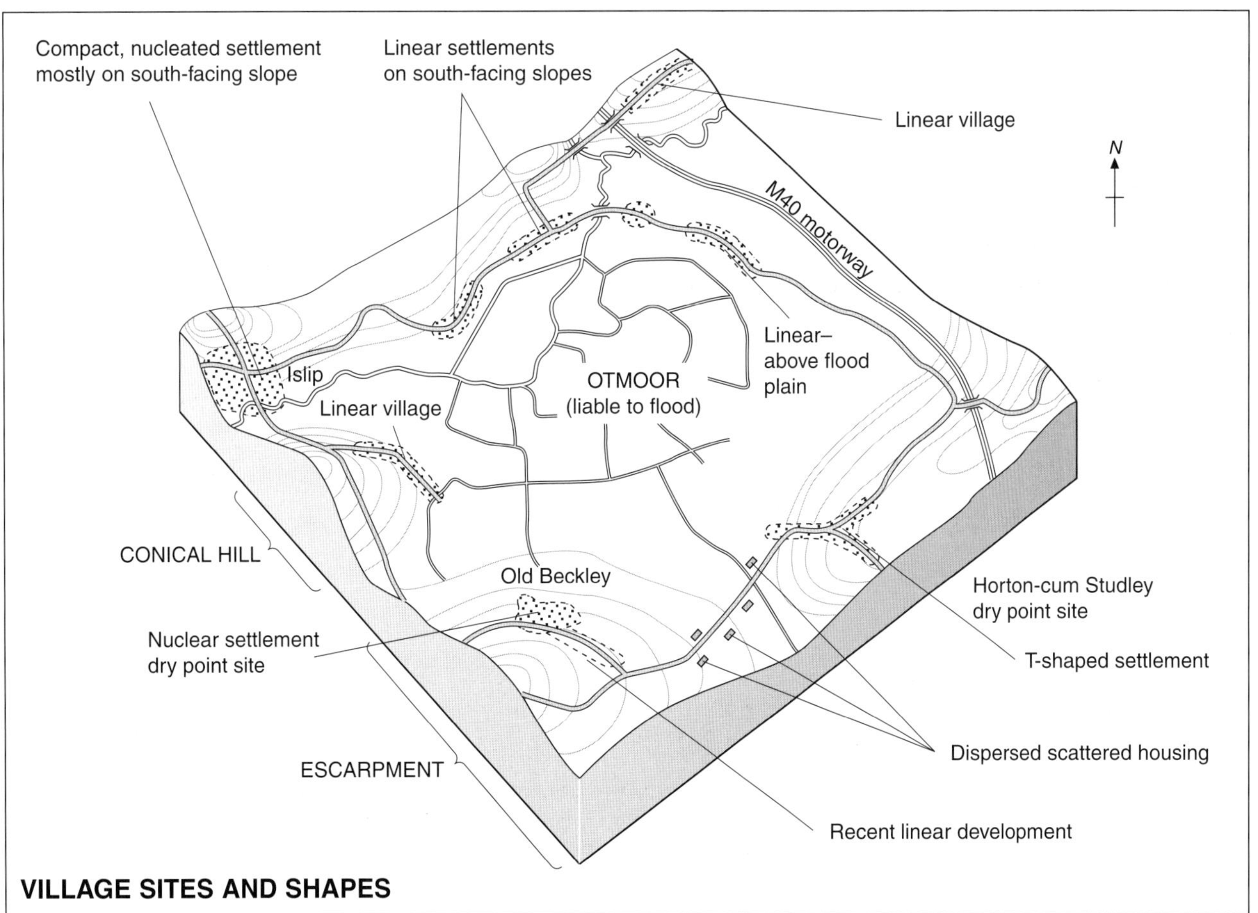

Central place theory (1)

SETTLEMENT PATTERNS

Central place theory attempts to explain the relative size and spacing of settlements. It was developed by Walter Christaller in 1933, based on his observations in southern Germany. In his theory there are a number of key terms:

- central place – a settlement such as a hamlet, village, or market town
- range – the maximum distance that people are prepared to travel for a good or service
- threshold – the minimum number of people required for a good or service to keep in business
- low order goods – necessity goods or convenience goods; bought frequently, such as bread and newspapers
- high order goods – luxury or shopping goods, bought or used infrequently, e.g. solicitors
- sphere of influence – the area served by a settlement; also referred to as a hinterland.

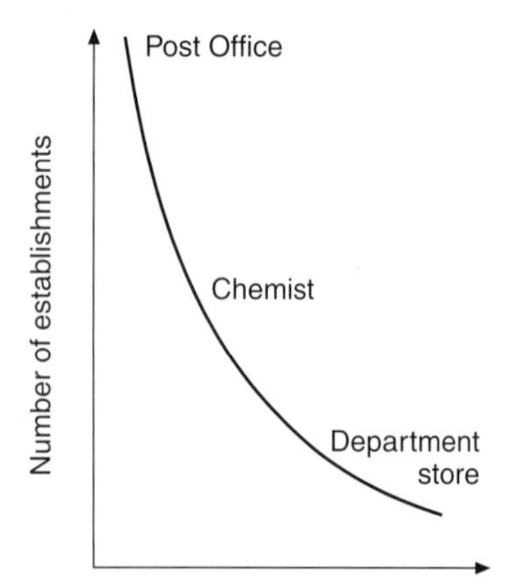

In central place theory there are a number of key assumptions.

- An **isotropic plain** is an area where there is no variation in relief, climate, or population density. It is the geographical equivalent of a pool table without any holes.
- **Rational behaviour** means that people minimise the distance in which they travel in order to obtain a good or service. Hence they visit their nearest central place. According to central place theory, the K3 model is developed as an optimum solution to the marketing of goods, that is, the minimum number of centres needed to serve the total population.

THE RELATIONSHIP BETWEEN SETTLEMENT SIZE AND FUNCTION

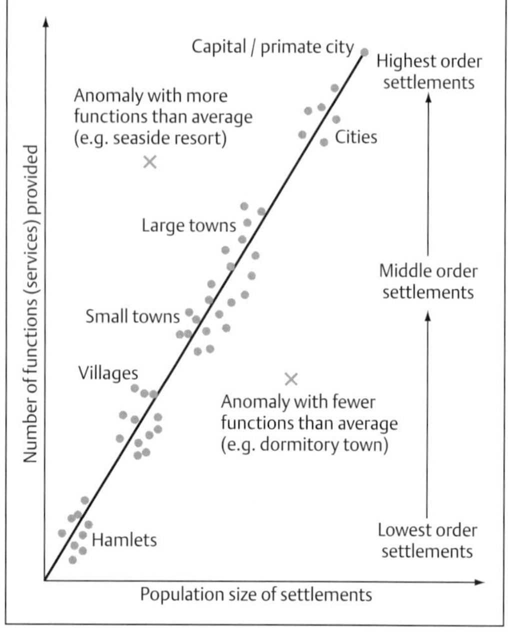

CHRISTALLER'S CENTRAL PLACES AND SPHERES OF INFLUENCE

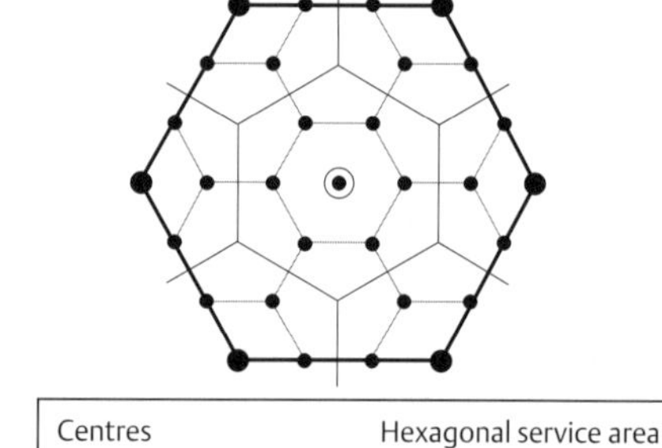

Centres	Hexagonal service area
Lowest order	Lowest order
Middle order	Middle order
Highest order	Highest order

CENTRAL PLACE PATTERNS

K = 3
The 'marketing principle'

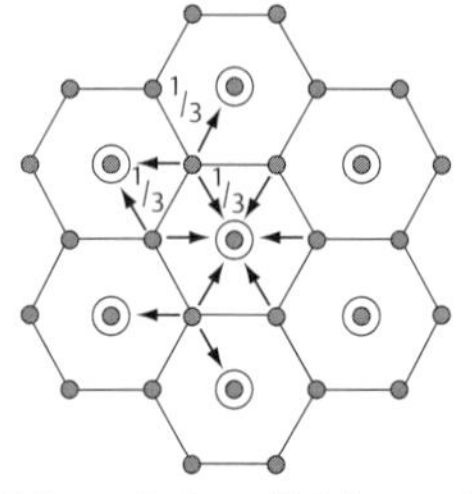

1 (population of higher order settlements) + (6 × $^1/_3$)
($^1/_3$ of population from each of 6 lower order settlements)

K = 4
The 'traffic principle'

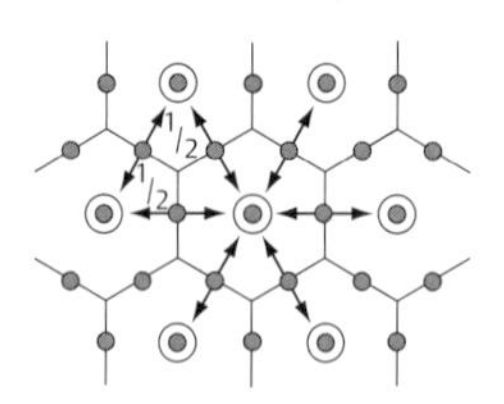

1 (population of higher order settlements) + (6 × $^1/_2$)
($^1/_2$ of population from each of 6 lower order settlements)

K = 7
The 'administrative principle'

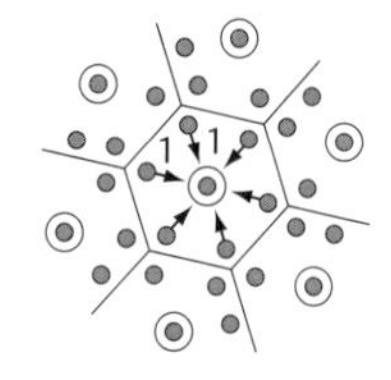

1 (population of higher order settlements) + (6 × 1)
(total population from each of 6 lower order settlements)

K = 3 gives the best choice for shoppers – it minimises the number of service centres needed to serve the whole population

K = 4 is the best network for transport – it minimises the number of roads needed

K = 7 is best network for administration – no settlement is split into different administrative areas

key
- Highest order settlements
- Lowest order settlements

Central place theory (2)

OTHER SETTLEMENT PATTERNS

Alternatives to Christaller's central place theory often deal with urban patterns rather than rural patterns.

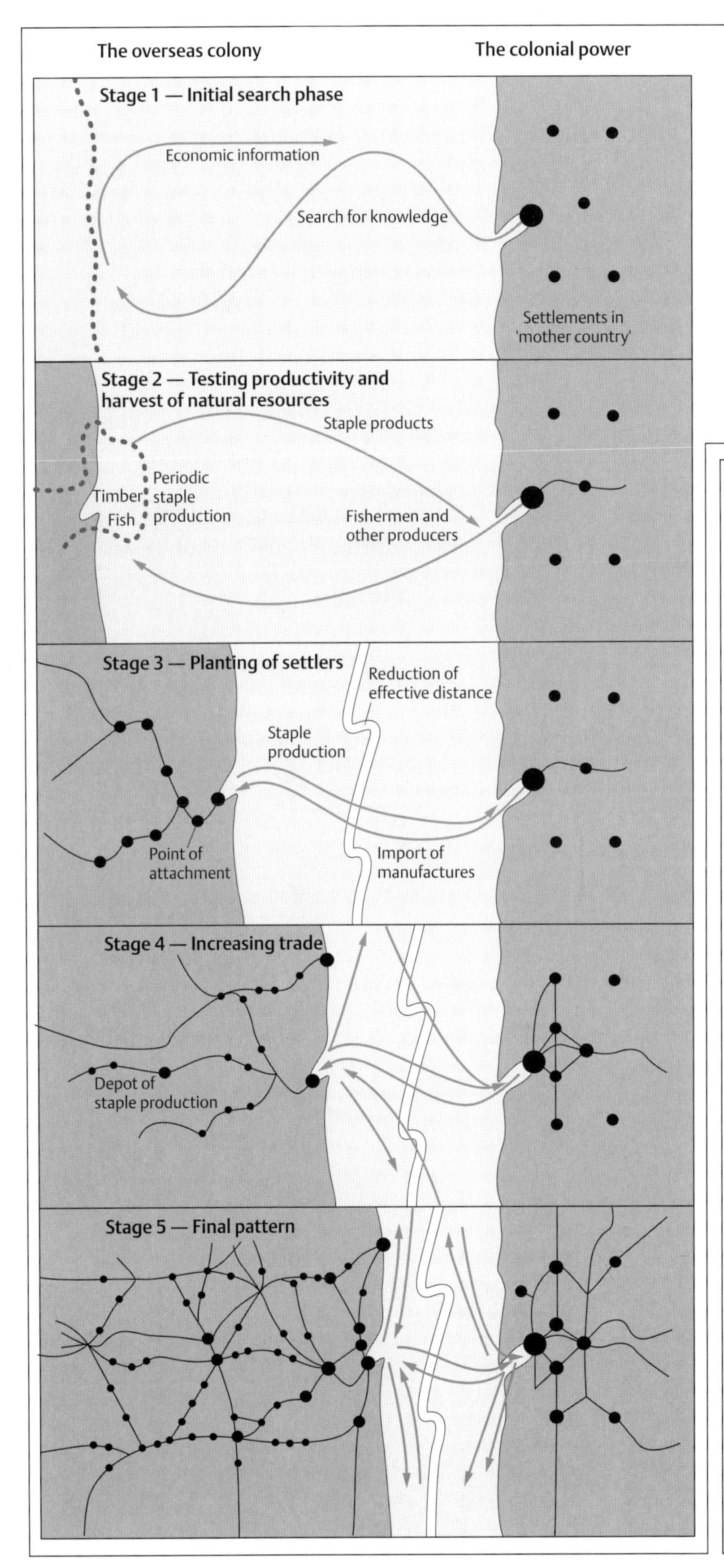

VANCE'S MERCANTILE MODEL (1970)

Vance's mercantile model shows the development of central places with time, due to colonial mercantile (trade) interaction between the colony and the empire. It is based on the case study of the north-east USA and adds a historical-geographical dimension to the study of central places. Vance's model stresses the importance of external influences. The hierarchy evolves from the top down, with large seaboard cities acting as initial centres of commerce. In the USA, Boston was the focal point for change.

THE LÖSCH MODEL (1954)

Lösch modified Christaller's model by producing many more hexagons. He rotated the hexagons to produce a landscape very different to that of Christaller. He claimed that the size of the central place varies with distance from the main city, namely, they get larger away from the centre. A sectoral pattern emerges, with a distinct city-rich city-poor distribution.

'City rich' and 'city poor' sectors in Löschian landscape

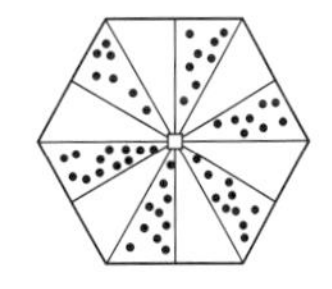

Distribution of all centres in one sector

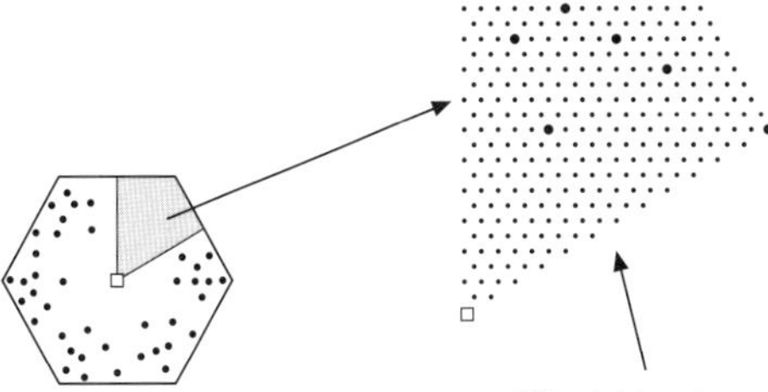

Distribution of large cities in Löschian landscape

- □ Metropolitan centre
- • Large towns or cities
- · Other cities

'City-rich' sector may develop along transport lines, e.g. the M4, leading to regional agglomeration

Change in rural areas

DEFINITION OF RURAL

Less densely populated parts of a country which are recognised by their visual 'countryside' components. Three criteria are used:

- economic - a high dependence on agriculture for income
- social and demographic - the 'rural way of life' and low population density
- geographical - remoteness from urban centres

TYPES OF RURAL AREA

- Extensive land use where there is little demand for land, e.g. Highlands and Islands of Scotland.
- Intensive land use in the developing world as a result of overpopulation.
- In the West, high land values and negative externalities in urban areas has led to the 'suburbanisation' of rural areas.

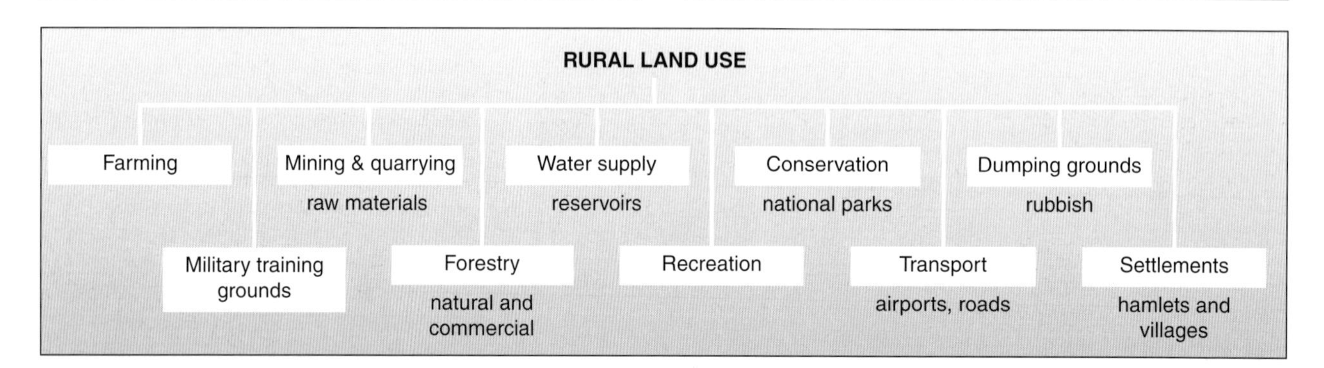

THE RURAL-URBAN FRINGE

Characteristics of changing communities in the rural-urban fringe include:

- **segregation** of rural areas into large blocks of one class or price of housing
- **selective immigration** of mobile middle-classes who live and work in distinct and separate social and economic worlds
- **commuting** of middle class workers

Rural neighbourhood — Agricultural village — Small town — Rural-urban fringe community — Suburban community — Small city — Metropolitan city

Rural-urban continuum

- **collapse of geographical and social hierarchies** as the community becomes more outward-looking and the 'squirarchy' is replaced by a class-based structure polarised around housing segregation.

CLOKE'S MODEL OF RURAL SETTLEMENTS

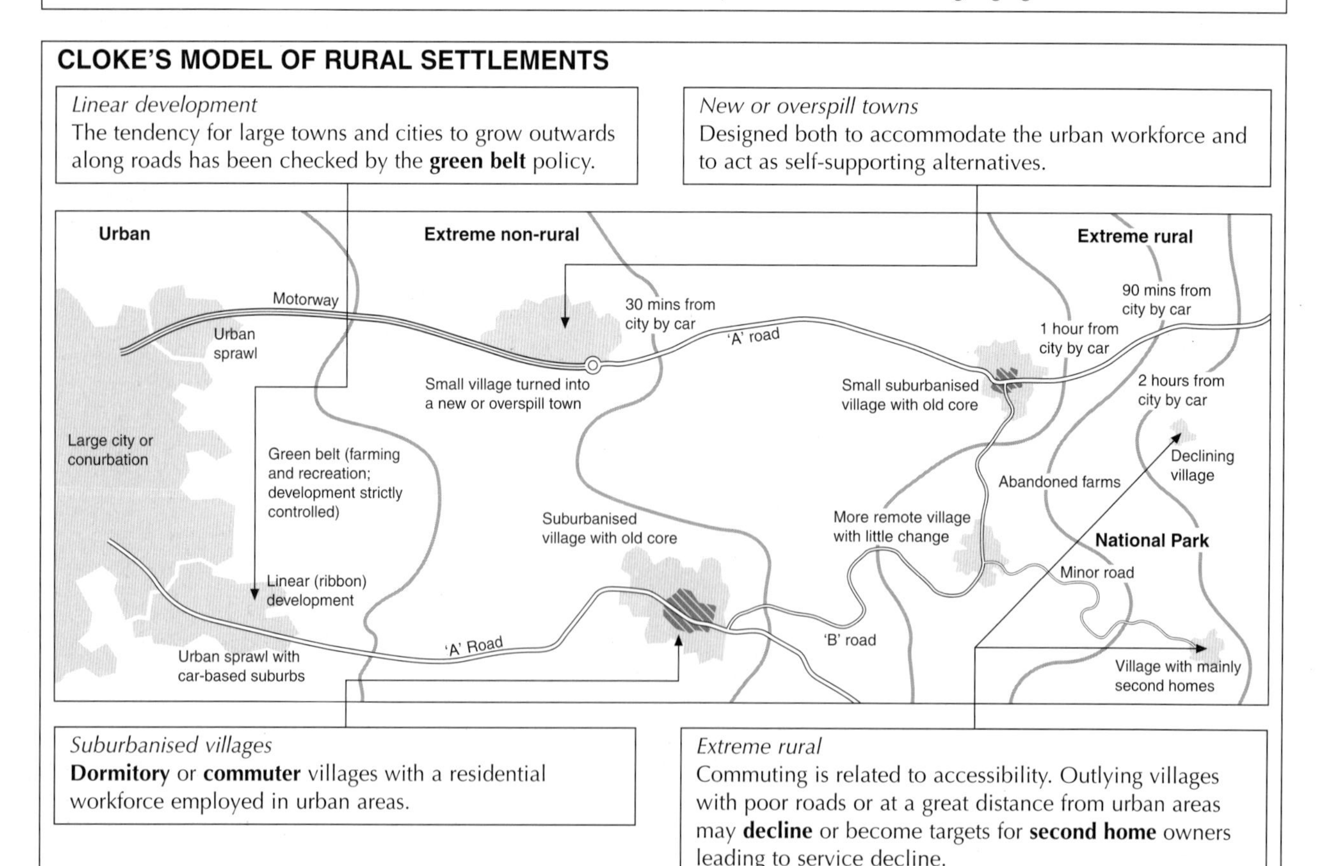

Linear development
The tendency for large towns and cities to grow outwards along roads has been checked by the **green belt** policy.

New or overspill towns
Designed both to accommodate the urban workforce and to act as self-supporting alternatives.

Suburbanised villages
Dormitory or **commuter** villages with a residential workforce employed in urban areas.

Extreme rural
Commuting is related to accessibility. Outlying villages with poor roads or at a great distance from urban areas may **decline** or become targets for **second home** owners leading to service decline.

Case study: the Gower peninsula

The Gower Peninsula is in south-west Wales, to the west of Swansea. It is an Area of Outstanding Natural Beauty forming the north-westerly limit of the Bristol Channel. In the past 30 years the rural area of Gower has seen dramatic changes in the social and economic characteristics of its population. Processes of repopulation and depopulation are apparent in the villages of the Gower. Growth and decline are based on closeness to Swansea, accessibility, planning decisions, and natural beauty.

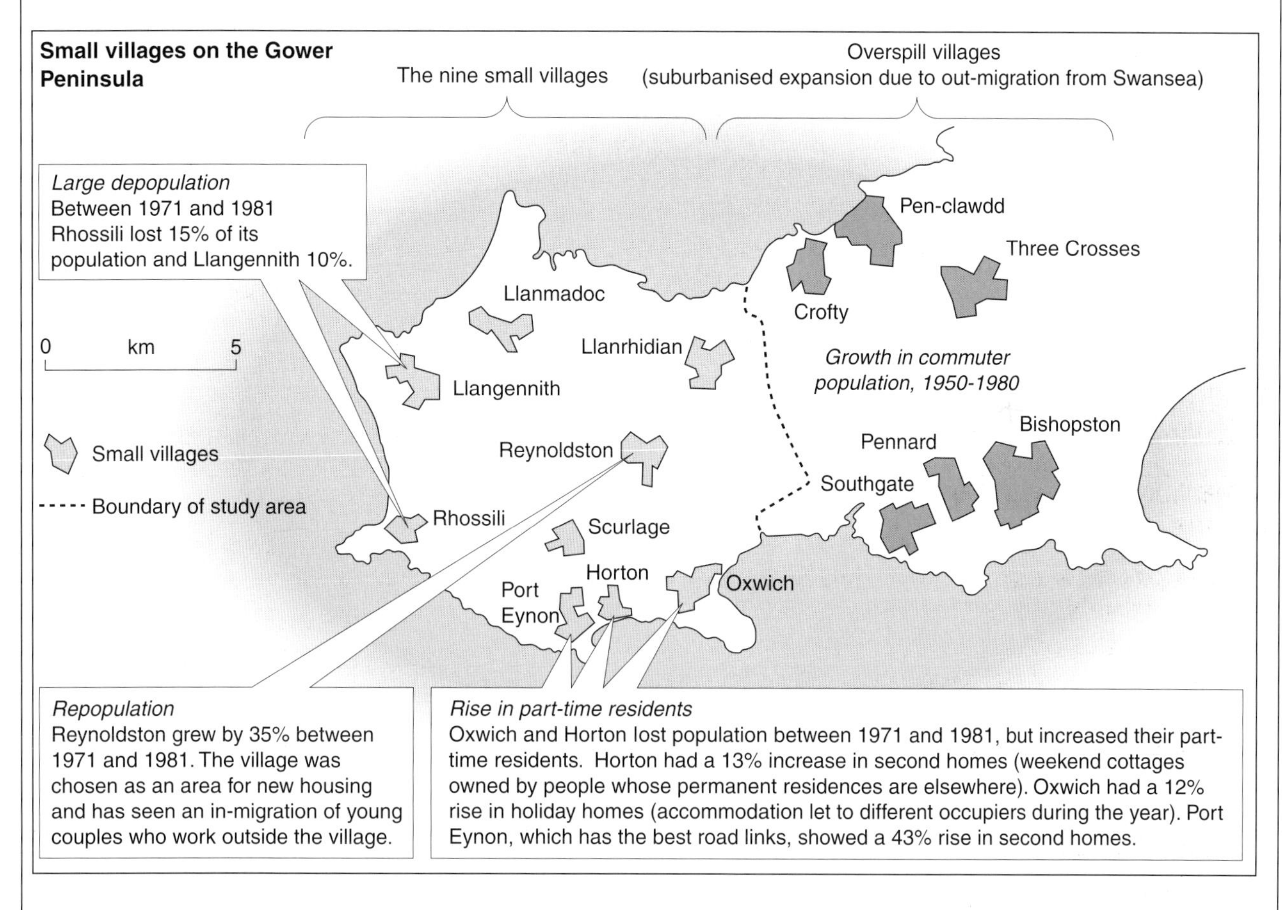

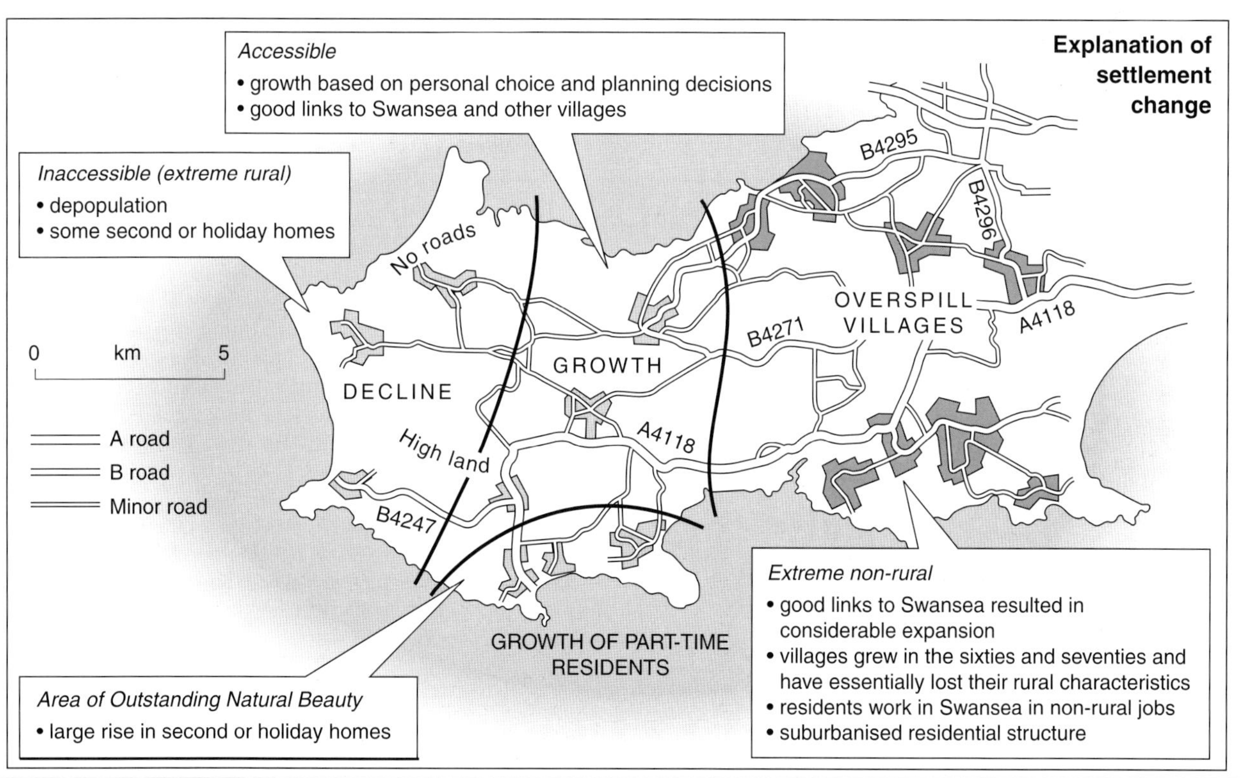

Village change

Many villages have grown at alarming rates and have lost their original character, form, and function. These are often described as dormitory, commuter, or suburbanised villages. Further away, villages without good transport links or beyond the distance of commuting, have tended to retain more of their original character. However, although the size of the population may not change much, the composition does. Younger people move out to be replaced by older ones. These villages may become something of retirement villages, although lacking in many of the functions that such people require.

Rural change

Characteristics	Original village	Suburbanised village
Housing	Detached, stone-built houses with slate or thatch roofs; some farms, most over 100 years old; barns.	New, mainly detached or semis; renovated barns or cottages; expensive planned estates, garages
Inhabitants	Farming and primary jobs; labouring or manual jobs	Professional / executive commuters; wealthy with young families or retired
Transport	Bus service; some cars; narrow, winding roads	Decline in bus services as most families have one or two cars; better roads
Services	Village shop, small junior school, public house, village hall	More shops, enlarged school, modern public houses and/or restaurant
Social	Small, close-knit community	Local community swamped; village may be deserted by day
Environment	Quiet, relatively pollution free	More noise and risk of more pollution; loss of farmland and open space

SECOND HOMES

Second homes are defined as a specialised form of holiday accommodation in remote rural areas beyond the travelling limit for weekend recreation. In Britain, second homes account for about 2% of all properties and are growing at a rate of about 15 000 annually.

There are a number of key questions regarding second homes:
- Are they a justifiable call on resources given that up to four million people in the UK live in sub-standard housing?
- What are the effects of second homes on local communities?
- Are they a justifiable use of public money?

Advantages	Disadvantages
• They are a rational alternative for the economic development of backward rural areas. For example, land can be sold off at higher prices to urbanites and business may increase for local trade. • Second home owners are not a burden on local services which are aimed at the permanent population. However, the taxes from second homes helps to support these services. • The competition between second home owners and young people may not be as great as generally assumed. Most second-home owners require an isolated home without electricity or a bath, so that they can renovate it.	• House and land prices escalate. Property-owners are generally very willing to sell land to the second home market because of greater profitability. • Weekenders do not buy their provisions at local stores, yet they demand piped water and mains sewerage, the cost of which is mostly borne by the local population. • There may be conflict between the permanent and the temporary populations for services. • Roads become congested and there is an increase in the number of accidents. • The local environment may deteriorate due to increased visitor pressure. • New buildings may not fit into the local surroundings.

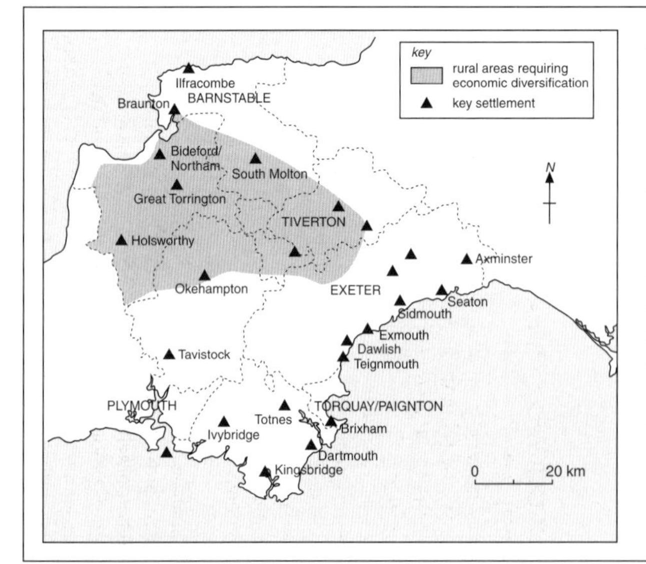

KEY SETTLEMENTS IN BRITAIN

The key settlement policy adopts some of the principles of **central place theory** by locating key services or facilities in a selected settlement e.g. a school or a clinic. It assumes that people will be able and willing to travel to the selected settlement. It is based on the assumption that concentration of facilities is more economic than their dispersal.

Key settlements vary widely. Some are mostly service centres; some are largely associated with public investment such as schools, hospitals, or council estates, while others are concerned with private housing developments, industrial development, or employment growth. The concept has close links with that of **threshold** – a certain facility is not economic below a critical population size.

Rural deprivation

There are a number of similarities between remote rural areas and inner cities. These include:

- decreased demand for labour
- unemployment
- low wages
- inappropriate skills
- a decline in services
- limited new investment
- population decline
- gentrification
- reduced morale
- inaccessibility
- a high dependency ratio
- dereliction and blight
- high cost of public services.

Unlike inner cities where these features are concentrated spatially, in rural areas they are spread over a large part of the country.

RURAL SETTLEMENT - THRESHOLDS AND FUNCTIONS

Percentage

100
90
80
70
60
50
40
30
20
10
0

0–99 | 100–199 | 200–299 | 300–499 | 500–999 | 1000–2999 | 3000–9999

Parish population bands

key

Bus service
Village halls
Schools
Pubs
Post Offices
Permanent shops

RURAL DEPRIVATION IN THE 1990s

The Rural Development Commission's 1991 survey of rural services revealed that:

- 39 % of parishes had no shop
- 40 % had no post office
- 51% had no school
- 29% had no village hall
- 73% had no daily bus service
- fewer than 10% had a bank or building society, day-care centre, a dentist or a daily train service.
- only 40% of parishes had a primary school and these usually had a population of over 500 people
- only 16% of parishes had a permanent GP surgery (they are most common in parishes of more than 1000)
- only 5% of parishes had an out-of-school childcare group and a nursery
- 61% of settlements with a population of under 300 had one or more weekly mobile shops – about a quarter of people in rural areas still depend on a local shop for their everyday needs.

DECLINING TRANSPORT IN RURAL AREAS

For residents in rural areas don't own a car there are possibilities but these do not provide the same service, comfort, and guarantee as one's own transport. Options include:

- subsidised fares
- integrating transport with other services such as post buses (taking passengers on post vans)
- car sharing
- mobile services
- concentration of services in key settlements.

Small settlements have many advantages – such as standards of behaviour, social cohesion, and environmental attributes. There are, however, many social problems and these are closely linked to economic problems. In addition, many local authorities lack sufficient funding to provide adequate services.

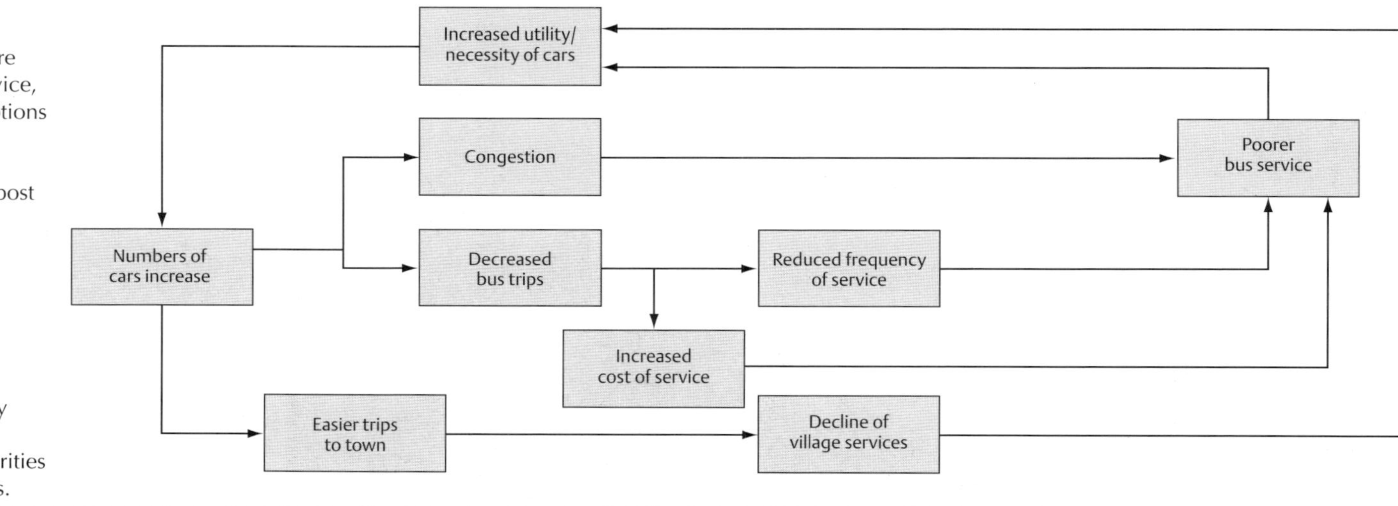

Land use models (1)

MODELS OF URBAN FORM

The growth of cities in the nineteenth and early twentieth centuries produced a recognisable form with many common features. It included a central commercial area, a surrounding industrial zone with densely packed housing, and outer zones of suburban expansion and development. Geographers have spent a lot of time modelling these cities to explain 'how they work'.

Every model is a simplification. No city will 'fit' these models perfectly but there are parts that can be applied to most cities in the developed world. All models are useful in as much as they focus our attention on one or two key factors.

BID-RENT THEORY

The concept of bid rent is vital to models of urban land use. Bid rent is the value of land for different purposes, such as commercial, manufacturing and residential purposes. Land at the centre of a city is most expensive for two main reasons – it is the most **accessible** land to **public transport**, and there is only a small amount available.

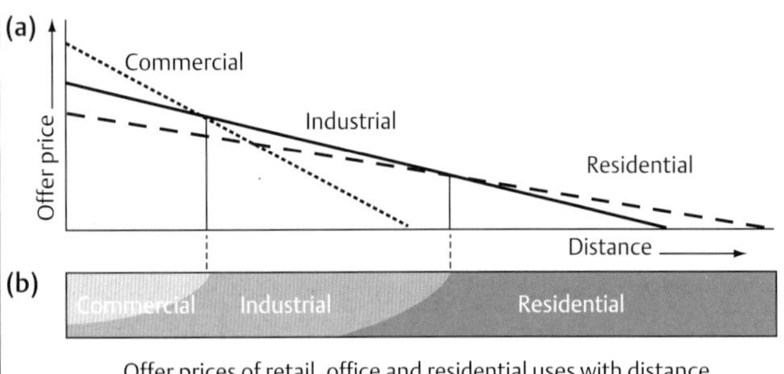

Offer prices of retail, office and residential uses with distance from the city centre:
(a) section across the urban value surface
(b) plan of the urban value surface

Land prices generally decrease away from the central area, although there are secondary peaks at the intersections of main roads and ring roads. Change in levels of accessibility, due to private transport as opposed to public transport, explains why areas on the edge of town are often now more accessible than inner areas.

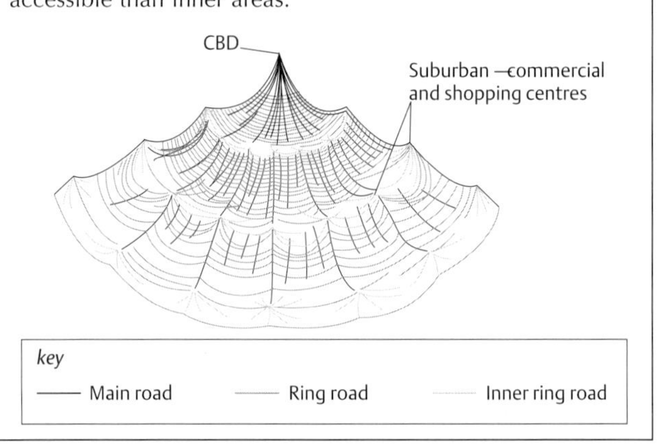

BURGESS' CONCENTRIC MODEL (1925)

The most simple model is that of Burgess. He assumed that new migrants to a city moved into inner city areas because they were the cheapest type of housing, and were closest to the sources of employment. With time, residents move out of the inner city area as they become wealthier. This was often the **second generation migrants**.

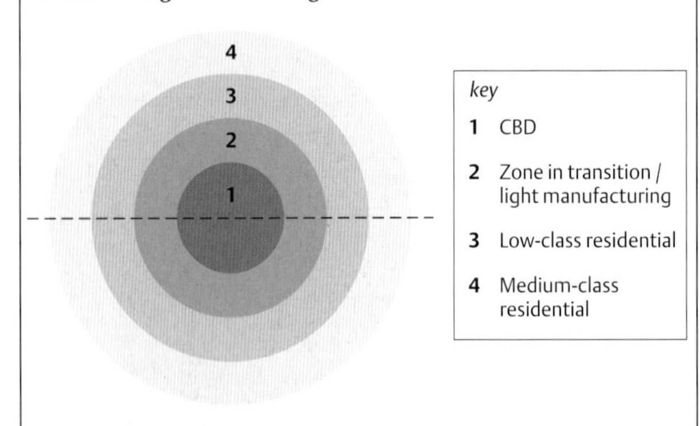

HOYT'S SECTOR MODEL

Homer Hoyt (1939) emphasised the importance of transport routes and the incompatibility of certain land uses. Sectors develop along important routeways, while certain land uses, such as high-class residential and manufacturing industry, deter each other.

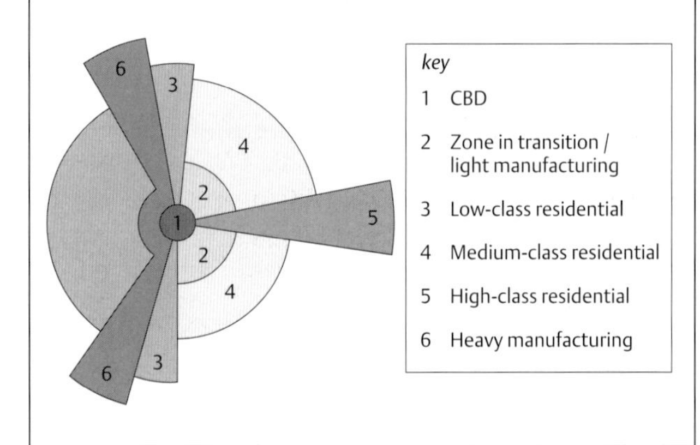

MANN'S MODEL FOR A UK CITY

One model designed specifically for British cities is Mann's concentric circles and sectors. He found that there was an east-west split in most British towns, with the west end being wealthier and most industry being located in the east. This is due to the fact that many industrial areas developed in the east end of towns, especially if prevailing winds and rivers were carrying pollution eastwards. As well as the east-west contrast, there is an inner-city outer-city contrast. Size of house and socio-economic status both increase with distance from the city.

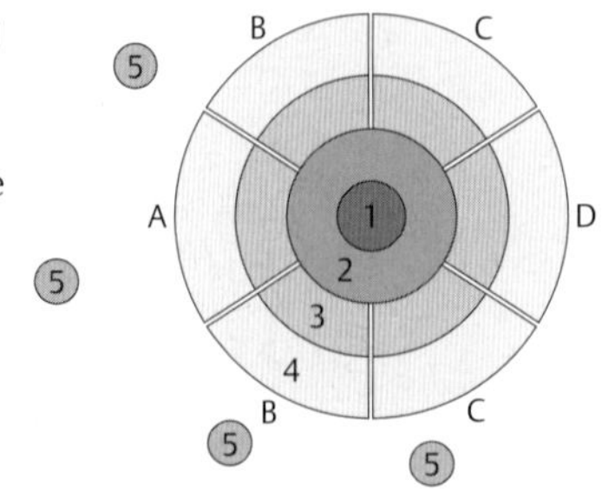

key
1 City centre
2 Transitional zone: small terraced houses in sectors C and D, larger by-law houses in sector B, large old houses in sector A
3 Pre - 1918 housing
4 Post - 1918 residential areas, with post—1945 development mainly on the periphery
5 Commuting distance 'villages'

A Middle-class sector
B Lower middle-class sector
C Working-class sector and main municipal housing areas
D Industry and lowest working-class sector

Land use models (2)

HARRIS AND ULLMAN'S MULTIPLE NUCLEI MODEL (1945)

The multiple nuclei model of Harris and Ullman shows that cities do not have a single centre but that there are many pre-existing centres. As cities grow they incorporate some of these centres. In addition, new centres out of the heart of the city may be planned for commercial, industrial, and residential purposes. Although this model was based upon North American cities some parallels can be drawn with cities in the UK, or example, as Oxford grew, it incorporated Cowley, Headington, and Botley. Residential developments such as Blackbird Leys and Barton are new nuclei around the city.

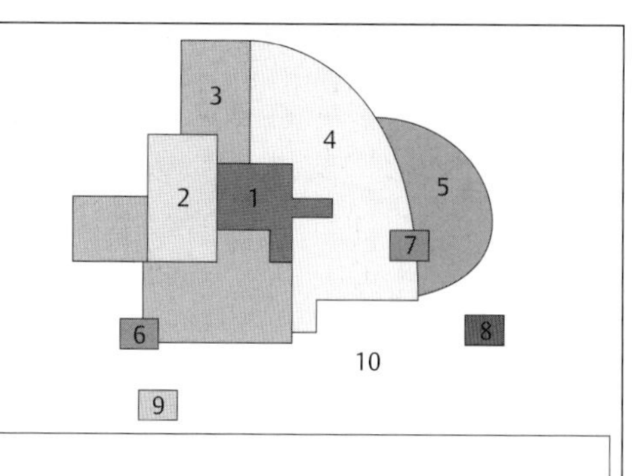

key

1 CBD
2 Zone in transition / light manufacturing
3 Low-class residential
4 Medium-class residential
5 High-class residential
6 Heavy manufacturing
7 Outlying business district
8 Residential suburb
9 Industrial suburb
10 Commuter zone

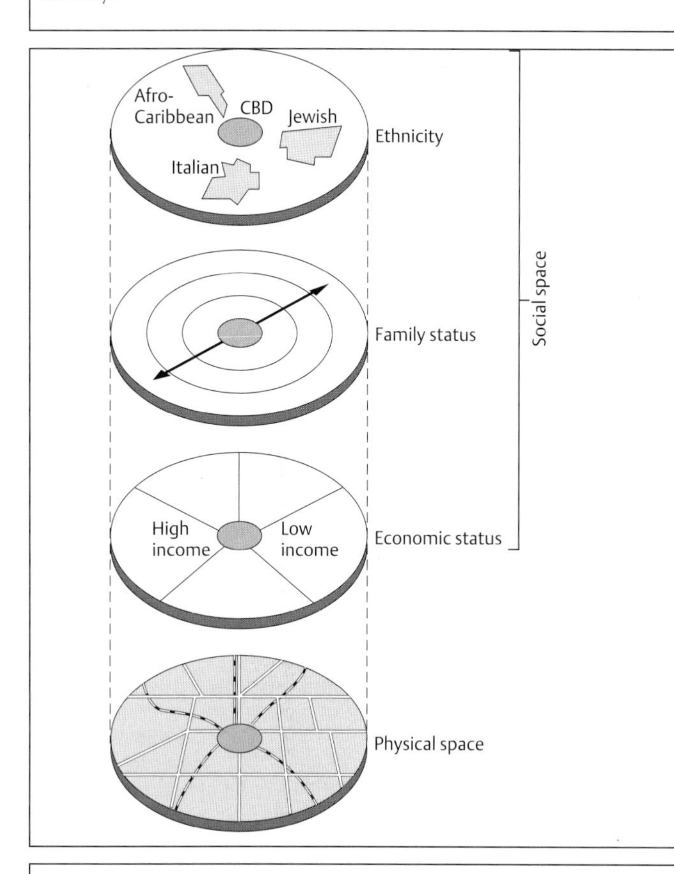

MURDIE'S ECOLOGICAL MODEL OF URBAN FORM

Murdie's model of urban ecological structure brings together a number of factors, such as the physical space, the economic, social, ethnic environments, and variations in the family life-stage. The result shows us that cities are indeed complex.

The model:

- identifies social, economic and ethnic areas
- the 'residential mosaic' is superimposed on the physical space
- economic status has a sector pattern; family status has a concentric pattern; ethnic groups occur in clusters

SOCIO-ECONOMIC, DEMOGRAPHIC AND HOUSING CHARACTERISTICS

The diagram shows how the age-structure of the population, the socio-economic status of the population, and the housing environment can be combined to produce nine categories. These include young, white-collar (service employment) workers living in owner-occupied housing, and elderly manual workers living in rented accommodation. Some of the combinations are more likely than others. They include:

Type 1 Inter-war owner-occupied housing
Type 2 High status owner-occupied
Type 3 Post-war semi-detached owner-occupied
Type 4 Post-war detached owner-occupied
Type 5 Student bed sitters
Type 6 Inter-war council estates/inner city council flats
Type 7 Post-war council estates/inner city high rise flats
Type 8 Privately rented low status
Type 9 Rooming houses.

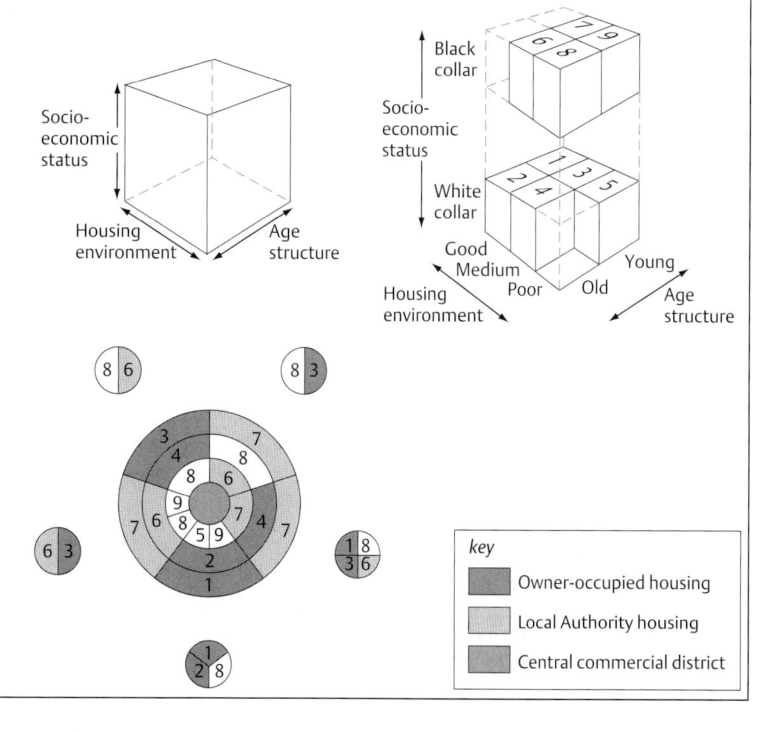

Filtering and gentrification

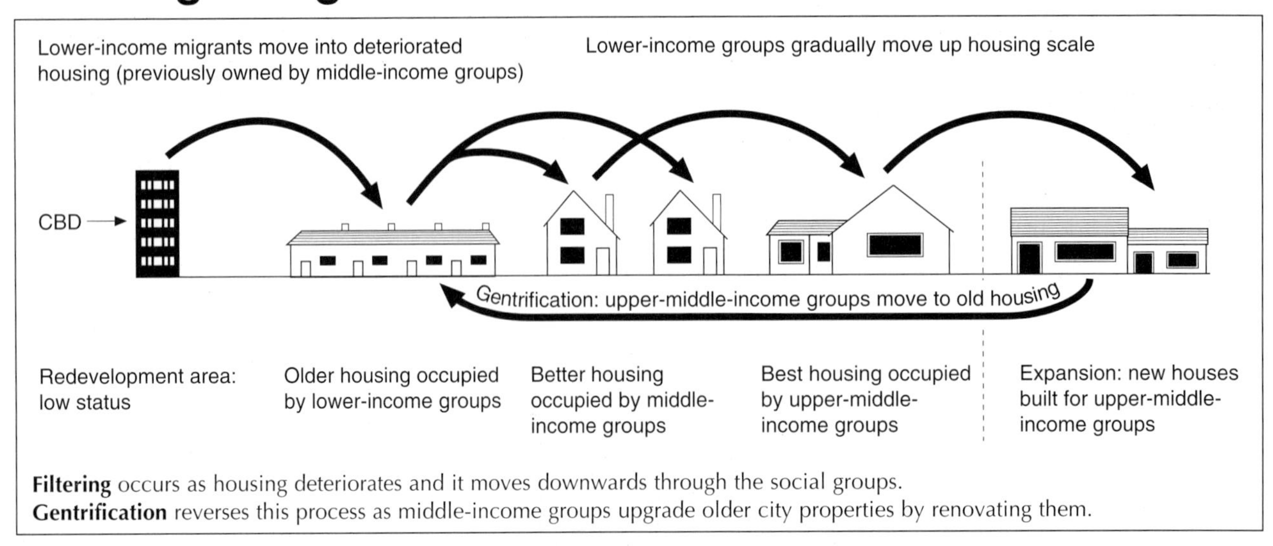

Filtering occurs as housing deteriorates and it moves downwards through the social groups.
Gentrification reverses this process as middle-income groups upgrade older city properties by renovating them.

FAMILY LIFE-CYCLE MODEL

Middle-income life-cycle		
1	Semi-detached	Childhood
2	Rented room/bedsit in crowded Victorian house	Pre-parenthood
3	Starter home (owned)	Child-rearing
4	Family home (owned)	Primary-age children
5	Family home (owned)	Adolescent children
6	Retirement bungalow (owned)	Grand-parenthood/ elderly

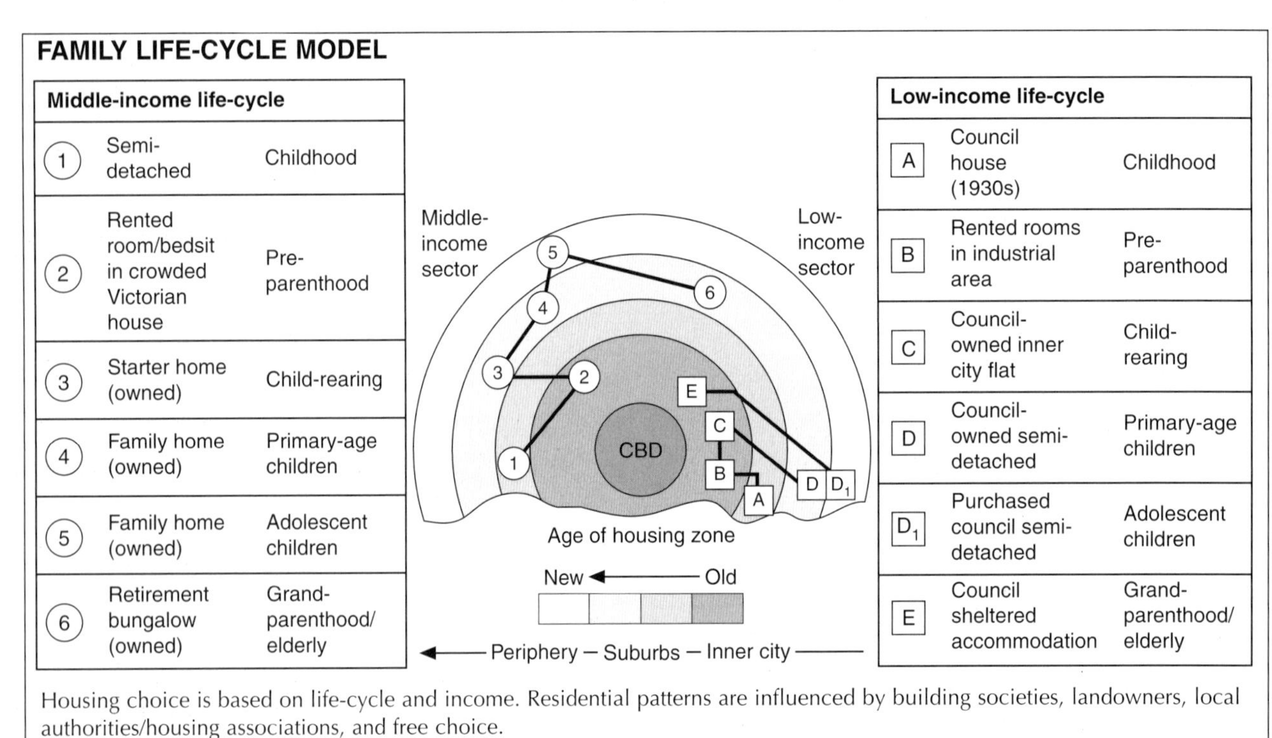

Low-income life-cycle		
A	Council house (1930s)	Childhood
B	Rented rooms in industrial area	Pre-parenthood
C	Council-owned inner city flat	Child-rearing
D	Council-owned semi-detached	Primary-age children
D_1	Purchased council semi-detached	Adolescent children
E	Council sheltered accommodation	Grand-parenthood/ elderly

Housing choice is based on life-cycle and income. Residential patterns are influenced by building societies, landowners, local authorities/housing associations, and free choice.

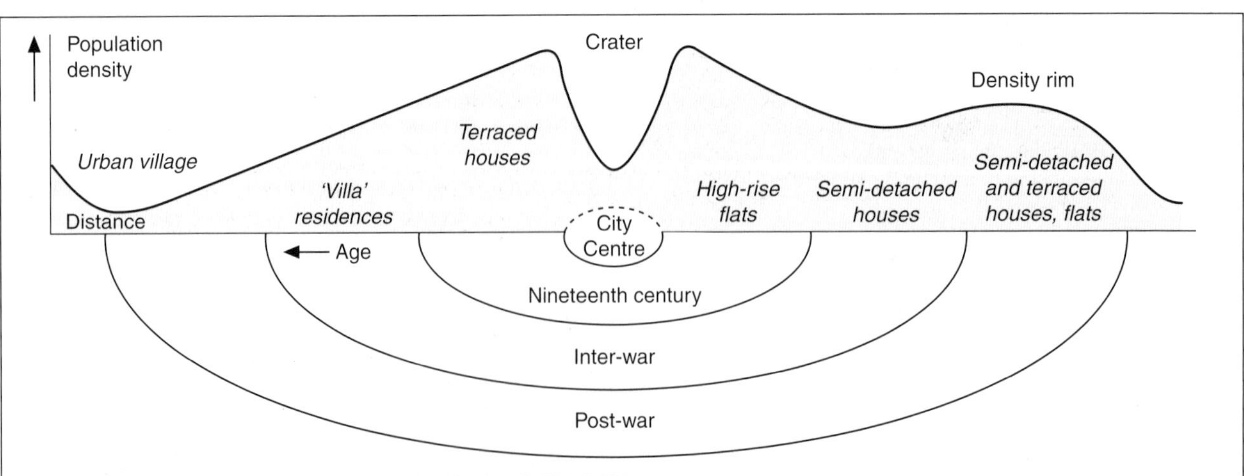

THE DENSITY CURVE WITH A HIGH-DENSITY RIM

There are low residential densities in the city centre. Peak residential densities are beyond the central area, in the inner city. The rise in densities in the outer suburbs is due to the development of council housing estates in the last thirty years.

The central business district (CBD)

The central business district (CBD) is generally at the heart of the city and is the focus for the urban transport system. It is the centre of the city's commercial, social, and cultural life. It possesses a number of clearly defined characteristics.

CHARACTERISTICS OF THE CBD

Absence of manufacturing
- a few specialised activities such as newspapers

Concentration of offices
- central location needed for clients and workforce
- offices tend to locate in zones

Multi-storey development
- high land values encourage buildings to grow upwards
- the real area of the CBD should therefore be measured in floor space rather than ground space

Vertical zoning
- land use changes within multi-storey blocks
- shops and services occupy ground floors

Low residential population
- high bid rents mean that few people live in the CBD
- there may be some luxury flats, especially in large cities

Pedestrianisation
- since the 1960s urban traffic management has limited the movement of vehicles within the CBD
- pedestrianisation has made shopping safer but some town centres have lost character

Concentration of retailing
- accessibility attracts shops with wide ranges and high threshold populations
- department stores and high threshold chains dominate in the centre
- specialist shops occupy less accessible sites

Comprehensive redevelopment
- the clearance of sites for complete rebuilding
- sometimes the CBD is extended into the inner city, causing conflict with residents
- redevelopment can also shift the centre of the CBD, causing some businesses to decline

CASE STUDY: THE CBD OF SWANSEA

The competition for highly accessible sites in the CBD gives rise to a hierarchy of land uses, with a variation in the intensity of land and building use between the innermost part of the CBD - the core - and its surrounding frame.

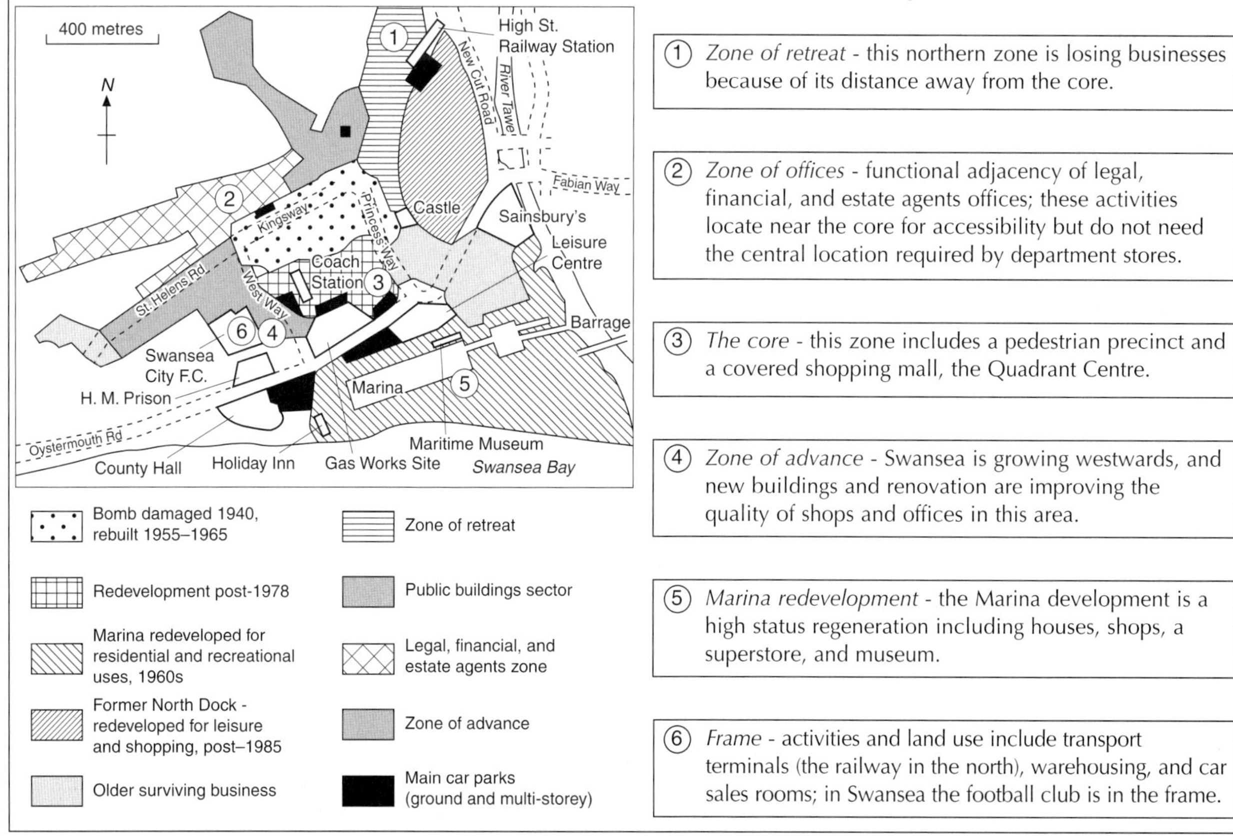

① *Zone of retreat* - this northern zone is losing businesses because of its distance away from the core.

② *Zone of offices* - functional adjacency of legal, financial, and estate agents offices; these activities locate near the core for accessibility but do not need the central location required by department stores.

③ *The core* - this zone includes a pedestrian precinct and a covered shopping mall, the Quadrant Centre.

④ *Zone of advance* - Swansea is growing westwards, and new buildings and renovation are improving the quality of shops and offices in this area.

⑤ *Marina redevelopment* - the Marina development is a high status regeneration including houses, shops, a superstore, and museum.

⑥ *Frame* - activities and land use include transport terminals (the railway in the north), warehousing, and car sales rooms; in Swansea the football club is in the frame.

Inner cities

PROCESSES AFFECTING INNER CITIES

- lack of services
- unemployment
- drugs and crime
- urban decay
- high population density
- road congestion and pollution
- lack of green space.

Ageing

- relates to housing, social services, infrastructure, industrial base, population
- the result is that population is more dependent on social and medical services; housing is in need of renovation or replacement; transport systems are outmoded; urban infrastructure in need of repair.

Changing land uses

- new construction related to the activ ities of the central business district, new roads, universities, hospitals
- slum clearance associated with tower blocks and moving the inner city poor to the periphery
- urban policy can also change land use with respect to social class as middle class housing replaces working class areas.

Changing social structure

- minority groups such as blacks or new immigrants concentrate in parts of the area
- high concentrations of ethnic minorities may be due to internal (defence, support, preservation) or external factors
- population decline and housing markets which 'lock in' the poor
- gentrification may reverse this.

CASE STUDY: ETHNICITY - ASIANS IN LEICESTER

Ethnic concentrations are a characteristic of the inner city. Ethnic minorities are groups who are culturally differentiated from the majority population. Visibility is due to language, religion, and race. Concentration in the inner city is due to 'uncontested filtering' into housing made available by the out-migration of the indigenous population (very similar to the Burgess model). In Leicester, immigrants from India and Pakistan and from other British cities were drawn by the following pull factors:

- jobs in industry and textiles and expanding engineering plants
- kinship ties and a resilient economy which drew Asians from other British cities

There were three clear waves of migration:

- early post-war male migration (1950-60); the main motive was economic and the move was seen as short term
- family reunion (1960-68); this saw families moving from Asia to join with families, and was a period of house buying and voluntary segregation to retain cultural identity
- East African Asian migration; more middle class than original migrants, and this has led to internal segregation.

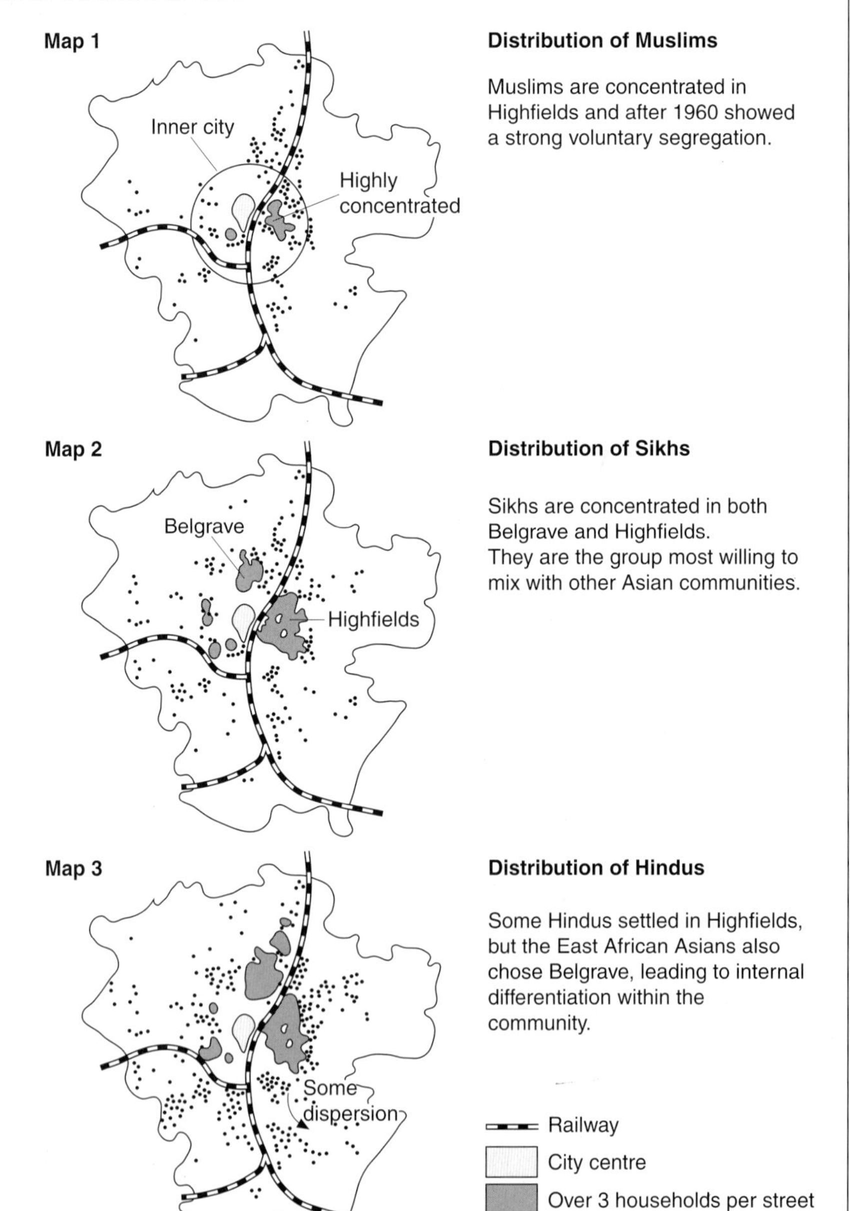

Distribution of Muslims

Muslims are concentrated in Highfields and after 1960 showed a strong voluntary segregation.

Distribution of Sikhs

Sikhs are concentrated in both Belgrave and Highfields.
They are the group most willing to mix with other Asian communities.

Distribution of Hindus

Some Hindus settled in Highfields, but the East African Asians also chose Belgrave, leading to internal differentiation within the community.

CASE STUDY: URBAN REDEVELOPMENT IN GLASGOW

Slum clearance (1957-74)

- comprehensive redevelopment of the tenement areas (Govan, Gorbals, Royston) of the inner city, which were cleared by bulldozers
- existing communities were broken up and forced to relocate to bleak peripheral estates

Peripheral council housing (1952 to the early 1970s)

Problems - post-war housing was high density (700 persons per acre) and poorly maintained, with a lack of basic amenities (over 50% of all houses with no bath) and infested by vermin, especially rats

Policy - 500 000 people were dispersed to new towns like East Kilbride and Cumbernauld and peripheral council housing at Castlemilk (8500), Pollock (9000), Drumchapel (7500), and Easterhouse (10 000)

Evaluation - estates lacked amenities; displaced and divided communities; poor design and rushed construction led to problems of damp and vermin

TOP-DOWN
Government-generated initiatives which attempt to change social or economic structure.

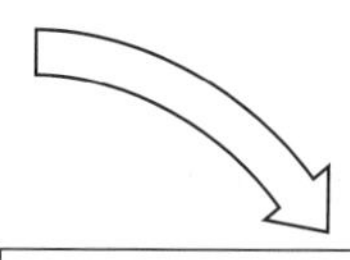

The GEAR project (1976-87)

Problems - Glasgow Eastern Area Renewal (GEAR) was a response to the mistakes of slum clearance and peripheral estates

Policy - modernisation rather than demolition with newly-built housing (2000 private homes) combining with existing housing stock

Evaluation - succeeded in attracting 300 new factories and improving housing (1200 homes were rehabilitated); stopped out-migration of residents; although 'job-rich', most jobs are taken by commuters

Transport

- Glasgow's ambitious transport policy included the construction of one of the UK's few 'urban freeways', the M8
- improved communications led to an increase in commuting and the loss of some good quality inner city housing

Slum clearance high
Deprivation (1991)
Planning Priority Area
Drumchapel
Maryhill
Possilpark
Royston
Easterhouse
M8
Govan
Gorbals
Haghill
GEAR
Pollok
Castlemilk
0 km 4

Deprivation

- despite planning policies, multiple deprivation still exists, especially in the peripheral housing schemes
- many argue that long-term unemployment cannot be solved by urban planners
- others argue that newer policies like GEAR ignore social problems

The Govan initiative (1987-94)

Problems - factory closure; decay and decline of housing stock; environmental damage by M8 motorway

Solutions - small-scale developments including start-up units for businesses, environmental improvements (landscaping), education and training for resident workforce

Success/failure - rebirth of local shipyard providing jobs for local workers; 'bottom-up' approach served the needs of the community

BOTTOM-UP
Locally-based initiatives including small-scale social action.

FACTORS AFFECTING THE MODERN WESTERN CITY

Economic

Deindustrialisation - the global shift of manufacturing growth from the west to the developing world

Sectoral change - the switch in employment and investment from manufacturing to services

Decentralisation of industries out of congested cities to greenfield sites and industrial estates on the periphery

Unemployment - growing long term unemployment and a predominance of part-time, less skilled, and female workers

Social and demographic

Contraction of urban population as middle classes move to suburbs

Rural-urban conflict - the change in rural life as villages are colonised by urban commuters

Funding - the contraction of urban population due to out-migration of middle classes has led to problems in funding public services

Traffic congestion - the rise in commuting has put tremendous pressure on transport systems and caused increased pollution

Urban decay - poor housing coupled with high unemployment creates areas of poverty and deprivation

New towns and green belts

CASE STUDY: THE GREATER LONDON PLAN

Until the mid-1960s the basis for planning in the South-East was the 1944 Greater London Plan ('The Abercrombie Plan'). The plan was set up to solve a number of problems:

- London was too large. Too many of the UK's jobs were centred on London.
- Other areas were suffering unemployment. London was congested. Many dwellings were slums.

New Towns

New Towns Act, 1946

- New Towns were set up to provide alternatives to London in terms of housing and employment
- eight New Towns were created around London - with target populations of between 25 000 and 80 000
- 28 settlements were expanded to take another 535 000 migrating Londoners ('expanded towns').

Green belts

Green Belt Act, 1938

- a zone of land around London within which building is controlled
- set up to stop the sprawl of London and the merging of neighbouring towns, to protect farmland, and to restrict harmful activities on rural-urban fringe
- 25 km wide; has many towns within it, which can only expand by infilling the spaces between existing buildings.

NEW TOWNS - SUCCESS OR FAILURE?

- First wave New Towns were close enough to London to allow daily commuting
  - they helped to relieve the housing problem
  - but added to congestion
- Second wave New Towns were much bigger and were built further out and have become independent growth poles, e.g. Milton Keynes.

GREEN BELTS - SUCCESS OR FAILURE?

- Green belts have succeeded in protecting many rural areas from urban sprawl.
- Green belts have caused many problems, such as:
  - increased commuting as urban dwellers relocate outside the green belt
  - house prices inside the Green belt rise
  - much of the green belt is poor quality and not worth preserving
  - market forces have seen planning regulations relaxed, e.g. the construction of the M25.

HOW NEW TOWNS AND GREEN BELTS WORK

The city cannot grow so New Towns are built outside the Green Belt to house workers and their families

Many people commute from the New Town across the Green Belt to the city for work

Original city

Expanded towns are existing towns which are enlarged

Green Belts prevent urban sprawl (spread) and provide land for recreation and farming

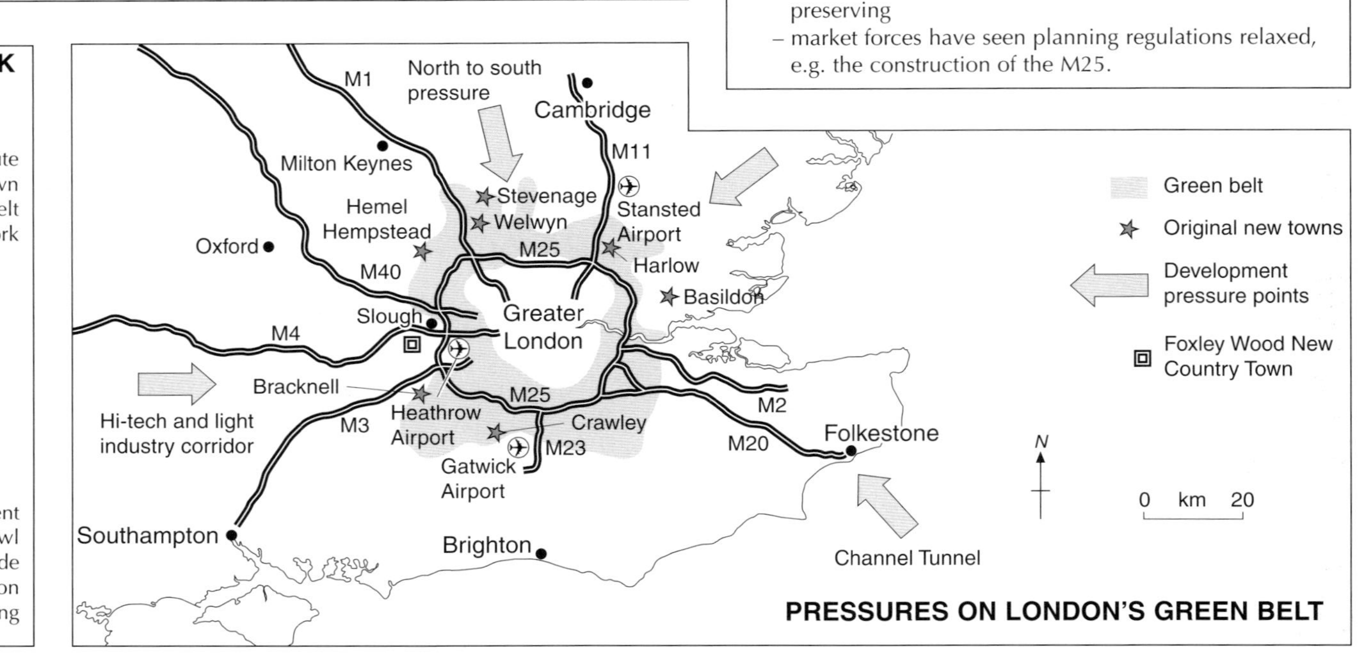

PRESSURES ON LONDON'S GREEN BELT

Supercities, rank-size rule

There are an increasing number of very large cities with populations of over 5 million people. These are known as 'supercities'. In most cases these are 'primate cities'. This means that they dominate a country in terms of size, location of manufacturing, investment, and power. Very often, supercities are ports with important trade functions or centres of former empires. However, being big is not always a good thing. The quality of life in many large cities is very poor. There are severe problems with:

- housing
- health
- employment
- education
- water and sanitation

THE QUALITY OF LIFE IN SUPERCITIES

	Population (millions)	Murders per 100 000	% of income spent on food	Persons per room	% of houses with water/ electricity	Tele-phones per 100 people	% of children secondary school	Infant deaths per 1000 live births	Noise levels (1-10)	Traffic flow mph in rush hour	Quality of life score
Tokyo	28.7	1.4	18	0.9	100	44	97	5	4	28.0	81
Mexico City	19.4	27.8	41	1.9	94	6	62	36	6	8.0	38
New York	17.4	12.8	16	0.5	99	56	95	10	8	8.7	70
Sao Paulo	17.2	26.0	50	0.8	100	16	67	37	6	15.0	50
Osaka	16.8	1.7	18	0.6	98	42	97	5	4	22.4	81
Seoul	15.8	1.2	34	2.0	100	22	90	12	7	13.8	58
Moscow	13.2	7.0	33	1.3	100	39	100	20	6	31.5	64
Mumbai	12.9	3.2	57	4.2	85	5	49	59	5	10.4	35
Kolkata	12.8	1.1	60	3.0	57	2	49	46	4	13.3	34
Buenos Aires	12.4	7.6	40	1.3	86	14	51	21	3	29.8	55

THE RANK-SIZE RULE

The rank-size rule states that the population size of a given city tends to be equal to the population of the largest city divided by the rank of the given city. Thus:

Pr = P1/r

where **Pr** is the population of city ranked r
P is the population of the largest city
r is the rank of the city r.

There are a number of patterns of settlement order. Where two or more cities are larger than their predicted size, the pattern is referred to as a **binary pattern**. According to Christaller's central place theory there ought to be a series of **levels** or **steps**, coinciding with the levels in the settlement hierarchy, such as conurbations, cities, towns, and so on.

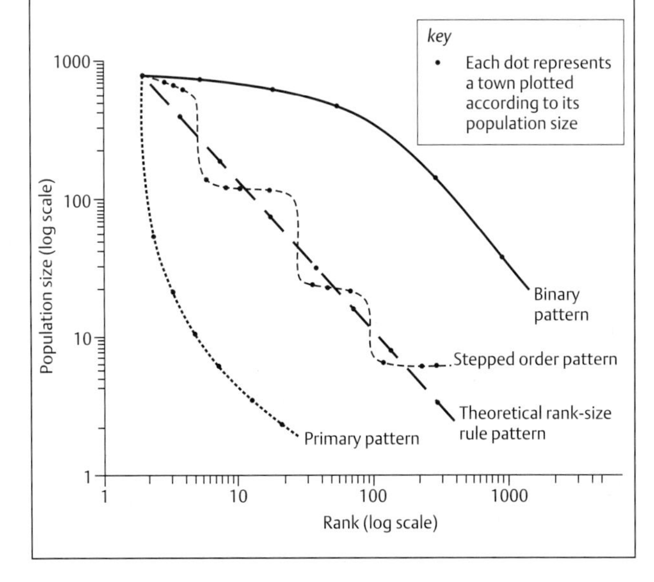

URBAN PRIMACY

A primate city is one with more than twice the population of the next largest city. Frequently, it is much more; as much as four times more in some developing countries where they are a common feature. Normally it is the capital city. They are a common feature of rapid urbanisation. A city achieves primacy through a number of ways:

- **colonial inertia** – developed by colonial powers to function as centres of trade and administration.
- **export dependency** – gateway ports dominate the urban system.
- **urban bias** – inequalities between urban and rural areas are increased by the development of manufacturing in the urban area, resulting in rural to urban migration.

Primacy has its positive factors. It allows economies of scale to be achieved, that is, it is more efficient to provide services to people that are concentrated in an area rather than spread out, and a large skilled workforce is attractive to industrialists. But there are problems associated with urban primacy (and large cities in general):

- social – housing shortage, traffic congestion, crime, illiteracy, pollution
- economic – including urban and regional inequalities, unemployment
- environmental – congestion, poor air quality, contaminated water
- political – over-concentration of power in the primate city.

LEDC cities: Cairo

Since the 1950s Cairo has expanded at a rapid pace. The population now exceeds 13 million. This growth is a result of both natural increase and migration from the countryside. The old city of Cairo is ill-equipped to cope with such a high-density population.

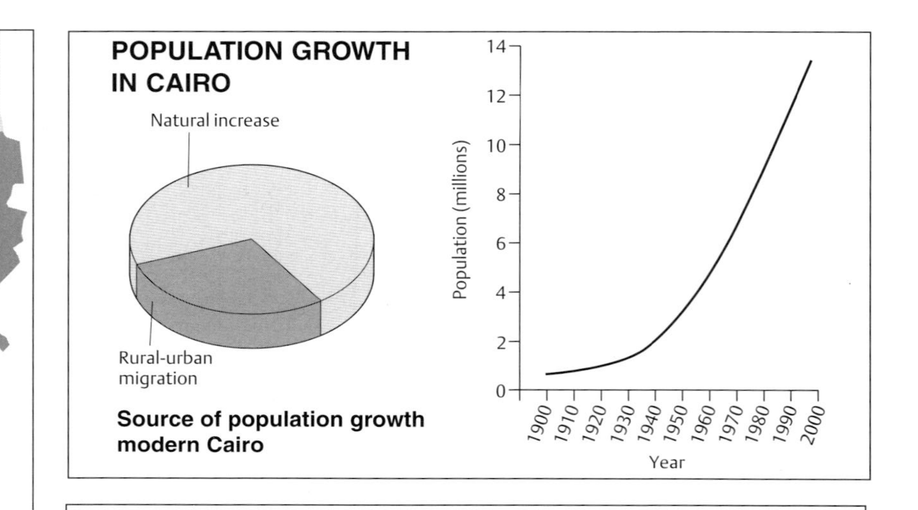

PROBLEMS

Housing
- informal housing (self-help, shanty, temporary) accounts for 80% of Cairo's housing
- 2 – 3 million people live in the Cities of the Dead (graveyards)
- Population density reaches over 30 000 per km^2

Traffic congestion

Lack of jobs and low salaries

Water and air pollution
- Vehicle fumes – smog
- Groundwater pollution

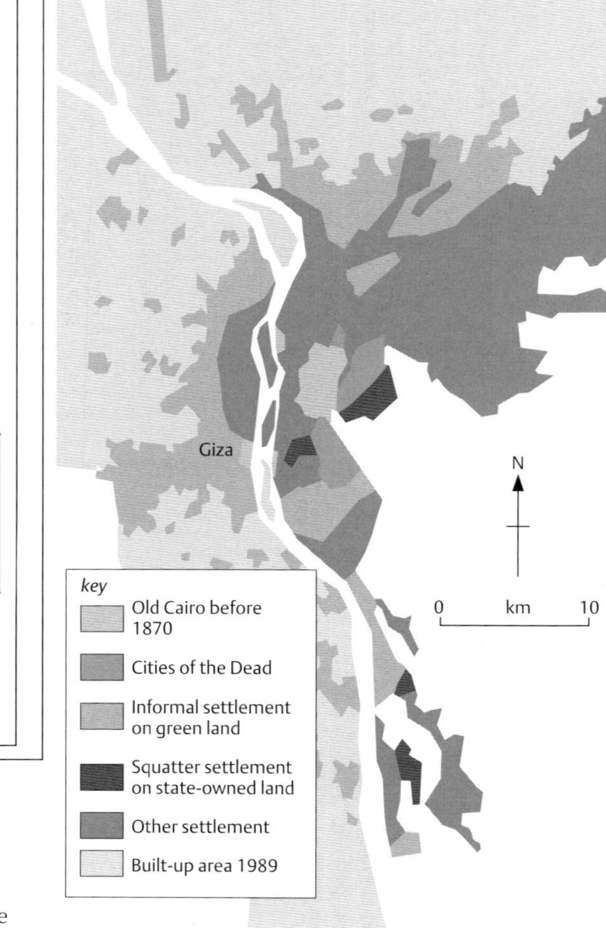

- Leaking sewers
- Rotting buildings

Environmental hazards
- in 1992, 30 000 buildings collapsed in an earthquake.

POPULATION GROWTH IN CAIRO

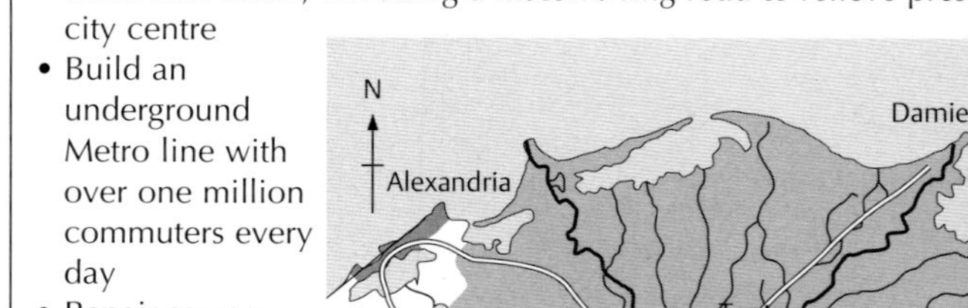

Source of population growth modern Cairo

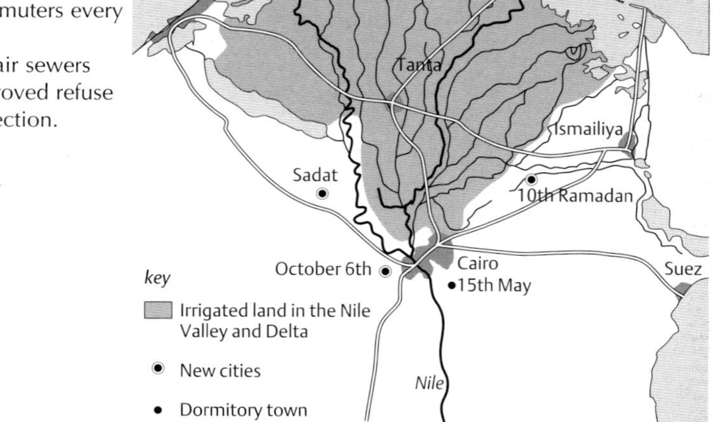

SOLUTIONS

- New satellite (edge) and dormitory towns (without main services and functions) such as 10th Ramada City and 15th May City. It has been necessary to create these to keep pace with the rapid population growth
- Build new roads, including a massive ring road to relieve pressure on the city centre
- Build an underground Metro line with over one million commuters every day
- Repair sewers
- Improved refuse collection.

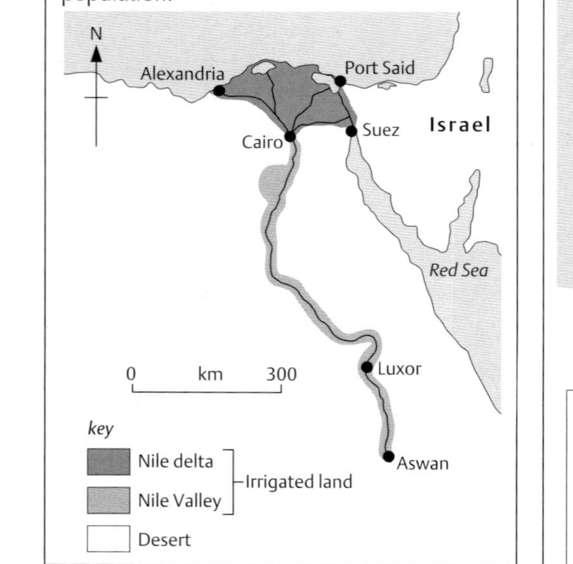

Agricultural systems and factors

Agriculture is the harvesting of crops and animal products for human and/or animal consumption and for industrial production.

CLASSIFYING AGRICULTURE

The following are not exclusive categories but indicate a scale along which all farming types can be placed.

Arable: the cultivation of crops, e.g. wheat farming in East Anglia.
Pastoral: the rearing of animals, e.g. sheep farming in the Lake District.

Commercial: products are sold to make a profit, e.g. market gardening in the Netherlands.
Subsistence (or **peasant farming**)**:** products are consumed by the cultivators, e.g. shifting cultivation by the Kayapo indians in the Amazonian rainforest.

Intensive: high inputs or yields per unit area, e.g. battery hen production.
Extensive: low inputs or yields per unit area, e.g. free range chicken production.

Nomadic: farmers move seasonally with their herds, e.g. the Pokot, pastoralists in Kenya.
Sedentary: farmers remain in the same place throughout the year, e.g. dairy farming in Devon and Cornwall.

FACTORS AFFECTING AGRICULTURE

Physical factors

Climate

Precipitation
- type
- frequency
- intensity
- amount

Temperature
- growing season (> 6°C)
- ground frozen (0°C)
- range of temperatures

Soil

Fertility
- pH
- cation exchange capacity
- nutrient status

Structure

Texture

Depth

Pests — Vermin, locusts, disease, etc.

Slope — Gradient

Relief — Altitude

Aspect — Ubac (shady) or adret (sunny)

Human factors

Political

Land tenure/ownership
- ownership, rental, share-cropping, state-control

Organisation
- collective, co-operative agribusiness, family farm

Government policies
- subsidies, guaranteed prices, ESAs, quotas, set-aside

War
- disease, famine

Economic

Farm size
- field size and shape

Demand
- size and type of market

Capital
- equipment, machinery, seeds, money, 'inputs'

Technology
- HYVs, fertilisers, irrigation

Infrastructure
- roads, communications, storage

Advertising

Social

Cultural and traditional influences

Education and training

Behavioural influences

Chance

Farming systems

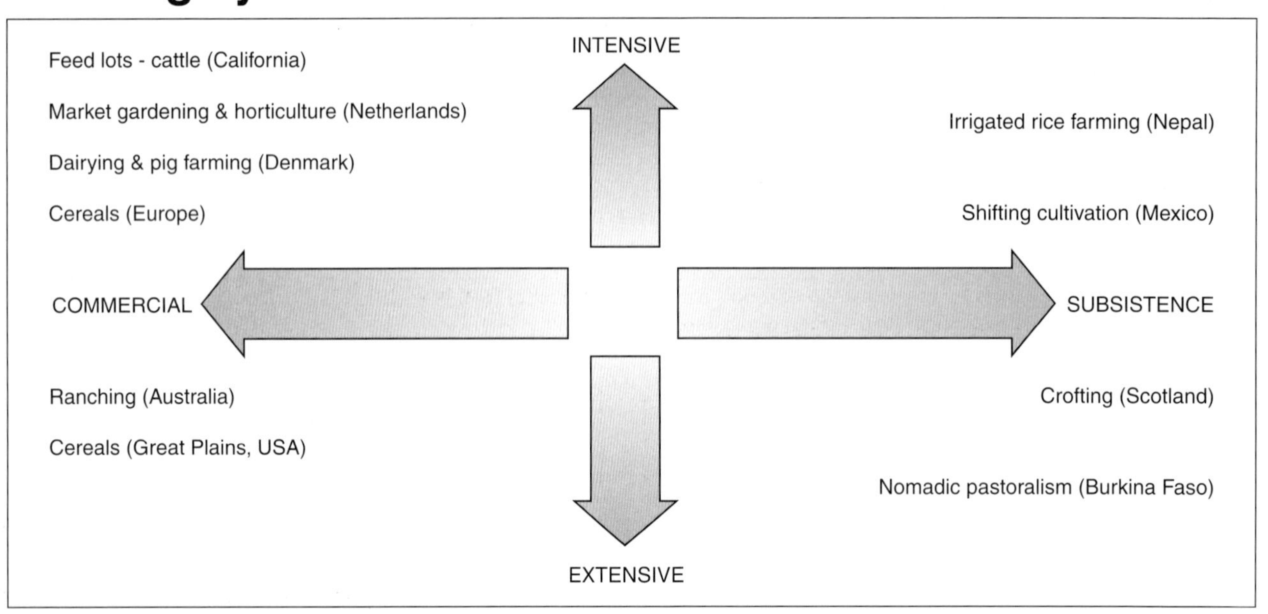

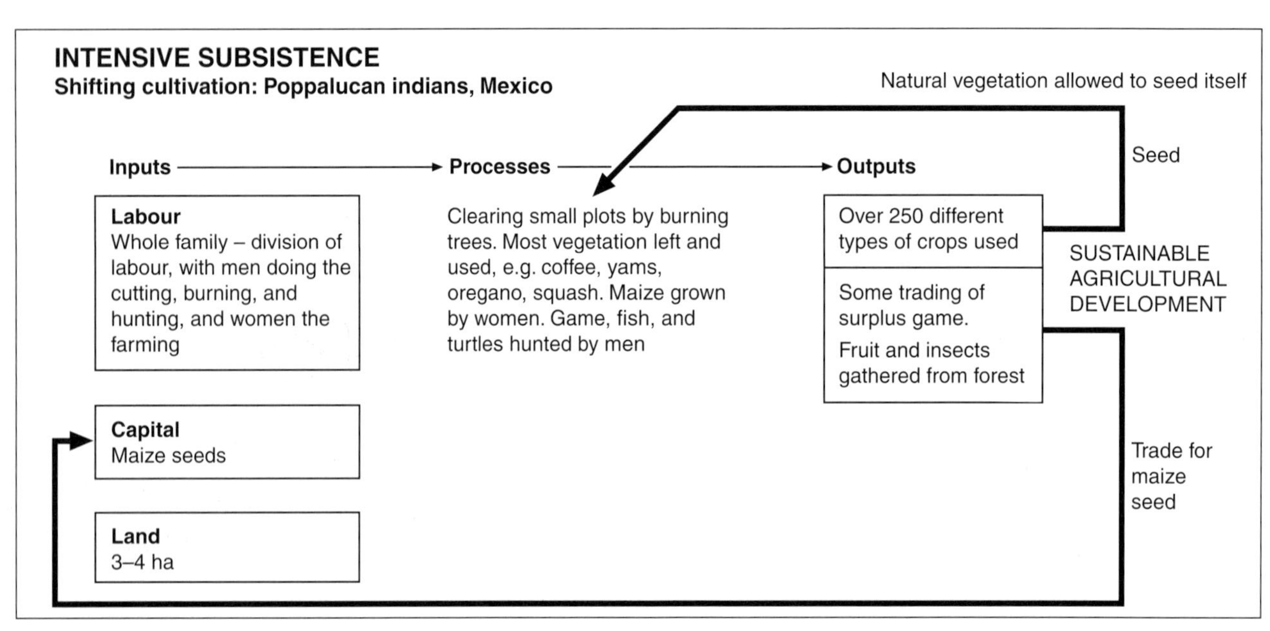

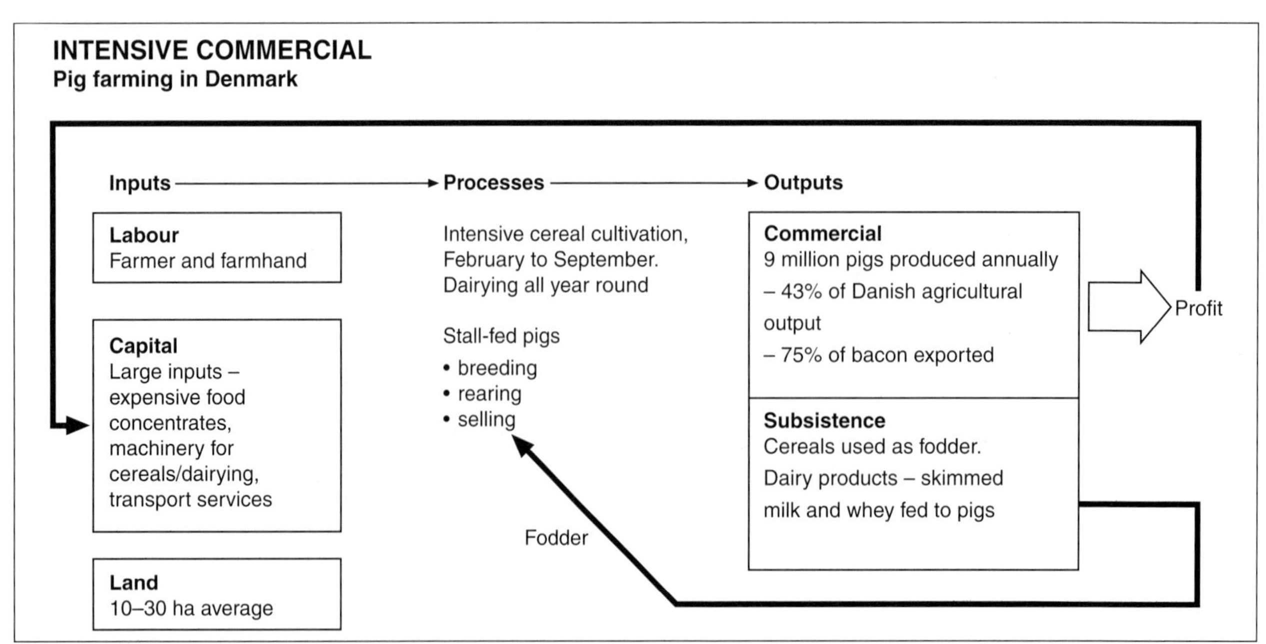

Agricultural ecosystems

Agricultural ecosystems can be compared with natural ecosystems in terms of productivity, biomass, nutrient cycling, and energy efficiency. Average productivity from agricultural systems is 650 g/m^2/yr, comparable with temperate grasslands or prairies. However, only about 0.25% of incoming radiation is utilised by crops, and of that less than 1% is harnessed by people through food.

ECOSYSTEMS

	Natural ecosystems	Agricultural ecosystems
Foodweb	Complex; several layers	Simple; mostly one or two layers
Biomass	Large; mixed plant and animal	Small; mostly plant
Biodiversity	High	Low - often mono-culture
Gene pool	High	Low, e.g. three species of cotton account for 53% of crop
Nutrient cycling	Slow; self-contained; unaffected by external supplies	Largely supported by external supplies
Productivity	High	Lower
Modification	Limited	Extensive - inputs of feed, seed, water, fertilisers, energy fuel; outputs of products, waste

UK NUTRIENT CYCLES

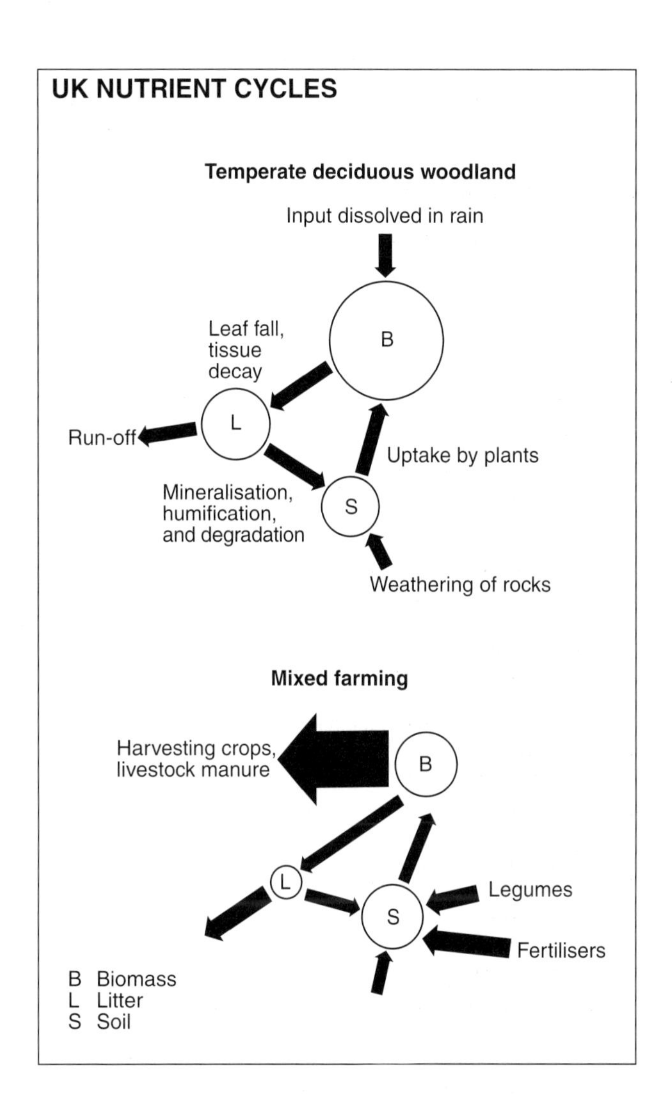

EFFECTS OF NOMADIC PASTORALISM

1. **Energy flow**
   Little change - cattle replace wild herbivores; limited killing of predators.
2. **Nutrient cycling**
   Little change - some concentrations of nutrients, in dung, may occur if herds remain in one place for a length of time. True nomadic movement returns and distributes nutrients over a wide area.
3. **Biological productivity**
   NPP low and variable - 150 g/m^2/yr in drier areas rising to 600 g/m^2/yr in wetter margins. Secondary productivity is low - hence farmers use milk, milk products, and blood, rather than meat.
4. **Ecological stability and modification**
   Over-exploitation of grass or over-concentration of herds removes vegetation, especially sweeter species, causing ponding of the surface, gullying, and desertification - the spread of desert conditions. This was due to climatic deterioration in traditional pastoral societies, but is increasingly common for economic, social, and political reasons: larger herds, shorter nomadic routes, and greater pressure around water sources, e.g. boreholes.

ENERGY RATIOS (ER)

$$ER = \frac{\textbf{Energy outputs}}{\textbf{Energy inputs}}$$

Shifting cultivation	65.0
Hunter-gatherers	7.8
UK cereal farm	1.9
UK allotment	1.3
UK dairy farm	0.38
Broiler hens	0.1
Greenhouse lettuces	0.002

Agricultural models

VON THUNEN'S MODEL

Johann Von Thunen's model of **locational or economic rent** (1826) suggests that land use and intensity of production declines with distance from a central market. High intensity market gardening, dairying, and horticulture predominate close to urban areas while extensive grain and livestock farming are located furthest away. Woodland was an important land use when Von Thunen developed his model and was found close to the urban area. Although his model is criticised for its simplicity and its assumptions (that farmers' sole aim is to maximise profits, i.e. **rational man**, and that physical conditions do not vary, i.e. an **isotropic plain**), aspects of his model can be observed at a variety of scales, from the individual farm up to land use in Europe. It is also important whenever transport is poorly developed, especially in developing countries.

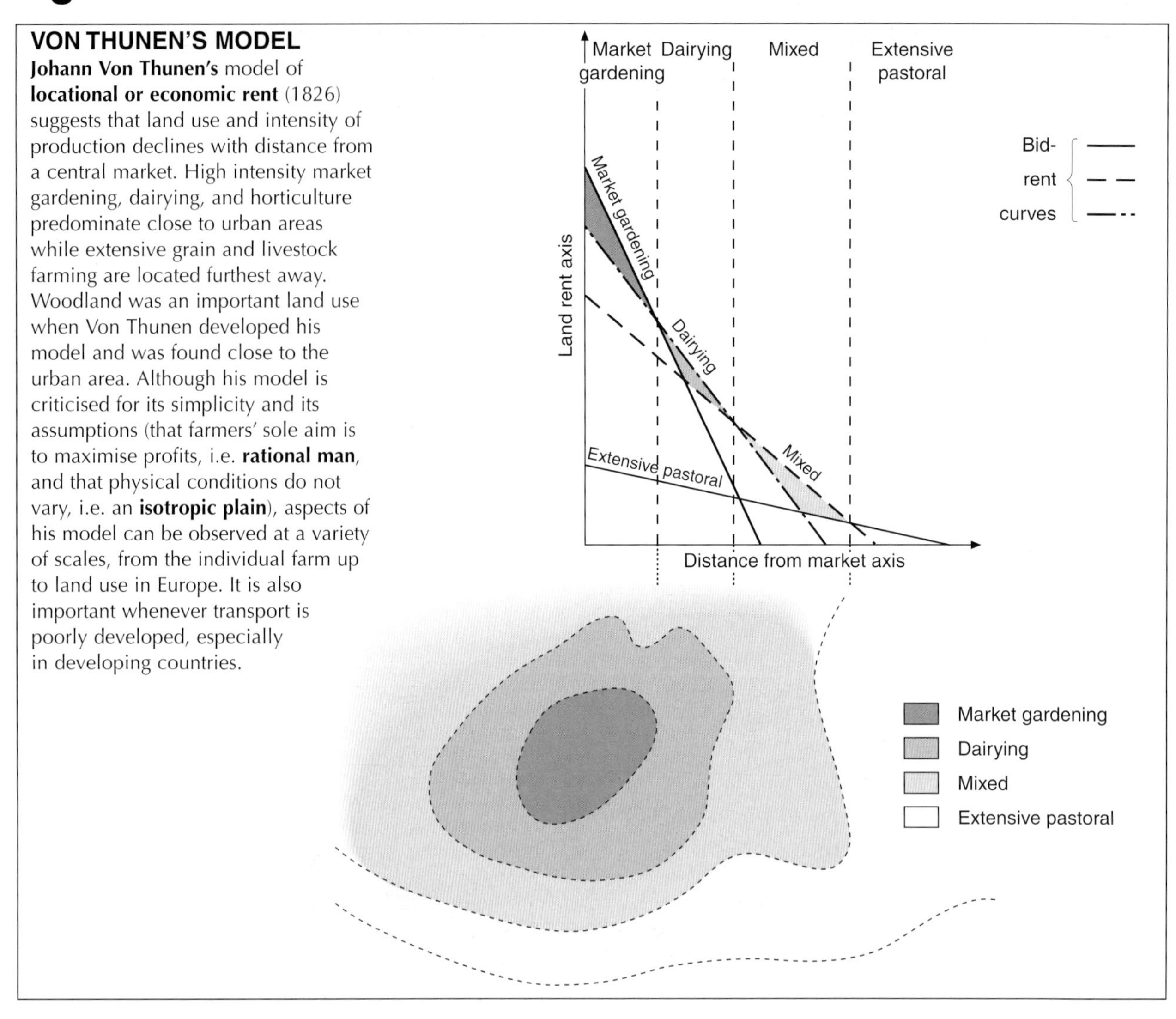

SINCLAIR'S MODEL

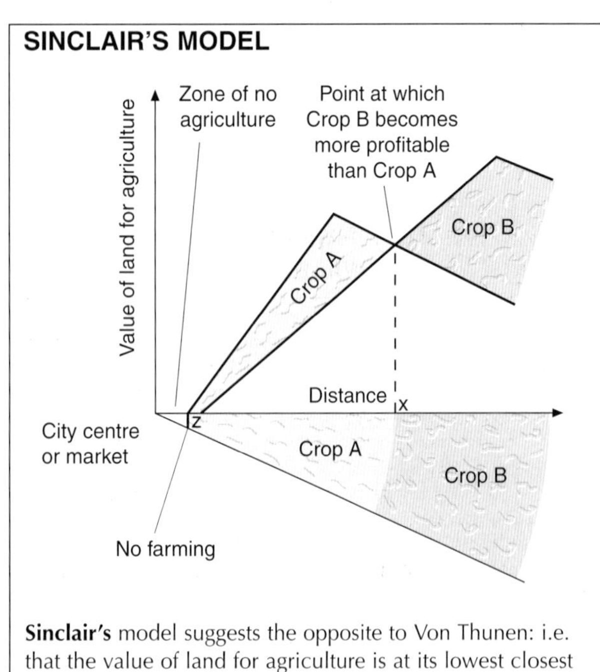

Sinclair's model suggests the opposite to Von Thunen: i.e. that the value of land for agriculture is at its lowest closest to urban areas. This is because the land is more valuable for **speculative developments**, e.g. for commercial, industrial, or residential uses. Beyond a certain distance, the land is used for agriculture as it loses its value for development.

HAGERSTRAND'S MODEL

Hagerstrand showed how **new innovations** and **techniques** were likely to be used only by a few people at first (innovators) before being adopted rapidly, although a few laggards would resist change. This meant the adoption of any technique followed an S-shaped curve.

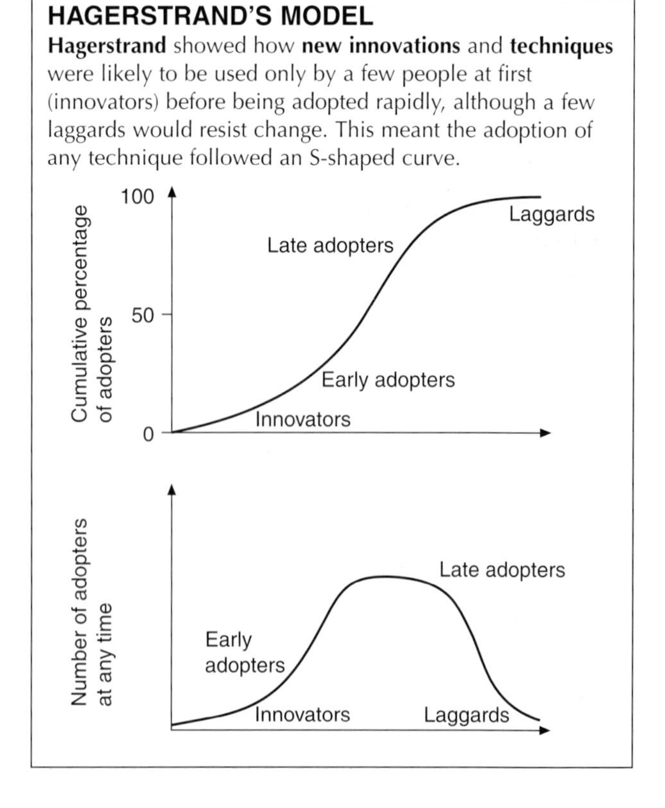

Agriculture in England and Wales

CEREAL FARMING IN EAST ANGLIA

Physical factors
- Cool summers and cold winters prevent the growing of arable crops. The temperature promotes growth.
- High rainfall, ≥ 2000 mm, is too much for arable crops and produces grass instead.
- Soils are often thin, infertile and easily eroded.
- Hilly land prevents the use of much machinery.

Human factors
- There is some limited support for sheep farmers, but it is not as great as for arable farmers.
- The area is quite remote and isolated, and far from the main markets.
- Decreased demand for red meat products, such as lamb, due to health reasons.
- Small amounts of capital inputs.
- The area suffered from nuclear fallout after the nuclear disaster at Chernobyl in 1986.

CEREAL FARMING IN EAST ANGLIA

Physical factors
- Warm summers, > 20°C, favour the ripening of crops.
- Low rainfall, <600 mm, but it falls mostly in Spring and Summer when plants need it.
- Fertile boulder clay is good for arable farming.
- Flat land allows the use of machinery.

Human factors
- Government (CAP) support for cereals.
- Good communications (M11, Felixstowe port).
- Increased demand for cereal products.
- Lots of capital inputs (money, seeds, fertiliser, and pesticides).

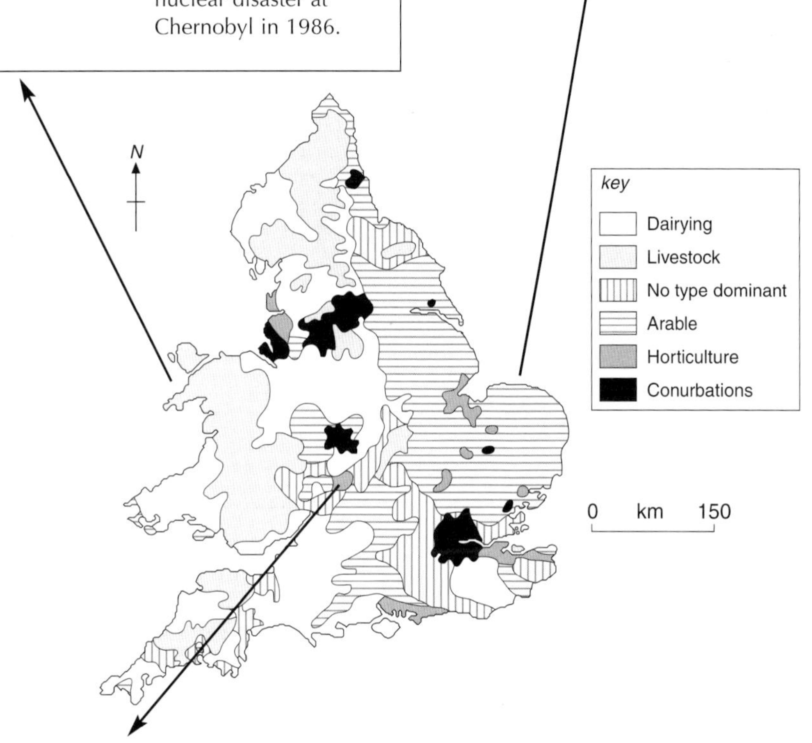

MARKET GARDENING IN THE VALE OF EVESHAM

Physical factors
- Much of the crops are under greenhouses so temperatures can be controlled. In other places, such as near Cheddar Gorge, south-west slopes are favoured because they are warmer.
- In greenhouses the water can be controlled. Some plants are grown in water - this is known as hydroponics.
- Soils are treated with lots of fertiliser.

Human factors
- Good communications (M5) and access to a large, wealthy urban market.
- Large demand for products such as strawberries, lettuces, tomatoes, raspberries, apples.
- Very high levels of capital inputs (money, seeds, fertiliser, and pesticides).

A COMPARISON OF FARMING IN EAST ANGLIA AND WALES

	East Anglia	Wales
Area of wheat ('000 ha)	297	12
Area of barley ('000 ha)	148	41
Area of rape ('000 ha)	32	1
Yield of wheat (tonnes/ha)	7.6	6.5
Yield of barley (tonnes/ha)	5.6	5.0
Yield of rape (tonnes/ha)	2.8	2.8
Number of sheep (000)	294	11 256
Number of cattle (000)	200	1340
Total income from agriculture (£ million)	463	232

The Common Agricultural Policy (CAP)

The **Common Agricultural Policy** was developed to achieve four main goals:

- To increase agricultural **productivity** and **self-sufficiency**.
- To ensure a **fair standard of living** for farmers.
- To **stabilise markets**.
- To ensure food was available to consumers at a **fair price**.

Three principles lay behind this:

- A single European agricultural market in which goods could move freely.
- Preferential treatment for European food.
- EU funding of the CAP.

The CAP achieved its main aim of increasing food supply by **guaranteed prices** and **intervention buying and storage**, i.e. a guaranteed market. This led to **intensification**, **specialisation**, and **concentration** of agricultural activities in the better suited areas.

The worst food crisis since 1974 has left the world with only 53 days supply of grain. The world is seriously short of food and up to 35,000 children die from hunger related diseases every day. Worldwide grain stocks are well below the FAO's minimum necessary to safeguard world food security. In 1987 there was over 100 days' worth of food supplies, but in 1995 there were just over 50, and by the end of 1996 it will be below 50 days'.

World food production has lagged behind food consumption since 1993. The drought of 1995 has led to the lowest harvest of food/head since the mid-1970s. The blistering summer of 1995, the hottest recorded in many parts of the northern hemisphere, destroyed millions of crops. The drought in Spain entered its fourth year and wheat yields slumped to less than half of their 1994 levels.

The European food mountain is fast being eroded. Its grain mountain has fallen from 33 million tons in 1993 to just 5.5 million tons in 1995. The only significant surplus is the wine lake, at 120 million litres.

Europe is not producing enough food, partly as a result of the 1992 CAP reforms which increased the amount of set-aside. Plans to reduce set-aside to 10% of land, rather than 15% in 1995, are an attempt to increase food production.

CAP REFORM

The CAP created huge surpluses - by the early 1990s, 33 million tonnes of cereal, 2200 million litres of wine, and 8 million tonnes of beef had to be stored. The cost of storing butter alone was over £5 million per week.

In 1992, the CAP was reformed. Surplus production was inefficient and costly to store, subsidies were excessively large, and intensive farming was harming the environment. Changes included reduction of price support, increased quotas, extensification of agriculture, and set-aside. However, by 1995-6, owing to the reforms of the CAP and a series of very hot summers, agricultural surpluses had been drastically reduced

REFORM OF THE CAP

The proposal	The winners	The losers
Freeze farm spending at £30.7 billion a year (half the EU's spending) between 2000 and 2006	Germany, The Netherlands, Denmark, and the UK are in the vanguard. They claim to have efficient farmers and pay much more into the EU than they get out.	France, Ireland, and 'Club Med' states oppose the freeze, not always on financial lines. Any country where farmers tend to take to the streets has worries.
Cut guaranteed prices in line with free market. Beef, cereals, and milk could come down by 30%	Big 'factory' farmers in Britain, France, eastern Germany, Denmark, Holland, and Sweden, those who can compete in a market where prices are not rigged.	Small hill farmers in Wales, the Scottish Highlands, Ireland and Spain. France is calling for cuts on cereals but fears its dairy farmers will not be able to compete.
Year-on-year cut in payments that compensate farmers for being exposed to world markets	Good for big farmers who will increase production in the free market. Direct payments will go up to compensate for cuts in guaranteed prices.	This horrifies the broadest spectrum of farmers. Small farmers will be forced to increase production to make up losses. Brussels suggests new funding to struggling areas.
Co-financing by member states and EU until nations take 25% of direct payments the EU now funds	A weapon to help Germany cut its £8 billion net contribution to the EU, which is backed by the UK and other countries that gain little from the CAP.	France is the most stubborn opponent, arguing that this is the beginning of the end of the 37-year-old policy. Ireland is also alarmed, for financial reasons.
CAP payments to big farmers—the millionaire grain barons	Hill farmers, smallholders, and southern producers will benefit. Despite previous reforms, the richest 4% of landowners still pocket 80% of CAP cash.	One of the few reforms opposed by Britain, this would cut payments to anyone receiving more than £100 000 per year, affecting 5500 farmers.
End of compulsory set-aside, helping clear way for cereal farmers to return to free market	Current plans would end curbs on grain production and are backed by countries with big cereal producers, such as the UK, Sweden, Denmark, Holland and France.	Governments in Germany, Austria, Greece, Ireland, and Portugal favour keeping a minimum guaranteed price coupled with a small set-aside provision.

Agriculture and environmental issues

SOIL EROSION

- Over one-third of arable land in the UK is at risk of soil erosion
- Sandy and sandy-loam soils with a slope angle of more than 3° are particularly vulnerable
- Soil losses are up to 250 t/ha in the South Downs, 160 t/ha in Norfolk and 150 t/ha in West Sussex.

The potential for soil erosion has increased considerably in recent years for a number of reasons:

- Spread of arable land use into pastoral areas
- Hedgerow removal
- Ploughing and draining of peaty soils
- Afforestation leaves bare ground between young trees
- Increased recreational pressure in rural areas.

Potential for erosion by rainfall

Low	< 700
	700–900
	900–1 100
	1 100–1 300
High	> 1 300

0 km 150

NITRATE POLLUTION

Increased levels of nitrates in ponds and streams can cause eutrophication, i.e. nutrient enrichment.

Up to 5 million people in England and Wales are supplied with water with high rates of nitrates. This is linked with:

- high rates of stomach cancer
- blue baby syndrome, due to oxygen starvation in the bloodstream

Solutions include:

- A change in land use to pastoral farming
- Less intensive arable agriculture
- Less use of fertiliser
- The use of cover crops in winter to absorb fertiliser.

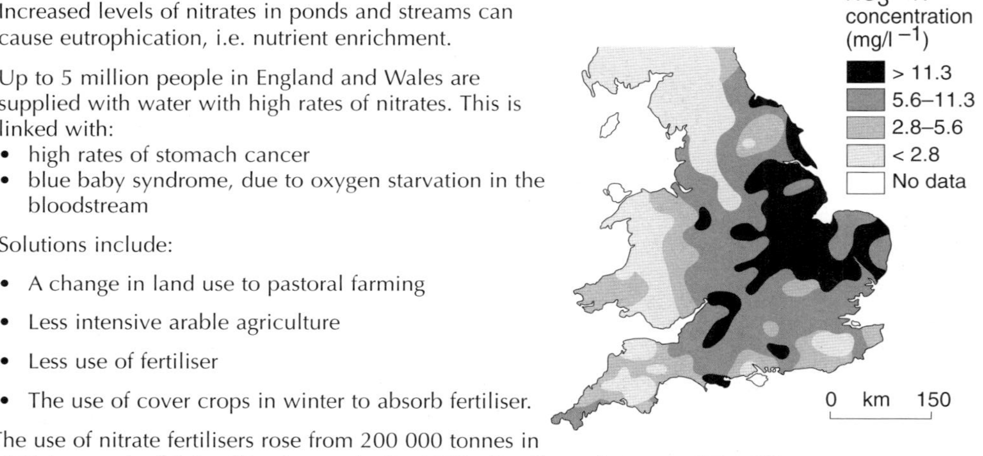

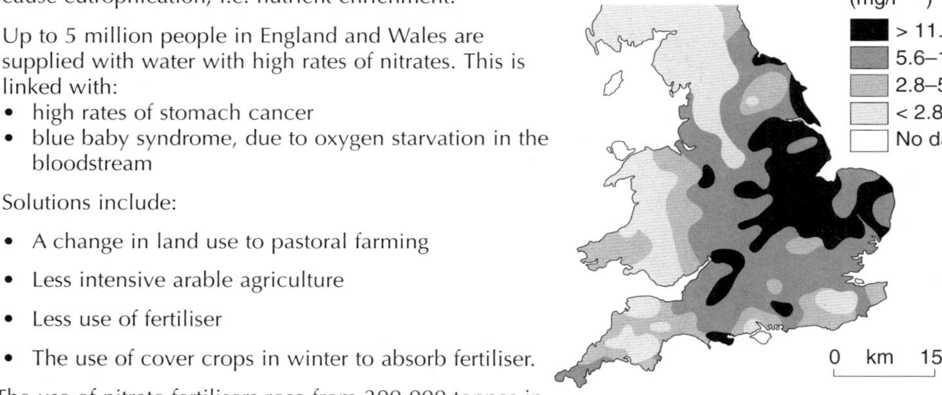

The use of nitrate fertilisers rose from 200 000 tonnes in 1945 to a peak of 1.6 million tonnes in the 1980s. It will cost between £50 million and £300 million each year to purify water that has become enriched with nitrates.

BSE

BSE (a disease in cows) and CJD (a disease in humans) belong to a rare group of diseases called spongiform encephalopathies. Most cases of CJD have occurred in places where BSE is more common. In 1996 10 cases of CJD were diagnosed.

The first case of BSE in Britain was in 1986. Most of the infection in cattle took place in the late 1980s and it peaked in 1992. BSE came into cattle when they ate meat that was infected with scrapie, a disease common in sheep. Cows that were fed on infected sheep tissue developed BSE. As these cows were then slaughtered, crushed, and fed back to other cows, some of these became infected.

Why did it affect the UK?

- Cattle carcasses in the UK were burnt at a relatively low temperature.
- Cattle in the UK derive up to 5% of their food from meat and bone meal.
- The government only offered a 50% grant for farmers to destroy infected animals. It is believed that this encouraged farmers to pass off sick animals as healthy and lengthened the period that humans were fed potentially infected beef.

As soon as other EU countries suspected that British animals might be spreading BSE they banned it. France and Ireland destroyed all animals in any herd that contained even one case of BSE. This did not occur in the UK. Up to 85% of beef herds and 40% of dairy herds remain unaffected by BSE.

To cost of BSE to Britain could be as high as £15 billion.

About 850 000 cows are killed every year. These are mostly cattle that have come to the end of their working lives and are used for products such as sausages, pate, pies, and glue.

Number of BSE cases (axis: 0, 50, 100, 200, 250, 161 200, 161 700)

Country	Number of BSE cases
Falkland Islands	1
Denmark	1
Canada	1
Oman	2
Italy	2
Germany	4
France	13
Portugal	31
Ireland	123
Switzerland	206
Britain	161 663

Reducing the environmental impacts

SET-ASIDE

The **set-aside** scheme was introduced on a voluntary basis in 1988 allowing farmers to take up to 20% of their land out of production and to receive up to £200 for each hectare set aside. The land could be left fallow, converted to woodland, or used for non-agricultural purposes. Reform of the CAP in 1992 reduced the amount of set-aside to a maximum of 15%, and further reform in 1994 reduced rotational set-aside to 12% and flexible set-aside to 15%. While many farmers took advantage of set-aside, others intensified production on the other land and made their least favourable land the set-aside! Between 1992 and 1993 the total area in England and Wales under cereals decreased by 400 000 hectares and there was a similar increase in the amount of set-aside.

ESAS

In 1985 the EU agreed to provide farmers with the means to farm **environmentally sensitive areas** in traditional ways which would preserve important biological and heritage landscapes. Less intensive, organic methods were favoured with increased amounts of fallow. By 1994, 10 500 farmers had signed or applied for ESA agreements, and payments during 1994/5 totalled about £25 million.

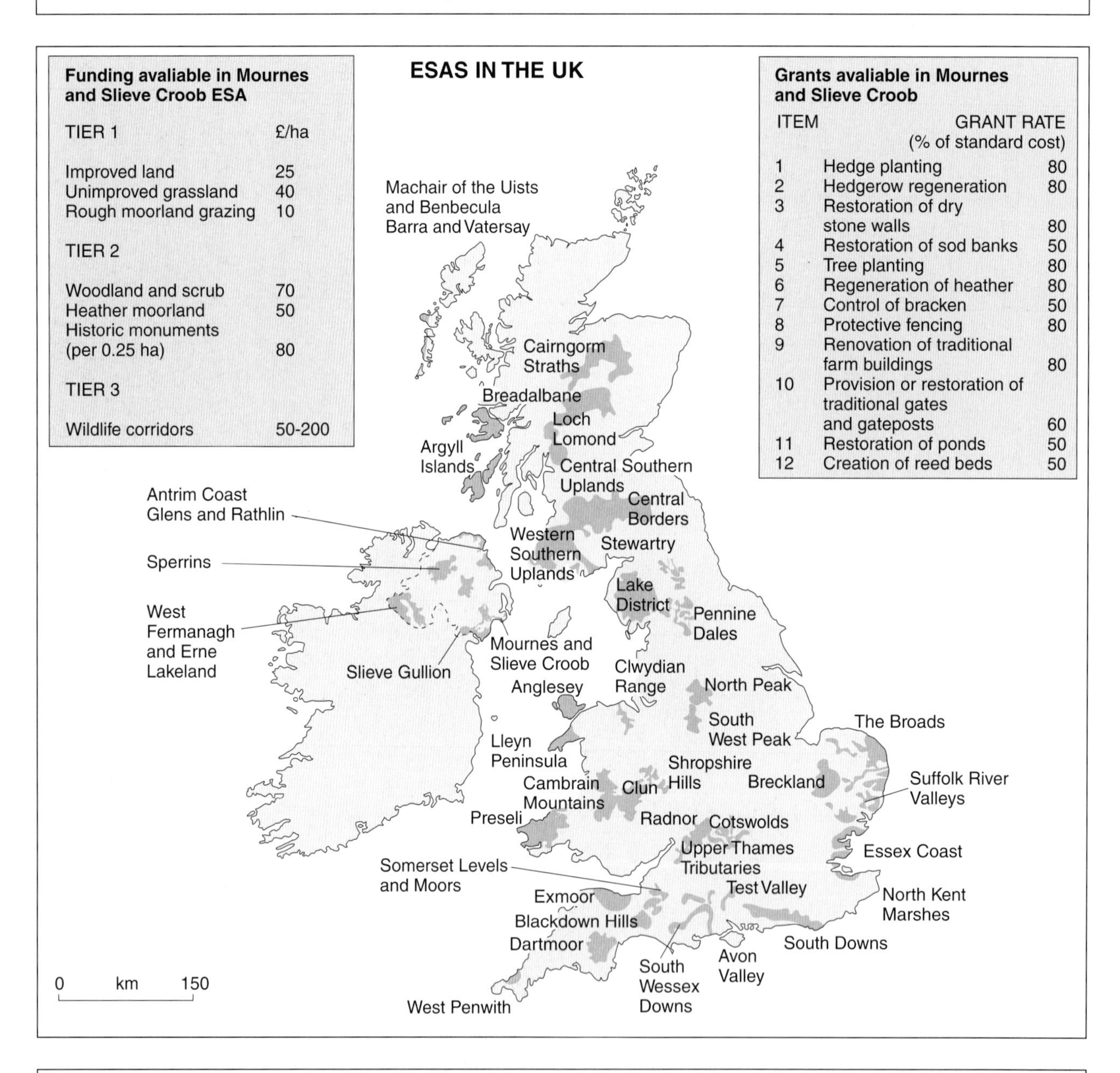

Funding avaliable in Mournes and Slieve Croob ESA

	£/ha
TIER 1	
Improved land	25
Unimproved grassland	40
Rough moorland grazing	10
TIER 2	
Woodland and scrub	70
Heather moorland	50
Historic monuments (per 0.25 ha)	80
TIER 3	
Wildlife corridors	50-200

Grants avaliable in Mournes and Slieve Croob

ITEM		GRANT RATE (% of standard cost)
1	Hedge planting	80
2	Hedgerow regeneration	80
3	Restoration of dry stone walls	80
4	Restoration of sod banks	50
5	Tree planting	80
6	Regeneration of heather	80
7	Control of bracken	50
8	Protective fencing	80
9	Renovation of traditional farm buildings	80
10	Provision or restoration of traditional gates and gateposts	60
11	Restoration of ponds	50
12	Creation of reed beds	50

Other schemes include:

- **Nitrate Sensitive Areas** to protect groundwater areas.
- **Habitat Schemes** to improve/create wildlife habitats.
- **Organic Aid Schemes** to encourage farmers to convert to organic production methods.
- The **Countryside Access Scheme** to grant new opportunities for public access to set-aside land and suitable farmland in ESAs.

Farming in LEDCs

The importance of agriculture

Agriculture remains the main source of **employment** for most people in LEDCs. However, its importance has declined in recent decades due to the growth of manufacturing and to decreased food prices. Nevertheless, it remains a vital part of many economies due to employment, **export earnings** and **food supply**.

Farming systems in more developed countries (MEDCs) and LEDCs are very different. Agriculture in MEDCs has more in common with manufacturing industry than it has with farming in LEDCs. For example, much of it is run by companies, and is **capital intensive** (costs lots of money), highly **mechanised**, **large-scale**, **market-orientated** (geared to consumer demand), and **government involvement** is crucial. By contrast, agriculture in LEDCs is typically **small-scale**, **labour intensive**, and **subsistence** by nature.

Over three-quarters of the world's population live in LEDCs, and in the poorest of these over 70% of the population are employed in agriculture.

The **global pattern** of agriculture in LEDCs can be divided into three main groups.
1 Tropical Africa, Iran, Iraq, and Cambodia - extensive farming, shifting cultivation, low yields, limited inputs, limited mechanisation, and a small proportion of irrigation.
2 Latin America - a small proportion of cultivated land, high proportion of grain, low but increasing crop yields, and limited use of high yielding varieties (HYVs) and fertilisers.
3 South and East Asia - intensive cultivation, especially of rice, high yields, and much use of HYVs.

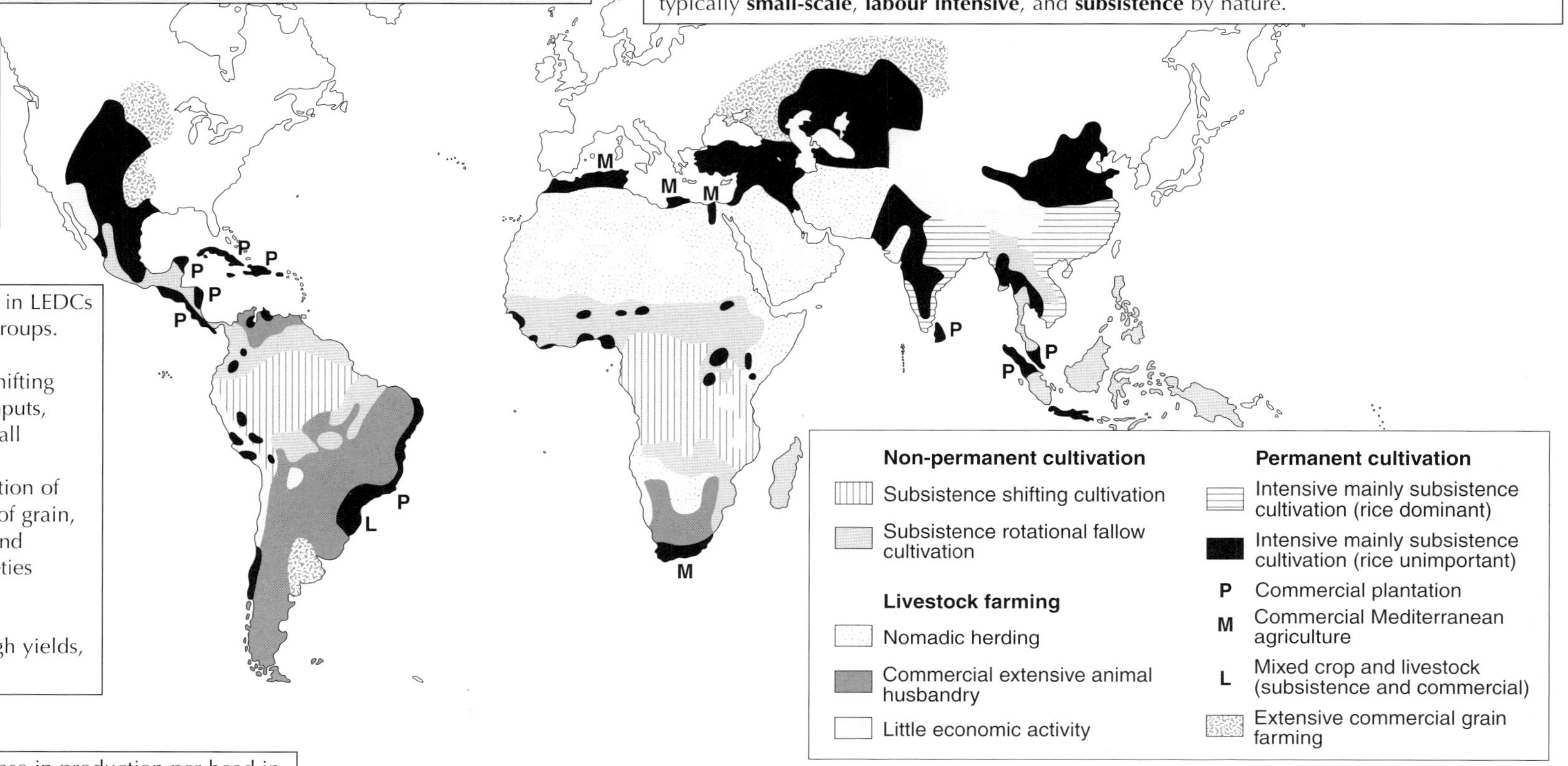

In LEDCs there has been a decrease in production per head in many countries. Reasons include:
- deteriorating environmental conditions
- poor farming practises
- over-population
- under-population, as in Rwanda, where there were not enough people to harvest crops from the fields
- the neglect of the agricultural sector by the government.

In parts of Africa, declining farmyields are widespread. At the heart of the problem is the fact that population growth exceeds agricultural production. Potential solutions are mostly related to intensification such as double cropping (two crops a year), irrigation, increased use of fertilisers, and greenhouses. But such developments are neither widespread nor even. For example, there is a very uneven pattern of fertiliser use with a large increase in Asia (especially China, India, and Bangladesh) but not in Africa.

The green revolution

THE PROBLEM

Population growth is more rapid than the growth of food production. In India, for example, by 2000 AD the population will reach 1 billion people and food production will need to increase by 40% to match demand. But much of India's land is of limited potential.

THE SOLUTION?

The **green revolution** is the application of science and technology to increase crop productivity. It includes a variety of techniques such as genetic engineering to produce higher yielding varieties (HYVs) of crops and animals, mechanisation, pesticides, herbicides, chemical fertilisers, and irrigation water.

HYVs are the flagship of the green revolution. During 1967-8 India adopted Mexican Rice IR8 which yielded twice as much grain as traditional varieties. However, it required large amounts of water and fertiliser. Up to 55% of India's crops are now HYVs and 85% of the Philippines' crops are HYVs. By contrast only 13% of Thailand's crops are HYVs.

THE SPREAD OF HYVS

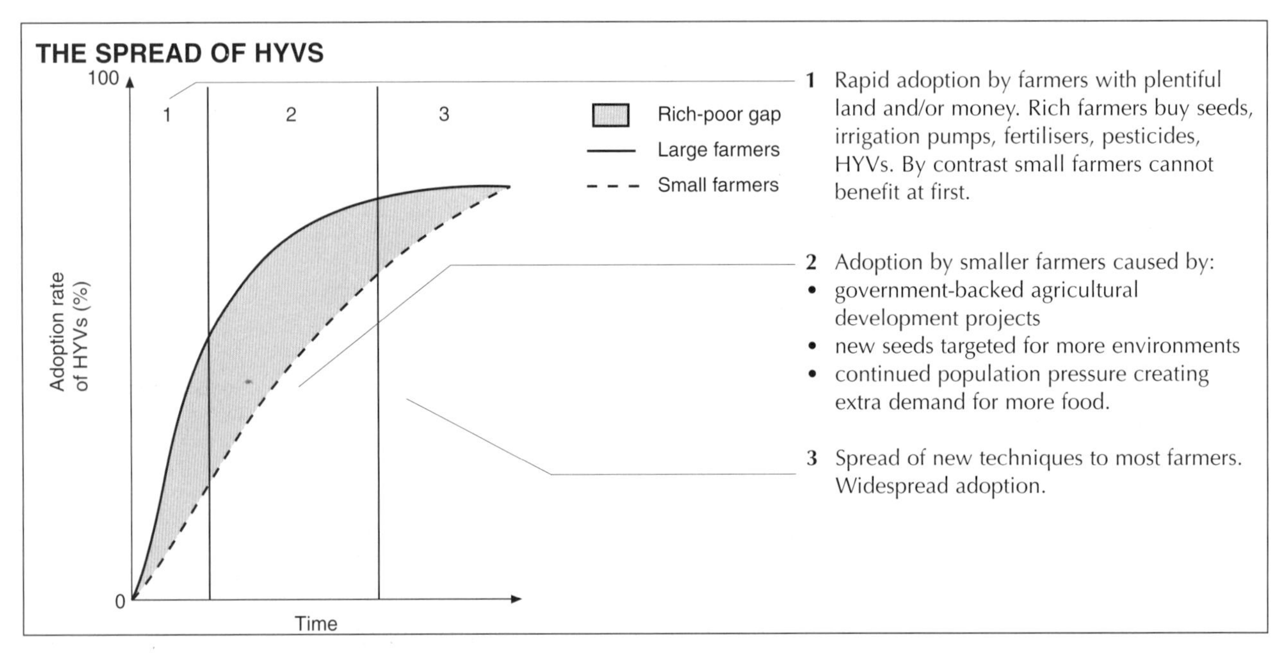

1 Rapid adoption by farmers with plentiful land and/or money. Rich farmers buy seeds, irrigation pumps, fertilisers, pesticides, HYVs. By contrast small farmers cannot benefit at first.

2 Adoption by smaller farmers caused by:
- government-backed agricultural development projects
- new seeds targeted for more environments
- continued population pressure creating extra demand for more food.

3 Spread of new techniques to most farmers. Widespread adoption.

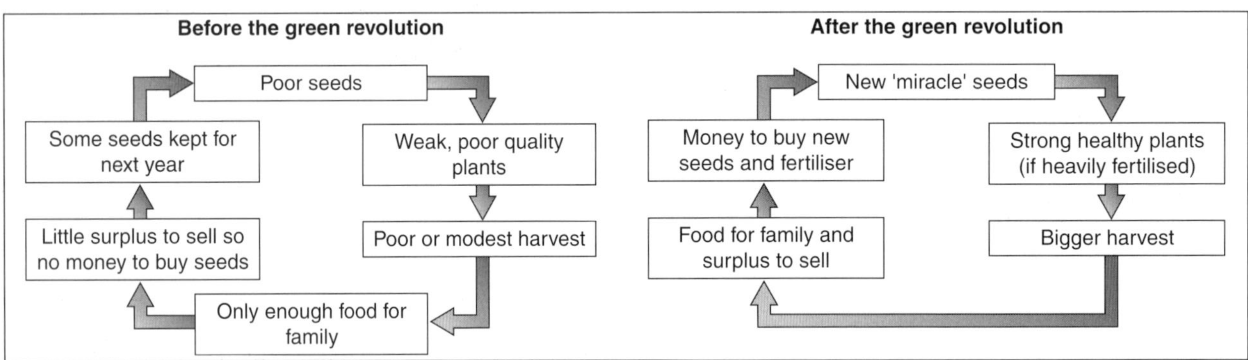

THE CONSEQUENCES

The main benefit is that more food can be produced:
- yields are higher
- up to three crops can be grown each year
- more food should lead to less hunger
- more exports create more foreign currency.

However, there are many problems:
- not all farmers adopt HYVs - some cannot afford the cost
- as the cost rises, indebtedness increases
- rural unemployment has increased due to mechanisation
- irrigation has led to salinisation - 20% of Pakistan's and 25% of Central Asia's irrigated land is affected by salt
- soil fertility is declining as HYVs use up all the nutrients; these can be replenished by fertilisers, but this is expensive
- LEDCs are dependent on many developed countries for the inputs.

Changes in South India: the effects of the green revolution	
Use of fertiliser	+138%
Human labour	+111%
Paddy rice	+91%
Sugar cane	+41%
Income	+20%
Subsistence food	-90%
Energy efficiency	-25%
Casual employment	-66%

Factors affecting industry

CLASSIFICATION

Primary industries The extraction of raw materials, e.g. mining, quarrying, farming, fishing, and forestry.

Secondary industries Manufacturing industries which involve the transformation of raw materials (or components) into finished products or semi-finished products, e.g. steelworks, the car industry, and high technology.

Tertiary industries These are concerned with providing a service to customers, e.g. transport, retailing, and medical and professional services.

Quaternary industries These provide information and expertise, e.g. universities, research and development, media, and political policy units.

CHANGING INDUSTRIAL STRUCTURE OVER TIME

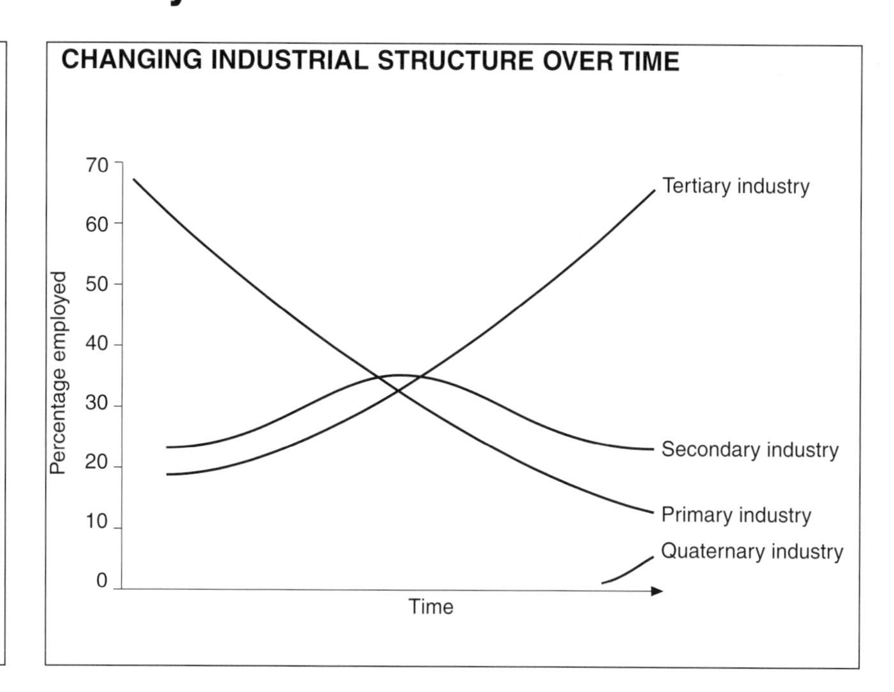

FACTORS INFLUENCING LOCATION

- Physical factors influenced industrial location in the nineteenth century.
- Old industrial regions were located where cheap energy and raw materials were found.
- Physical factors influenced mature industries like textiles, shipbuilding, and iron and steel.
- Transport costs and markets also influenced mature industries.
- In the 1990s, government policy and labour requirements have strong influences on industry.
- Industrial location is no longer country-specific and choices of location are global and strategic.

Physical factors

- Raw materials
- Energy/power
- Site/land
- Climate
- Natural routes
- Rivers
- Natural ports

Human-economic factors

- Transport
- Labour
- Technology
- Capital
- Industrial inertia
- Markets
- Product life-cycles
- Linkages between associated industries
- Government policies

DEFINITIONS OF KEY WORDS

Agglomeration economies Savings which arise from the concentration of industries either together or close to linked activities.

Greenfield site An industrial site located on the edge of an urban area in a place with no prior industrial use.

Transplant or branch plant An assembly plant owned and operated by a foreign-based company.

Industrial inertia The survival of an industry in an area even though the initial advantages are no longer relevant.

Transnational or multi-national corporation (TNC or MNC) A large, multi-plant firm with a worldwide manufacturing capability.

Break-of-bulk location A location which takes its advantage from a position where there is forced transfer of freight from one transport medium to another, e.g. a port or rail terminal.

Rationalisation A reduction in the production capacity of a multi-plant firm by factory closure.

Research and development (R&D) The branch of a manufacturing firm concerned with the design and development of new products; R&D employs highly skilled workers.

Classical location theory

All models of industrial location simplify the real world in order to illustrate one or more concepts which influence locational choice.

WEBER'S LEAST-COST LOCATION MODEL (1926)

Weber concentrated on *costs* to explain the *optimum location* for industry.

(i) Transport costs

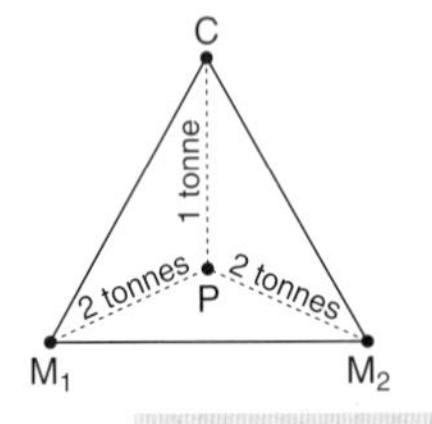

C
2 tonnes
P
1 tonne
1 tonne
M1
M2

C = Consumption point (market)
M_1M_2 = Raw materials
P = Optimum location

a. Weight-losing industry

Raw materials are heavier than finished product so transport costs are saved by locating close to the raw materials

Example: Iron and steel, which uses bulky raw materials

b. Weight-gaining industry

Finished product is heavier than raw materials so location is closer to the market

Example: Brewing, which gains weight through the addition of a ubiquitous raw material, water

The optimum location may shift if savings from labour costs or agglomeration outweigh the increased transport costs of moving.

(ii) Labour costs

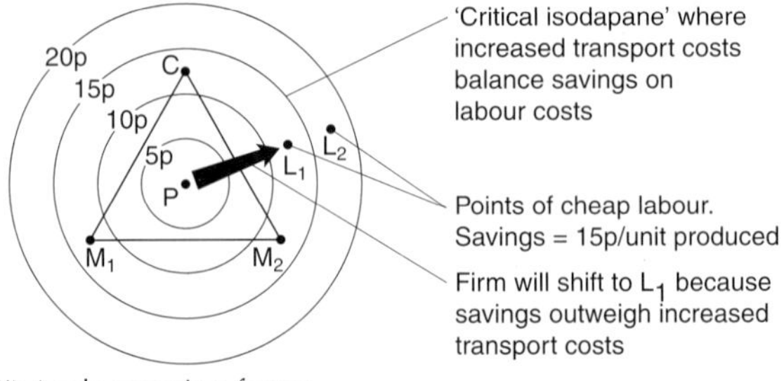

(iii) Agglomeration forces

Some firms can save money by locating together. This could be by sharing costs (industrial estates) or ideas (science parks).

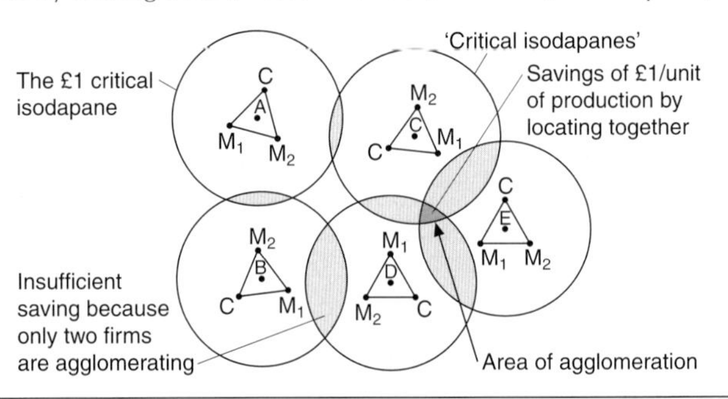

SPATIAL MARGINS OF PROFITABILITY

Smith and Rawstron also described sub-optimal location. Their *spatial margins* define whole areas of profitability. The model introduces the idea that firms may not choose the optimum location but rather be content with a satisfactory profit. It also focuses attention on the limits to freedom of locators.

(a) Variable sales, cost of production constant

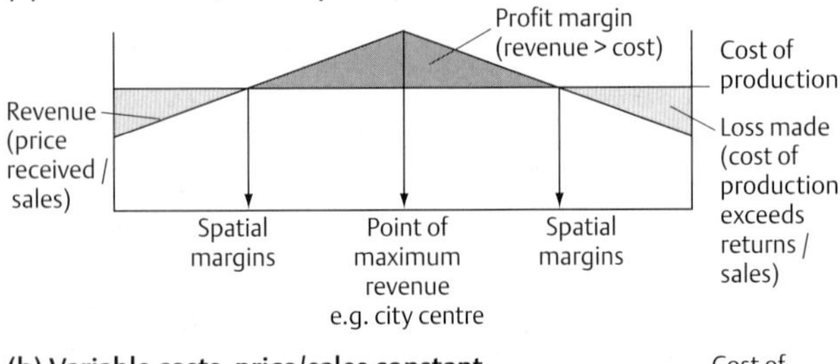

(b) Variable costs, price/sales constant

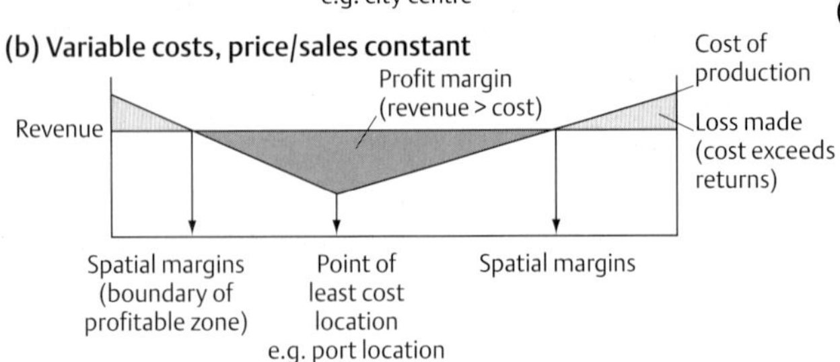

(c) Variable costs, variable revenue

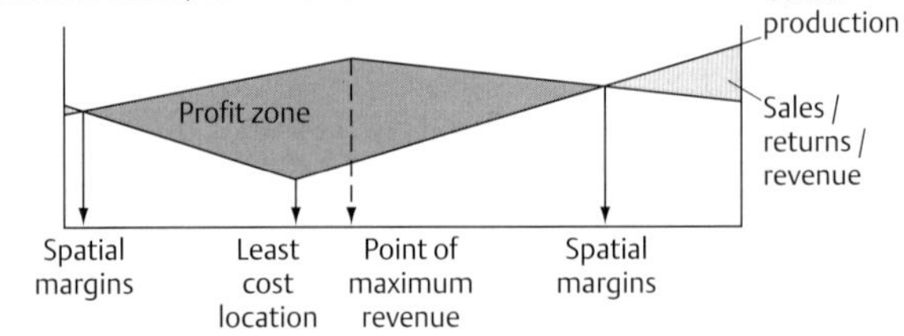

(d) Changes over time

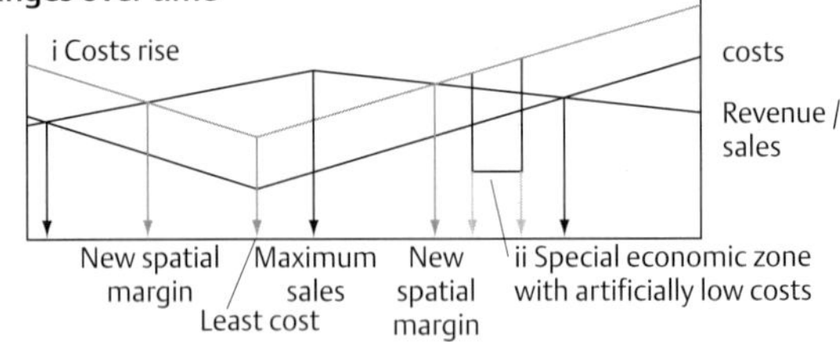

In situation i, the costs rise by a similar amount at all locations.

ii The government offers incentives, such as a tax break, reduction on rates, etc. so as to reduce costs and keep industry alive.

OTHER MODELS

Lösch's profit maximisation model (1954)

According to Lösch industries should locate as close as possible to their market. For many firms this means an urban location. In particular weight-gaining industries such as brewing and baking are attracted to sites close to the markets.

Pred's behavioural matrix (1967)

Pred's model relaxes Weber's assumption of 'economic man' acting with perfect knowledge. Sub-optimal locations are based on poor knowledge or lack of entrepreneurial flair. High ability and top quality information give an optimal location. Lack of ability and limited knowledge mean plants lose money and close.

New location theory

Over the last 30 years the influences on industrial location described by the classical models have changed in importance. In the 1990s, *labour* is the most important factor in the location of manufacturing industries. The term *spatial division of labour* is used to describe the way workforce can influence location on a national or global scale.

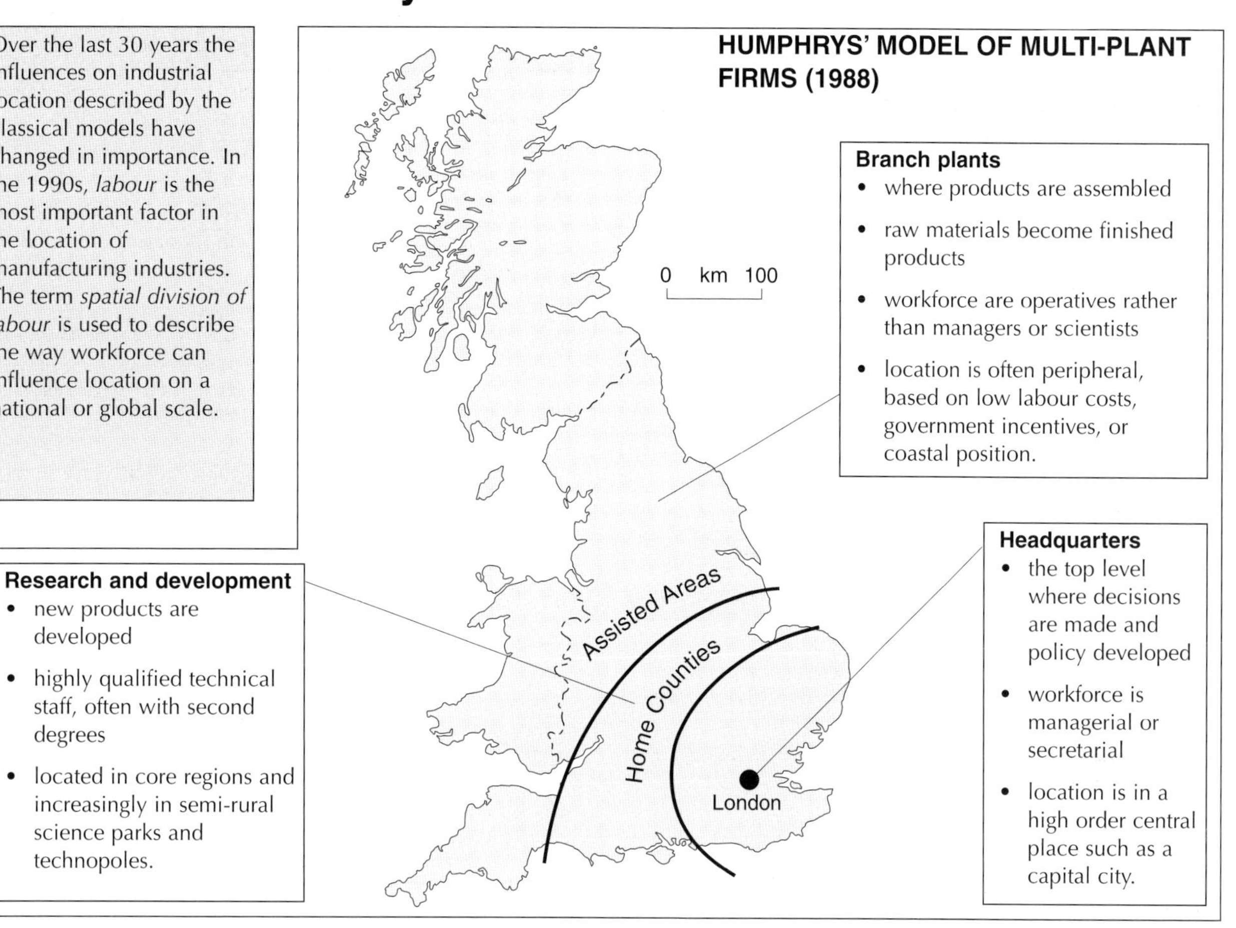

Branch plants
- where products are assembled
- raw materials become finished products
- workforce are operatives rather than managers or scientists
- location is often peripheral, based on low labour costs, government incentives, or coastal position.

Headquarters
- the top level where decisions are made and policy developed
- workforce is managerial or secretarial
- location is in a high order central place such as a capital city.

Research and development
- new products are developed
- highly qualified technical staff, often with second degrees
- located in core regions and increasingly in semi-rural science parks and technopoles.

VERNON'S PRODUCT LIFE-CYCLE MODEL (1966)

The product life-cycle model is associated with four stages of production which have clear implications both to the spatial division of labour and industrial location.

Manufacturing capacity shifts from core areas to peripheral areas and finally out of the country as products move through their life cycle.

1. *Infancy* - product is developed and is sold as a niche product to the core region; labour force is skilled but small.
2. *Growth* - demand expands and production techniques become standardised; labour force grows.
3. *Maturity* - little new research and development; production moves to peripheral areas where labour is cheaper.
4. *Obsolescence* - declining sales due to better or cheaper products from abroad; production moves to firms overseas or the product is superseded by new innovation.

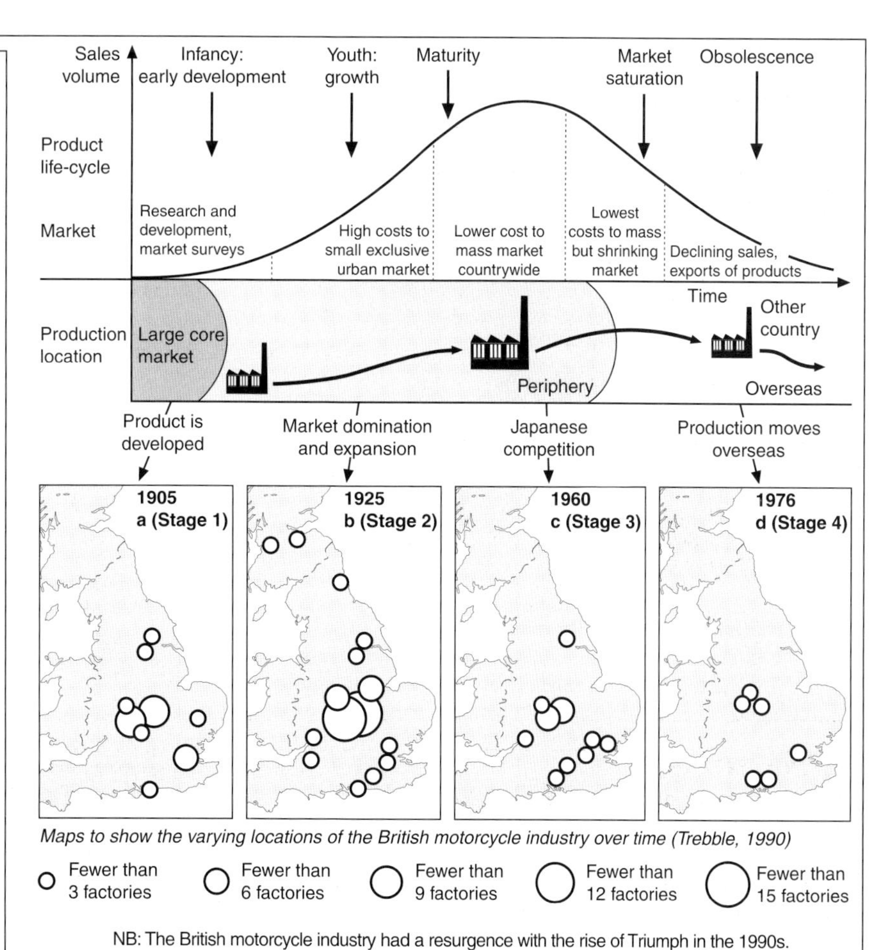

Maps to show the varying locations of the British motorcycle industry over time (Trebble, 1990)

NB: The British motorcycle industry had a resurgence with the rise of Triumph in the 1990s.

Deindustrialisation

In the period since the Second World War the UK has entered a new industrial phase. The characteristics of this *post-industrial society* are:

- a shift away from agriculture and manufacturing industries towards service industries
- the growing importance of large, multinational corporations
- a spatial division of labour within the growing economies of South-East Asia
- the decline of the older 'smokestack' industries.

One of the results of post-industrial change has been *deindustrialisation.*

Deindustrialisation is the long-term absolute decline in the manufacturing sector with respect to jobs and production.

The more mature industries like textiles, iron and steel, and shipbuilding are most likely to deindustrialise. The results can be plant closure, job losses, and regional decline.

TYPES OF DEINDUSTRIALISATION

There are two types of deindustrialisation:

Negative deindustrialisation

Plant closure due to inefficiency or obsolescence
↓
Labour is displaced and high unemployment results
↓
Services and active population migrate to more prosperous areas
↓
Regional problems

Positive deindustrialisation

Industries reduce plants and workforce to improve competitiveness
↓
Productivity is improved
↓
Displaced labour absorbed by the service sector or new manufacturing firms
↓
Regional prosperity

EXPLANATIONS OF DEINDUSTRIALISATION

1 Maturity

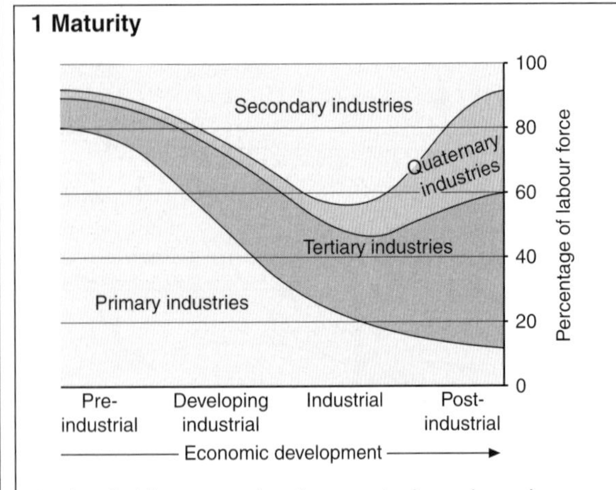

An inevitable progression from agriculture through manufacturing industry to the service industry, i.e. post-industrial change.

2 Overseas competition

Newly industrialising countries have the advantage of cheap labour, expanding national markets, and the newest technology. This has led to a *global shift* of manufacturing industry towards South-East Asia.

3 Management slow to innovate

In the boom period of the 1960s even inefficient plants could make a profit. This meant managers were unwilling to modernise and labour fought changes

4 Rationalisation

There are a number of options facing a rationalising industry:

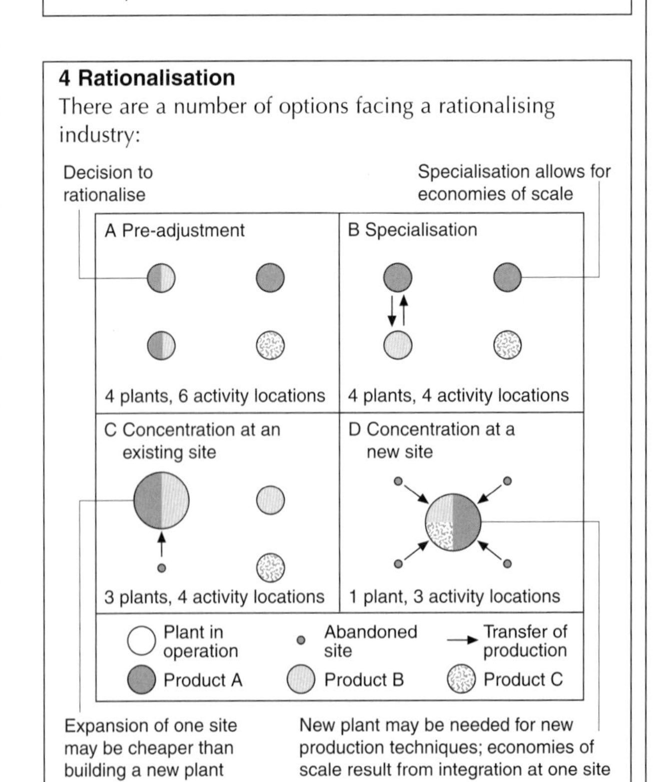

Reindustrialisation

Reindustrialisation is the development of new industries which has followed deindustrialisation in many regions of the developed world. Two processes are relevant to the expansion of manufacturing industry in the developed world: the inward investment of large, multinational corporations (MNCs), and the rise of small firms.

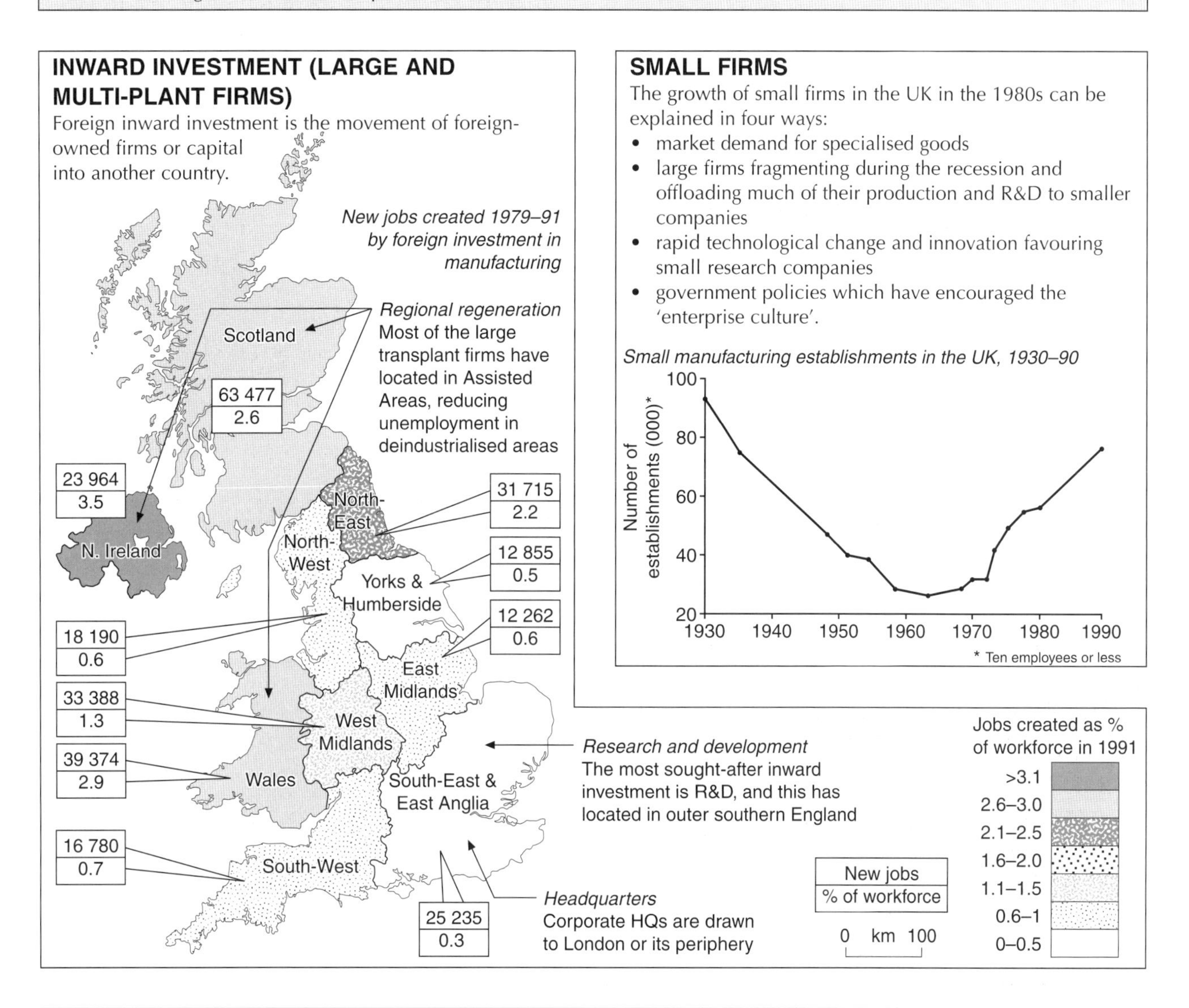

INWARD INVESTMENT (LARGE AND MULTI-PLANT FIRMS)

Foreign inward investment is the movement of foreign-owned firms or capital into another country.

SMALL FIRMS

The growth of small firms in the UK in the 1980s can be explained in four ways:

- market demand for specialised goods
- large firms fragmenting during the recession and offloading much of their production and R&D to smaller companies
- rapid technological change and innovation favouring small research companies
- government policies which have encouraged the 'enterprise culture'.

LARGE FIRMS VERSUS SMALL FIRMS

	Large firms	Small firms
Marketing	Comprehensive distribution and servicing facilities. High degree of market power with existing products	Ability to react quickly and to keep abreast of fast-changing market requirements (Market start-up abroad can be prohibitively costly)*
Management	Experienced managers able to control complex organisations and establish corporate strategies (Can suffer an excess of bureaucracy. Managers can become cautious planners rather than entrepreneurs)*	Lack of bureaucracy. Dynamic, entrepreneurial managers react quickly to take advantage of new opportunities and are willing to accept risk
Internal communications	(Internal communication often cumbersome; this can lead to missed opportunities and poor response to market change)*	Efficient and informal communication networks. Flexible and fast response to problems. Rapid change when faced with new opportunities
Qualified technical manpower	Ability to attract highly skilled technical specialists. Can support the establishment of a large R&D laboratory	(Often lack suitable qualified technical specialists. Unable to support R&D without external help from larger company)*

* Statements in brackets represent areas of potential disadvantage

The steel industry

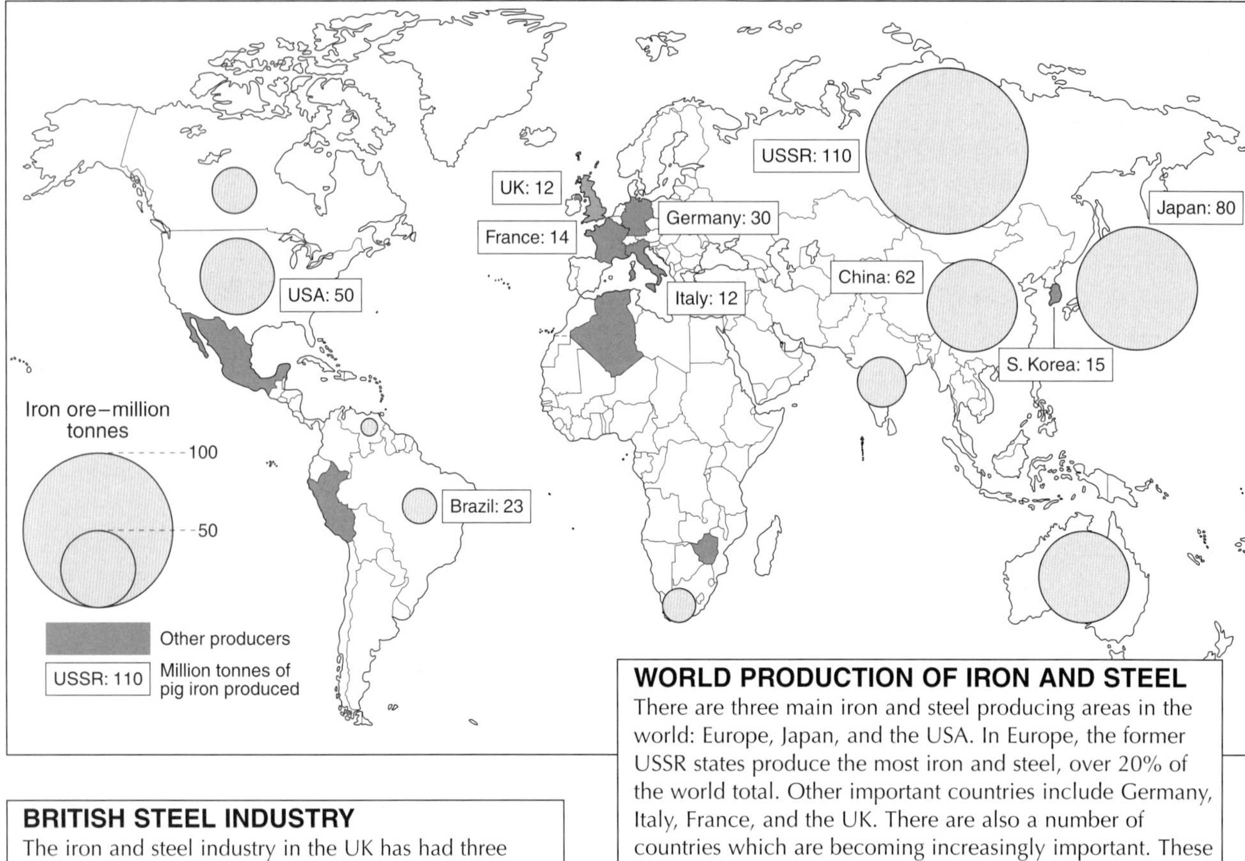

WORLD PRODUCTION OF IRON AND STEEL

There are three main iron and steel producing areas in the world: Europe, Japan, and the USA. In Europe, the former USSR states produce the most iron and steel, over 20% of the world total. Other important countries include Germany, Italy, France, and the UK. There are also a number of countries which are becoming increasingly important. These include China, India, Brazil, and Korea.

BRITISH STEEL INDUSTRY

The iron and steel industry in the UK has had three main stages. The location of the iron and steel works has changed with each of these stages.

1. Small-scale dispersed production - located close to raw materials, such as iron ore and charcoal, e.g. the Forest of Dean in Gloucestershire.
2. Coalfield locations - it took eight times more coal than iron ore to produce one tonne of steel. Therefore, the iron and steel industry moved to the coalfields to reduce transport costs, e.g. the South Wales coalfields.
3. Coastal locations - once the coal and iron ore reserves have gone, resources need to be imported. Coastal locations are better for the import of resources and the export of finished products, e.g. Teesside and Port Talbot.

- Many of the raw materials are imported from Poland, South Africa, the CIS, and Brazil.
- Steel plants have rationalised. They have become bigger and more competitive. However, smaller factories have closed down.
- Steel plants are now more mechanised - this creates unemployment.
- Plants are integrated - this means that different processes such as smelting and conversion to steel take place on the same plant.
- The demand for steel is decreasing. This is because there are other new materials such, as aluminium and plastics.
- There is increased competition between steel producers.
- There is a shift to Pacific Rim producers, that is countries in the East of Asia.

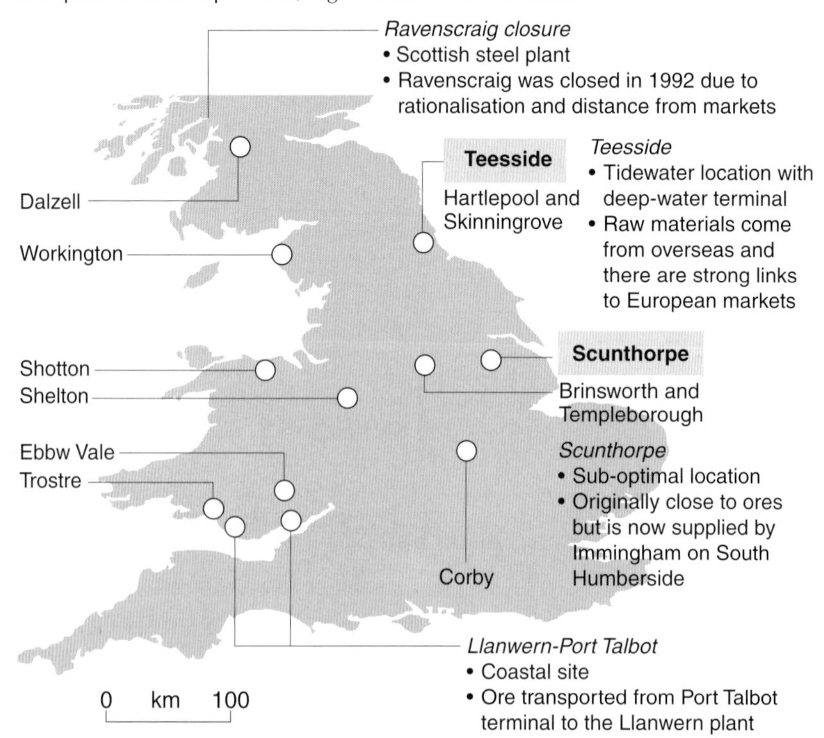

The car industry

The world car industry can be viewed at three levels: the global, the national, and the regional. Each level is related but has different locational characteristics.

1 GLOBAL LOCATION

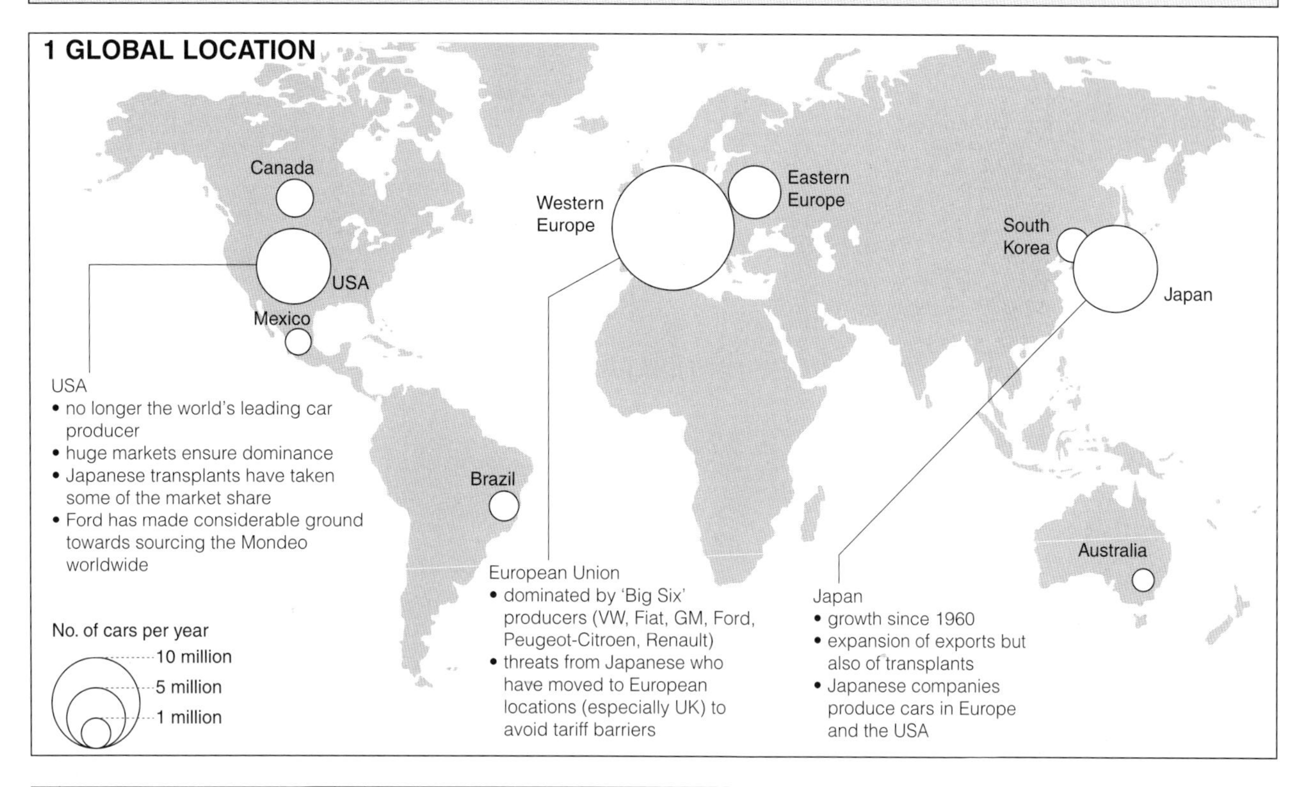

2 NATIONAL LOCATION - THE JAPANESE IN THE UK

The Japanese have shifted much of their production to Europe, especially the UK. There are two reasons for the move to Europe:

- to **avoid tariff barriers** and the restrictions of the Single Market
- to produce a car specifically for the **European market**

They have chosen the UK as their favoured location because of:

- the UK's **'open door' policy** - the government and most local authorities welcome Japanese investment
- **access to EU markets** - if the Japanese produce cars using 60% of their components sourced from Europe they avoid tariffs
- **regional assistance** - support available in intermediate and development areas can be a strong incentive
- **language and culture** - English is the universal business language
- **labour costs** - the UK has amongst the cheapest labour costs in Europe.

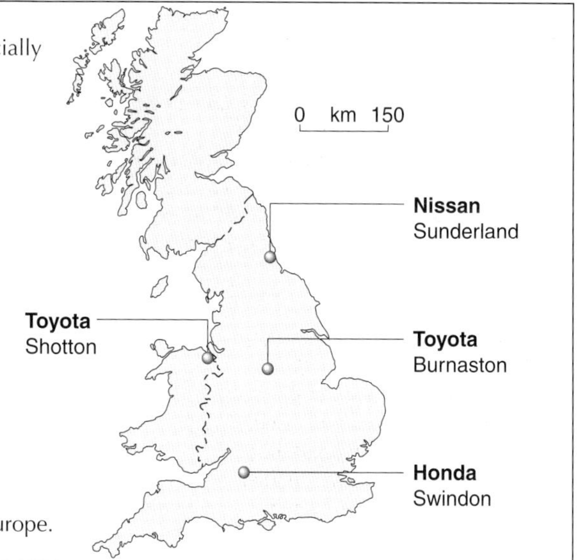

3 REGIONAL LOCATION - TOYOTA

- In 1989 Toyota invested £700 million in a car plant in Burnaston, Derbyshire.
- Toyota has 160 European-based suppliers at present, from ten countries in the EU.
- The plant operates **just-in-time production** (orders and receives parts as they are needed) and purchases parts and components worth £113 million from within 50 miles of the plant.

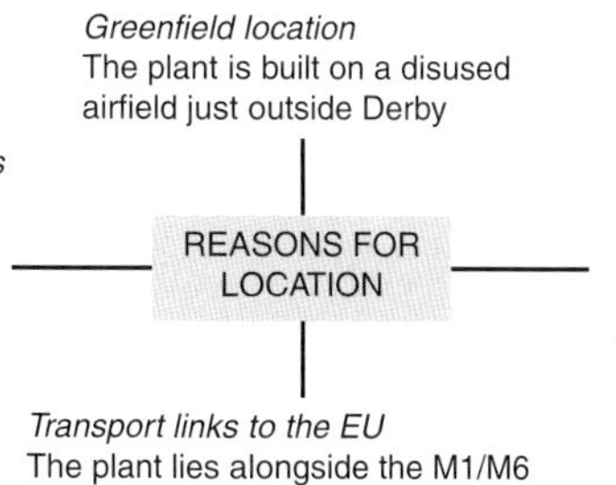

High technology industries

DEFINITIONS AND CHARACTERISTICS

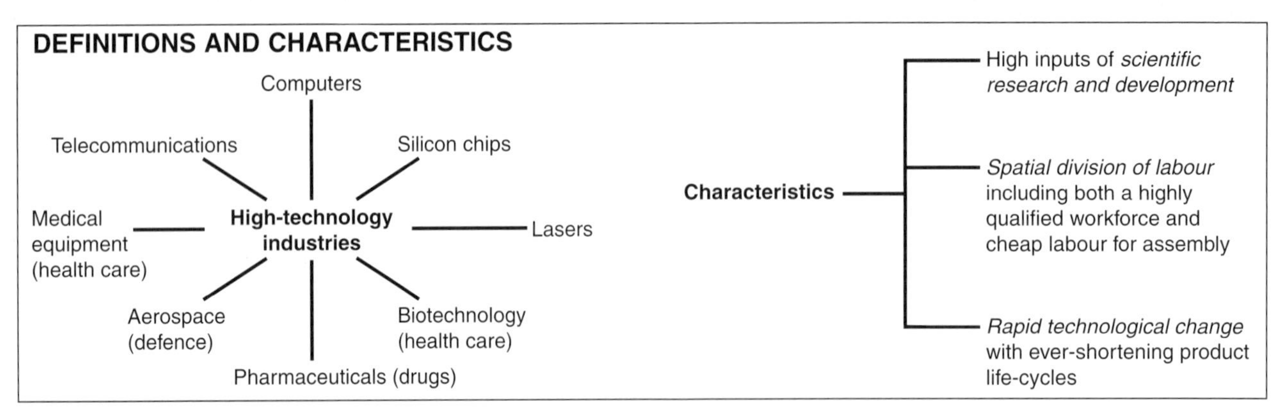

HIGH TECHNOLOGY INDUSTRIES IN THE BRITISH ISLES

The Emerald Corridor
- Ireland's growth axis between Dublin and Belfast
- improved motorway communications between Dublin and Belfast
- two international airports
- two major universities in Dublin, and one in Belfast
- centre of political power and decision making.

Silicon Strip
The M4 Corridor
- excellent communications, M4, M25, M5, Heathrow Airport
- concentration of universities, e.g. Bristol, Reading, London, and Oxford
- nuclear laboratories at Harwell
- Government research departments at Aldermaston (weapons), Bracknell (meteorology), and Farnborough (aircraft)
- concentration of head offices, e.g. Phoenix, Sun Life, and Nat West at Bristol.

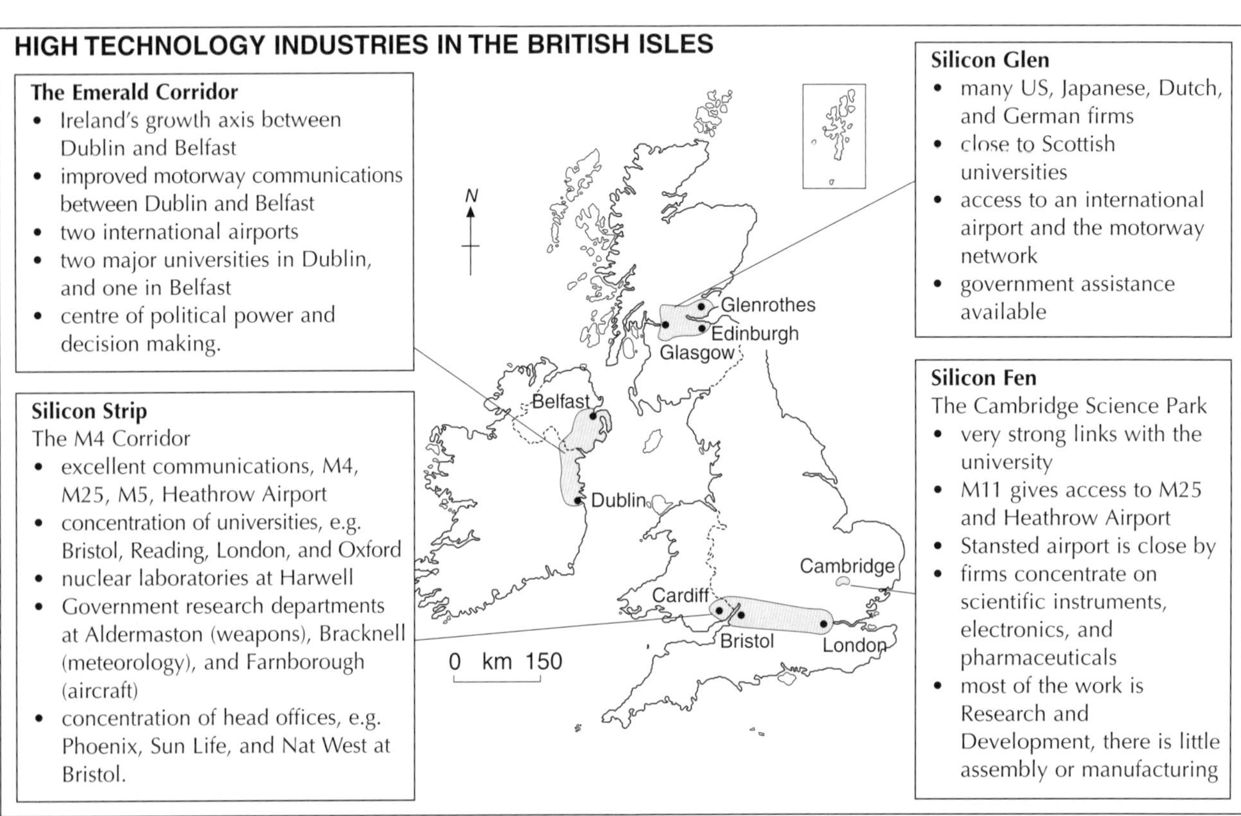

Silicon Glen
- many US, Japanese, Dutch, and German firms
- close to Scottish universities
- access to an international airport and the motorway network
- government assistance available

Silicon Fen
The Cambridge Science Park
- very strong links with the university
- M11 gives access to M25 and Heathrow Airport
- Stansted airport is close by
- firms concentrate on scientific instruments, electronics, and pharmaceuticals
- most of the work is Research and Development, there is little assembly or manufacturing

CASE STUDY: THE CAMBRIDGE PHENOMENON

Why Cambridge?
- prestige name and location
- attractive, landscaped sites
- links with the university
- highly qualified workforce
- links with related companies

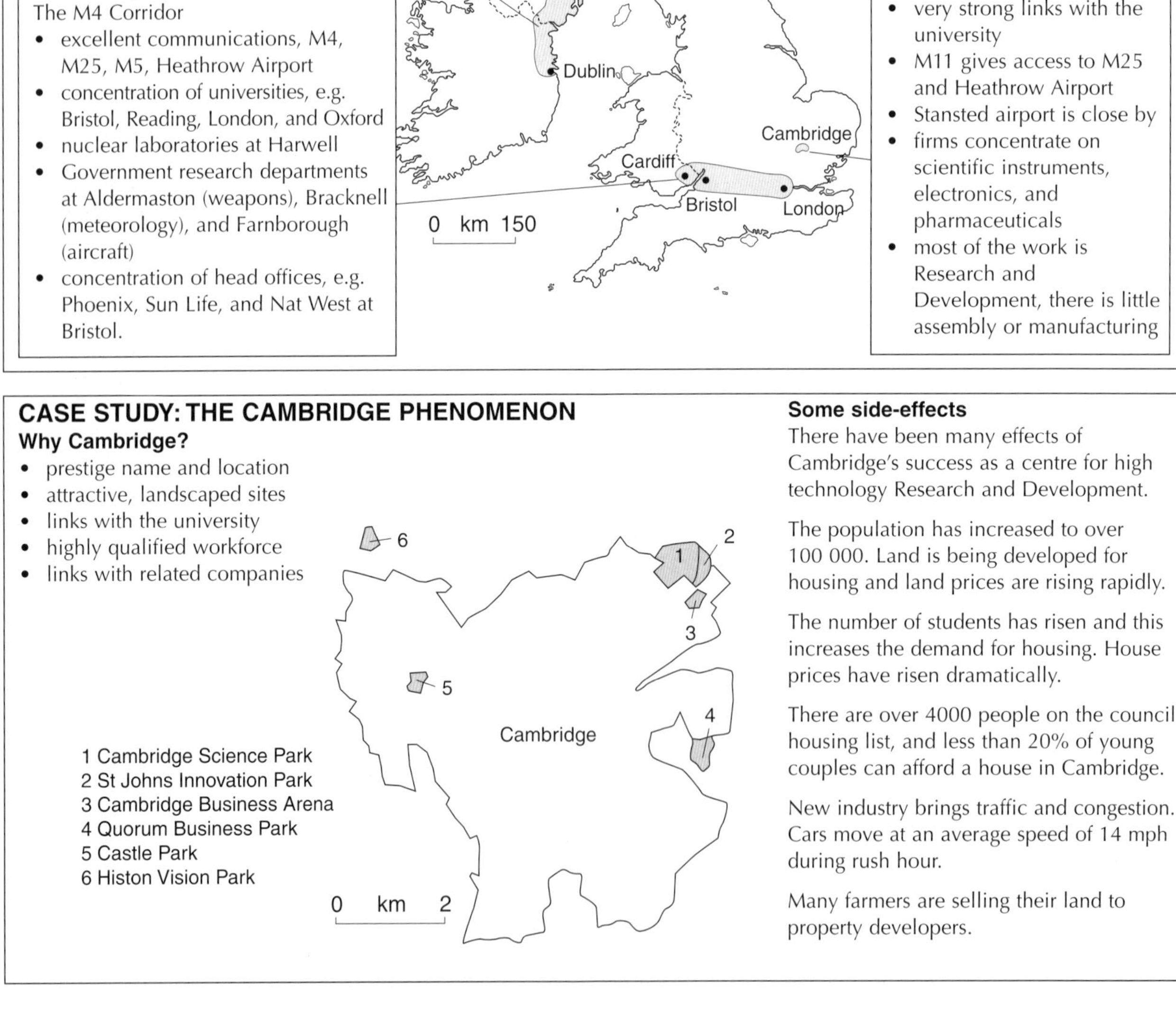

Some side-effects

There have been many effects of Cambridge's success as a centre for high technology Research and Development.

The population has increased to over 100 000. Land is being developed for housing and land prices are rising rapidly.

The number of students has risen and this increases the demand for housing. House prices have risen dramatically.

There are over 4000 people on the council housing list, and less than 20% of young couples can afford a house in Cambridge.

New industry brings traffic and congestion. Cars move at an average speed of 14 mph during rush hour.

Many farmers are selling their land to property developers.

Acid rain

THE DISTRIBUTION OF ACIDIFIED WATER

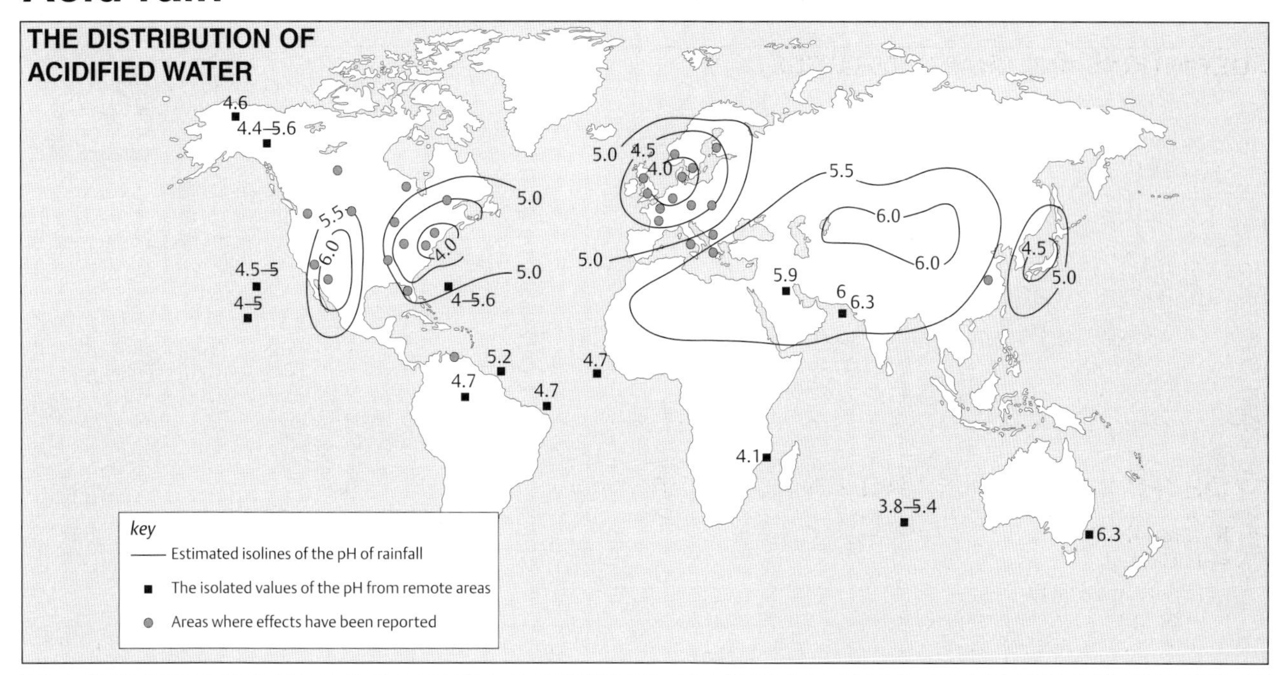

THE CAUSES

The major causes of acid rain are the sulphur dioxide and nitrogen oxides produced when fossil fuels such as coal, oil, and gas are burned. Sulphur dioxide and nitrogen oxides are released into the atmosphere where they can be absorbed by the moisture to become weak sulphuric and nitric acids, sometimes with a pH of around 3. Most natural gas contains little or no sulphur and causes less pollution.

Coal-fired power stations are the major producers of **sulphur dioxide**, although all processes that burn coal and oil contribute. Vehicles, especially cars, are responsible for most of the **nitrogen oxides** in the atmosphere. Some come from the vehicle exhaust itself, but others form when the exhaust gases react with the air. Exhaust gases also react with strong sunlight to produce poisonous ozone gas that damages plant growth and, in some cases, human health.

THE EFFECTS

Acidification has a number of effects:

- buildings are weathered
- metals, especially iron and aluminium, are mobilised by acidic water and flushed into streams and lakes
- aluminium damages fish gills
- forest growth is severely affected
- soil acidity increases
- there are links (as yet unproven) with the rise of senile dementia.

The effects of acid deposition are greatest in those areas that have high levels of precipitation (causing more acidity to be transferred to the ground) and those that have base-poor (acidic) rocks which cannot neutralise the deposited acidity.

DISPERSION AND DEPOSITION

Top of mixing layer
TURBULENCE LIMITED
Wind direction
Plume
ZONE OF TURBULENCE
Diffusion and dilution
Uptake of aerosols, gases, and particles
OXIDATION
Natural air turbulence
DISSOLUTION
1–2 km
DRY DEPOSITION
Emission source
Plume touches ground
WET DEPOSITION
Plume buoyant (1–2 km)
Mixing and diffusion (tens of km)
First dry deposition (5–25 km)
Dispersion of oxides and acids (hundreds, even thousands of km)

DRY AND WET DEPOSITION

Dry deposition typically occurs close to the source of emission and causes damage to buildings and structures. Wet deposition, by contrast, occurs when the acids are dissolved in precipitation, and may fall at great distances from the sources. Wet deposition has been called a 'trans-frontier' pollution, as it crosses international boundaries with disregard.

THE SOLUTIONS

Various methods are used to try to reduce the damaging effects of acid deposition. One of these is to add powdered limestone to lakes to increase their pH values. However, the only really effective and practical long-term treatment is to curb the emissions of the offending gases. This can be achieved in a variety of ways:

- by reducing the amount of fossil fuel combustion
- by using less sulphur-rich fossil fuels
- by using alternative energy sources that do not produce nitrate or sulphate gases (e.g. hydro-power or nuclear power)
- by removing the pollutants before they reach the atmosphere.

However, while victims and environmentalists stress the risks of acidification, industrialists stress the uncertainties. For example:

- rainfall is naturally acidic
- no single industry or country is the sole emitter of SO_2 and NOx
- more car owners have vehicles with catalytic convertors
- different types of coal have variable sulphur content.

Services*

SERVICES

The service sector includes all economic activities other than the production of goods. It includes insurance, banking, shipping, tourism, health care, refuse collection, entertainment, education, and retailing. Some services are provided by the State (e.g. NHS health care in the UK), while others are provided by the private sector (e.g. market research and advertising). Some jobs are extremely well paid (e.g. banking), others are very poorly paid (e.g. refuse collection). Many jobs are traditionally female (e.g. catering, cleaning), others male (e.g. transport).

TYPES OF SERVICES

A number of distinctions can be made:

1 Producer services are 'high order' activities, such as market research, management consultancy, financial, advertising, and legal functions, that are provided in a small number of highly developed metropolitan centres, generally capital cities, for other firms or organisations.
Consumer services are provided generally for people, e.g. health care, retailing, education, distribution, and refuse collection. These are more local in scale or 'low order'.

2 Non-basic services are provided for users in the local area.
Basic services are those which are provided for a market beyond the local economy, e.g. a national or a global market.

BANK
Barney's BANK
CITY BANK
NAT BANK

3 Private or **market** services are organised by independent companies, ranging from contract-cleaners and retailing, to banking and insurance.
State or **non-market** services are organised by the government, such as central and local government, state health care, and education.

An area which depends upon **consumer non-basic services**, such as a post office or a supermarket, will attract little extra growth, whereas a **producer basic service**, such as an international bank, will make large exports. This will contribute to the region's and the nation's economic growth.

SERVICE EMPLOYMENT IN THE UK

N
%
>75
70–74.99
65–69.99
≤64.99
Avg 72.9
Services are concentrated in the south east – the financial, legal, and administrative centre
0 km 150

SERVICE EMPLOYMENT IN EUROPE

≥70%
65–70
60–64.9
55–59.9
50–54.9
<50%
Average 63%
N
High rates of services are found in major cities and tourist areas.
0 km 1000

Multinational corporations (MNCS)

A multinational corporation (MNC) is an organisation that has operations in a large numbers of countries. Generally, research and development and decision-making are concentrated in the core areas of developed countries, while assembly and production are in developing countries and depressed, peripheral regions.

CASE STUDY: ICI

ICI (Imperial Chemical Industries) was formed in 1926 and has its headquarters in the UK. It employs about 130 000 people and has sales of about £6 500 million each year.

The corporation is a vast conglomerate that makes almost the complete range of chemicals and chemical-related products, including fertilisers, paints, pharmaceuticals, and plastics. Its sales and profits now depend on four main markets: the UK, Western Europe, North America, and Australia and the Far East.

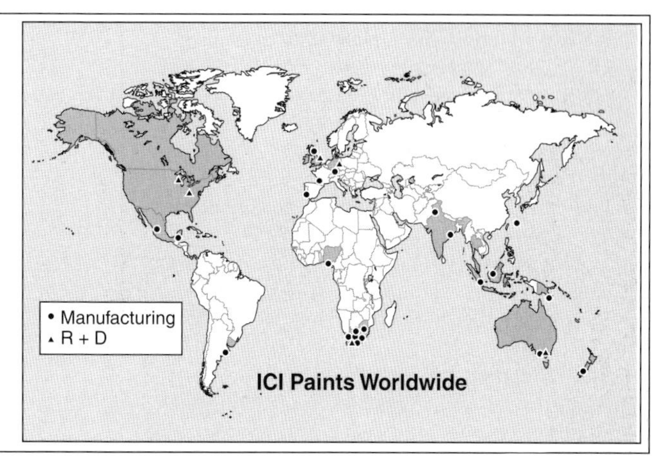

THE IMPACT OF MNCS ON THE HOST COUNTRY

advantages
- development of energy resources
- employment
- improvement of educational and technical skills
- provision of capital equipment
- development of resources and manufacturing
- investment and aid

disadvantages
- increased imports leads to increased national debt
- local labourers exploited
- few skilled workers are employed
- mechanisation reduces demand for labour
- local resources exported and cost of manufactured products is beyond country's range
- large proportion of profits go overseas

ECONOMIC POWER

The sheer scale of the economic transactions that MNCs make around the world and the effect they have on urban, regional, and national economies gives them tremendous power. Thus MNCs have become planned economies with vast internal markets.

- Up to one-third of all trade is made up of internal transfers of MNCs. These transfers produce money for governments through taxes and levies.
- Economic power comes from the ownership of assets.
- Over 50 million people are employed by MNCs.
- Although many governments in developing countries own their resources, MNCs still control the marketing and transport of goods.

POLITICAL POWER

- MNCs' political power may be limited due to governments have many sanctions, such as import quotas, and local content arrangements.
- Governments often promote the interests of their own MNCs rather than that of foreign ones. For example, it is very difficult for non-Japanese MNCs to operate in Japan.

THE CHANGING NATURE OF MNCS

Reduced demand and increased competition creates unfavourable economic conditions. In order to survive and prosper, MNCs have used three main strategies:

- **rationalisation** refers to a slimming down of the workforce, replacing them with machines.
- **reorganisation** includes improvements in production, administration, and marketing, such as an increase in subcontracting of production.
- **diversification** refers to firms that have developed new products.

MNCs are now:

- greatly slimmed down, employing more people indirectly through subcontracting
- integrating production, administration, and marketing on a global scale
- increasingly financially orientated.

The geographical consequences of these changes are varied.

- There are increases in business travel, telecommunications, and information processing.
- MNC workforces have become more international than ever before, e.g. ICI's foreign share of employment is over 50%.
- Greater integration of administration has led to congregation in particular 'world cities'. This allows increased access to finance, advice, information, telecommunications, and social contacts.
- Locational sifting of the workforce means that MNCs decentralise the part of production that requires repetitive tasks to low-wage, developing countries.

Retail developments

Retailing is a major component of European economies. It accounts for about 13 – 14% of GDP in most countries. Although the number of shops has fallen from 3.5 million in 1955 to under 2.5 million, the amount of floorspace has risen. Hence there are fewer, larger shops.

TRADITIONAL RETAILING

The retailing industry has changed from one dominated by small family firms to one in which large corporate organisations dominate. Traditionally, geographic accounts of retailing concentrated on the location and type of retailing outlet. A central place-type hierarchy was seen:

- low-order goods concentrated in neighbourhood stores and shopping parades
- high-order goods in high street shops, department stores
- out-of-town superstores and retail parks.

Central shopping area or **high streets** are characterised by departmental shops, chain stores, specialist shops, and, increasingly, by pedestrianised malls. Outlets sell mainly high-order goods that have a large range and threshold. The sphere of influence of central shopping areas is generally large. By contrast, **shopping parades** are clusters of shops. These usually include a supermarket, off-licence, newsagent, and other low-order outlets serving nearby residential areas. At the bottom of the hierarchy are **corner shops**. These are generally small, independent outlets with long opening hours, that sell a wide variety of products.

REGIONAL SHOPPING CENTRES

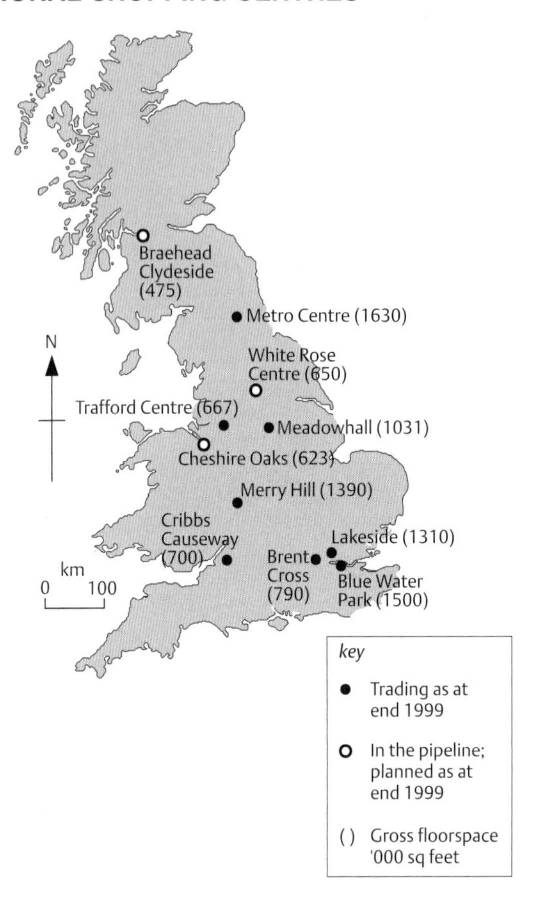

MODERN RETAILING

The retailing revolution has focused upon superstores, hypermarkets, and out-of-town shopping precincts. These are located on green-field suburban sites with good accessibility and plenty of space for parking and future expansion. The increasing use of out-of-town shopping centres, on a less frequent basis has led to the closure of many smaller shops which depended on frequent convenience trade.

Recently, however, government policies have tended to favour central shopping areas and neighbourhood schemes over out-of-town developments. Perhaps as a result recent trends are beginning to suggest a slow down in out-of-town developments and an increase in city-centre regeneration.

A COMPARISON BETWEEN OUT-OF-TOWN AND HIGH STREET RETAILING

High Street	Out-of-town
Tradition	Parking
Comparison	Safety
Convenience	Access
Access by public transport, cycle or by foot	Space*
Status*	Flexibility
Visibility*	Efficiency
Access to wider consumer market*	Lower land costs
Property value*	Joining the competition
*retailer specific	

FACTORS THAT AFFECT RETAILING

A number of factors explain the changes in the retailing:

- **demographic change** such as falling population growth, smaller households, and more elderly people
- **suburbanisation** and **counter-urbanisation** of more affluent households
- **technological change** as more families own deep freezers
- **economic change** with increased standards of living, especially car ownership
- **congestion** and **inflated land prices** in city centres
- **changing accessibility** of suburban sites, especially those close to ring-road intersections
- **social changes** such as more working women.

Two of the most important locational changes in the retailing industry are:

- the suburbanisation of retail activity
- the growth in large stores on the edge of town and out-of-town sites.

Distribution of services

LOCATIONAL FACTORS

There are a number of factors that influence the locational decision-making of service firms. These include:

- proximity to customers
- quality of telecommunications
- direct access to major motorways and an international airport
- recruitment of qualified personnel
- cost of office space and property
- access to information
- accessibility to other businesses
- image and prestige
- quality of environment
- distribution of customers
- changes in affluence and demand
- internal developments within service activities.

Services are over-represented to populations in large urban areas. The distribution of consumer (household) services and producer services varies significantly. Factors that explain the location of services in cities include:

- the need for information and knowledge
- prestige and attractiveness, favouring historical and cultural cores and/or attractive suburbs
- room for expansion
- lower wages and rents in the suburbs
- car accessibility and parking, favouring suburban locations.

Consumer services, and some producer services, tend to locate close to their customers (as seen in Witney, Oxon). Increasingly, however, some services have **decentralised** due to suburbanisation and counterurbanisation. This is especially true for less sophisticated producer services, the 'back-office' services. They include the routine functions such as personnel, finance, and accounts which do not need face-to-face meetings. These services are characterised by low wages and low office costs, and are able to trade services by means of telecommunications. Hence cheap suburban sites and cheap suburban labour are very attractive. These services have moved from city centres to concentrate in suburban and peripheral locations.

The location of services in Witney, Oxon

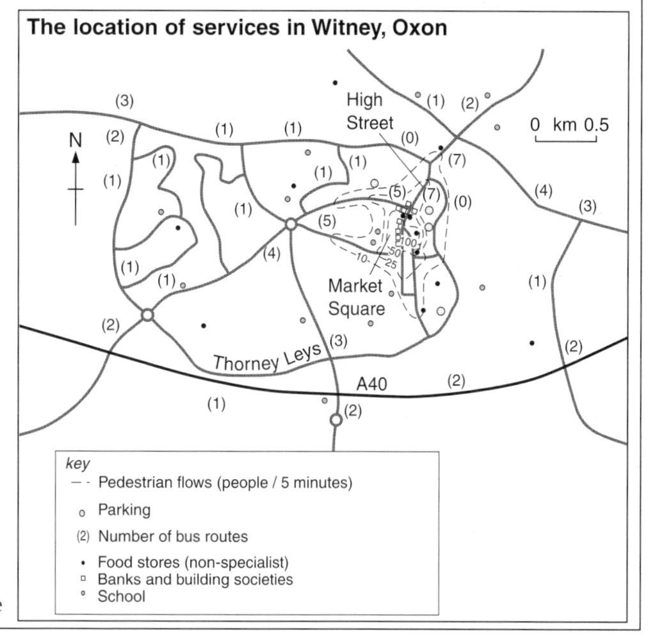

SERVICE CONCENTRATION IN LARGE CITIES

By contrast, high-value, sophisticated producer services are still concentrated in large urban areas. Producer services generally locate in the **CBD** due to excellent transport accessibility and the importance of having face-to-face meetings easily. Access to customers is relatively unimportant. The need for face-to-face meetings leads to a concentration of highly skilled, influential people in large urban areas, especially capital cities. The ever-increasing importance of specialised knowledge has reinforced the attractions of the CBD or economic core.

The location of large public institutions, such as hospitals and government offices, is more difficult to explain. They often require large spaces and easy accessibility. Once established they are hard to shift. The distribution of hospitals and clinics is also related to accessibility, need, and equality. These have a clear spatial element as well as social, cultural, economic, and psychological aspects.

Services concentration in central London

Inner London industrial areas
Central Business District
The City
West End: shops & entertainment
Specialized office areas
1 Insurance
2 Finance
3 Law
New office developments
Parks
Docklands
A Parliament
B St Pauls
C Bank of England
D Stock Exchange
E Lloyds' building
R Residential areas

Area of London
Redevelopment direction
Main assimilation directions:
The City
Decaying fringe
Soho
Mayfair
West End
London's Docklands
Canary Wharf
Isle of Dogs

Regional inequalities

There are large variations in the standards of living between different countries. There are also significant differences between regions of a country or a group of countries. The rich areas are often called the **core**, whereas the poorer areas are called the **periphery**.

In the European Union the core stretches from the South East region in the UK, through Belgium and Germany, to northern Italy. This area has been called the **hot banana**. The poorer regions include Greece, Spain, Portugal, and Ireland.

The richer areas, such as the South East in England, grow because:

- in the beginning they have **initial advantages**, such as more resources or a better location, or both
- as they grow, they develop **acquired advantages** such as a skilled labour force, a marketing network, more industries, and investment.

By contrast, the peripheral areas decline. Their younger, more skilled workers migrate to the core in search of work. The periphery falls further behind the core.

Regional inequalities can be measured in a number of ways, including:

- unemployment rates
- migration rates
- number of people owning their own home
- percentage of people staying on at school after GCSEs
- percentage of jobs in agriculture or manufacturing
- cost of land per hectare
- amount of regional aid from the government.

EXAMPLES OF REGIONAL DISPARITIES

Local: disparities between the inner city areas and rich suburbs.

National: Britain's North-South Divide, Italy's Mezzogiorno.

International: core and periphery in the European Union, e.g. Portugal's position in the EU.

Global: the developed west versus the less developed countries.

FINGLETON'S POSITIVE AND NEGATIVE FEEDBACK (1991)

There are two views of regional inequality and regional policy:

Negative feedback (neo-classical) suggests market forces are all that are needed to draw investment to peripheral areas and attract labour from core areas.

Positive feedback (cumulative causation) suggests continual divergence due to a region's comparative advantage which continues to attract investment and labour.

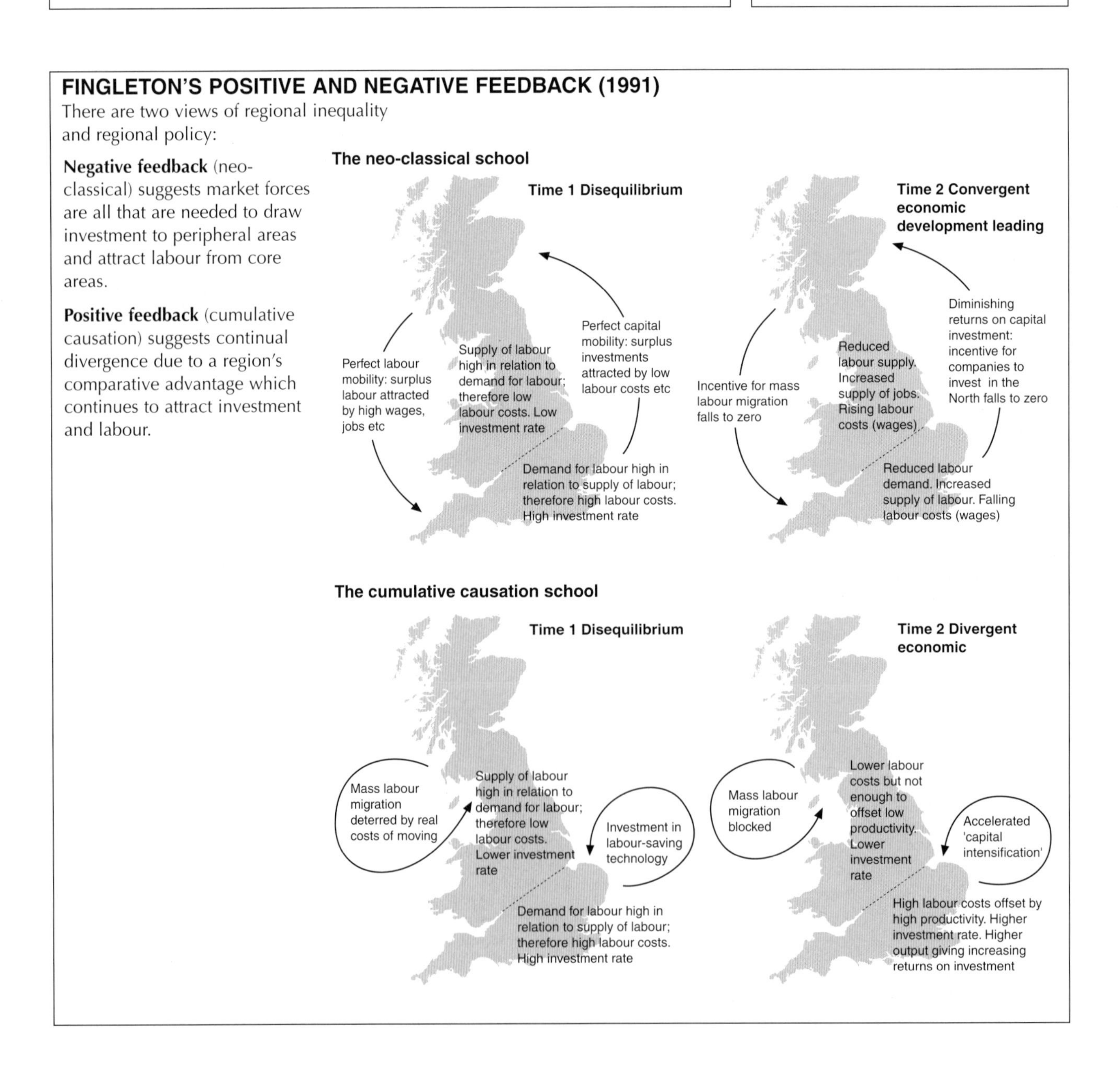

Regional inequalties in Italy

In terms of national growth, Italy has recently overtaken Britain as the fifth-biggest industrial power of the west. However, this expansion has not benefited the whole country.

Industrial triangle
- the cities of Milan, Turin, and Genoa enclose one of the EU's most affluent regions
- northern Italy has been compared with Japan and Taiwan in its rapid and sophisticated industrial expansion
- products include textiles, machine tools, leather goods, footwear, and fashion clothes.

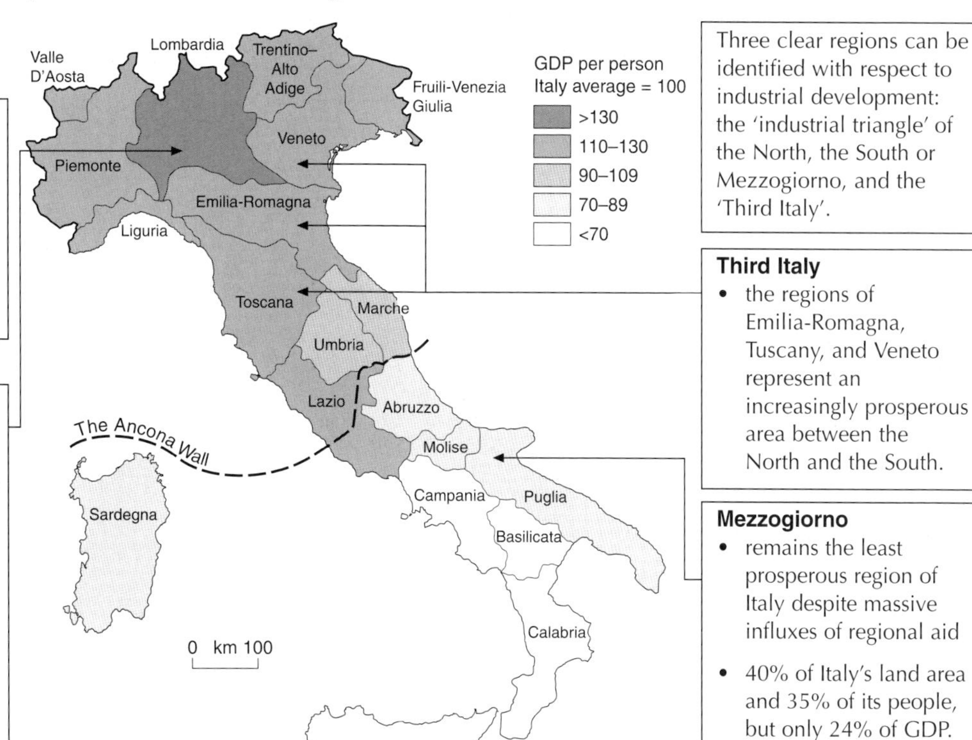

ITALY'S NORTH-SOUTH DIVIDE

Three clear regions can be identified with respect to industrial development: the 'industrial triangle' of the North, the South or Mezzogiorno, and the 'Third Italy'.

Third Italy
- the regions of Emilia-Romagna, Tuscany, and Veneto represent an increasingly prosperous area between the North and the South.

Mezzogiorno
- remains the least prosperous region of Italy despite massive influxes of regional aid
- 40% of Italy's land area and 35% of its people, but only 24% of GDP.

Regional Contrasts

	Lombardy	Mezzogiomo
GDP	139	62
Unemployment	3.4%	17.4%
Employment (agri)	3.2%	17.0%
Crude birth rate	8.5%	12.4%
Migration per year	+0.7%	-1.8%

Income inequalities

Percentage: 62, 60, 58, 56, 54, 52

Year: 1951 55 60 65 70 75 80 85 90 95

The South's income per capita as a percentage of the rest of Italy

THE MEZZOGIORNO

Four factors have been suggested to explain the different rates of development between the North and the Mezzogiorno.

EXPLANATION

Southernist view
the South was neglected by northern politicians

Rural decline
absentee landlords meant that low paid tenant peasants had little incentive to improve the land

'Two nations'
cultural differences between 'modern' North and 'backward' South

Peripherality
the South's distance from the core markets of the EU

In 1950 the *Cassa per il Mezzogiorno* (Fund for the South) was set up to raise southern living standards. Two main policies were pursued:

- land reform, which broke up the large semi-feudal estates of the South creating about 120 000 new, small farms
- growth pole strategy, which legislated that 60% of all state investment went to the South with investments in large steel and chemical plants ('cathedrals in the desert').

Big industrial ventures in the Mezzogiorno have failed because:
- they were capital-intensive projects built without reference to the needs of the local economy or to the markets for steel and petrochemicals, which were in the North or outside Italy
- the OPEC price-hike, which quadrupled the price of oil in 1973, raised the price of raw materials for the new industries
- there is overcapacity in the European steel market.

There were some successes, but growth pole strategy was abandoned in 1986.

New policy is to attract private investment through grants and subsidies. Fiat's plant in Melfi has resulted from this policy.

Regional inequalties in China

China has important regional differences in physical geography, economic well-being, and employment opportunities.

Physically, the west of China is dominated by the rugged Tibetan plateau, reaching over 3500 m in height, whereas the east of China is characterised by lower altitudes, fertile river valleys, and deltas.

The physical contrasts are mirrored by important social and economic inequalities. The western and interior regions remain peripheral and underdeveloped, whereas the south-east and east coast regions are more developed and form China's core.

The eastern, coastal parts of China account for about:

- 66% of industrial production
- 80% of export earnings
- 90% of foreign capital.

In particular, ports and industrial-agricultural areas have experienced rapid economic growth.

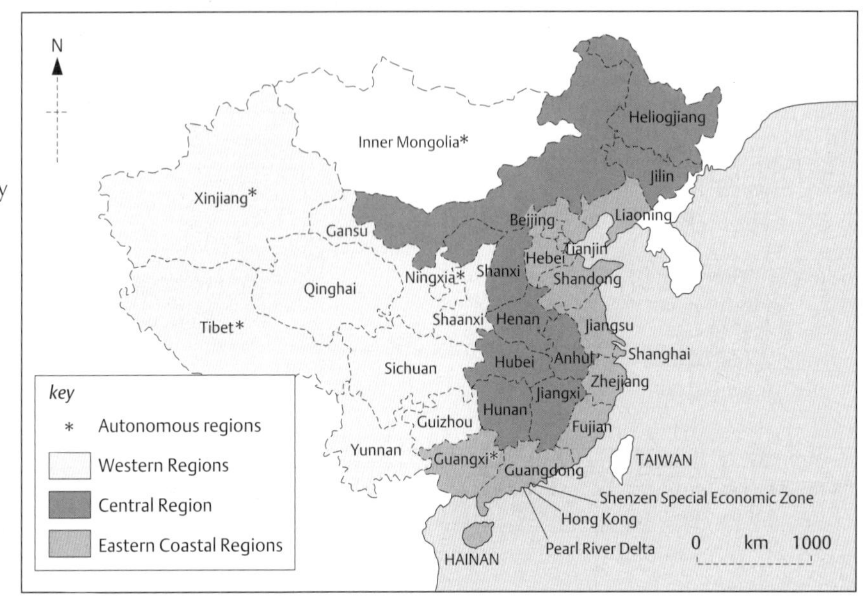

Until 1976 China was a centralised economy. Since then, however, it has followed a more open, free market. Special Economic Zones have attracted foreign investment and infrastructural developments have increased their accessibility and attractiveness to investors.

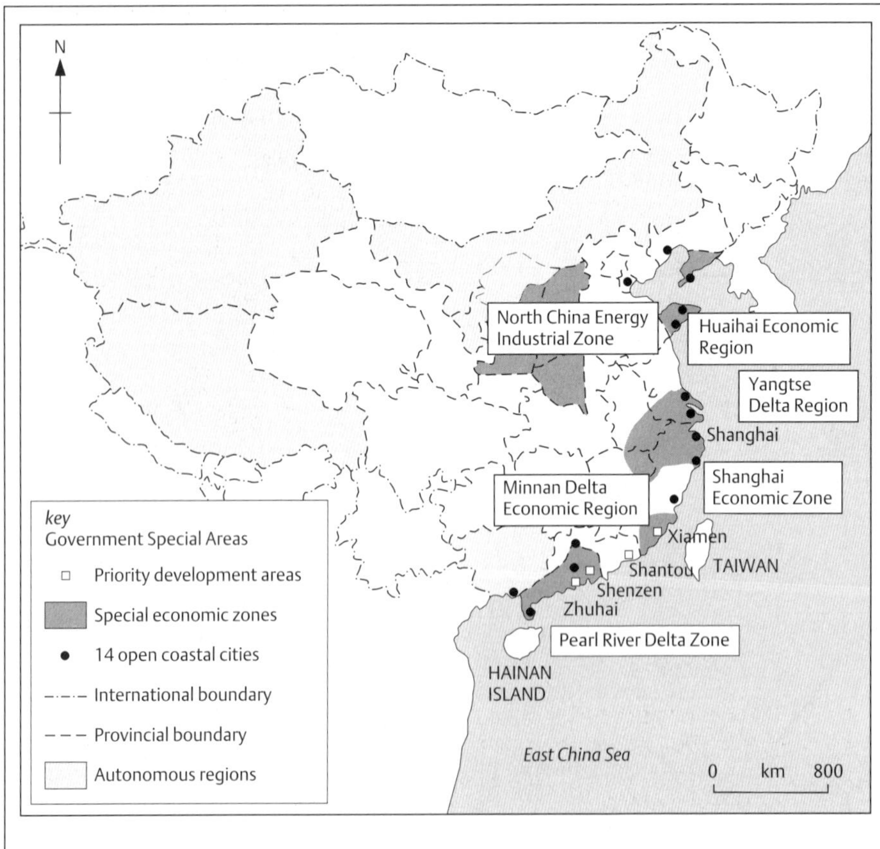

THE SPECIAL ECONOMIC ZONES

Shenzen, Zhuhai, Shantou, Xiamen, and Hainan are Special Economic Zones. They attract foreign investment through:

- tax incentives
- reduced import and export tariffs
- cheap labour
- improved communications and transport facilities.

In 1984 the Chinese government allowed a further 14 coastal cities to establish areas of economic development to attract foreign investment, especially in high technology industries. Larger open zones were announced in 1989, including the Yangtze River Delta near Shanghai, and in 1992.

Following the economic reforms of the late 1970s government policy favoured the coastal areas at the expense of the western interior provinces. By 1993 it was evident that the coastal economy was booming but the interior economy was declining. Thus, most recently, the government is promoting western and inland provinces at the expense of the south-east.

Regional policy in the UK

There has been a long history of regional policy in the UK. The policies put forward have attempted to encourage the growth of manufacturing in the deindustrialising regions of the North. The map of regional assistance has changed between 1970 and the present. The changes reflect a shift in political ideology in addition to the changing fortunes of the UK's regions.

In the UK there is a significant north-south divide. The south is wealthier than the north, and there are more jobs in service industries, and less unemployment. By contrast, the north has higher unemployment, and there are more jobs in manufacturing industries, and agriculture.

Region	Unemployment rates %	GDP in relation to European average
UK	6.2	102
North East	9.1	84
Yorkshire and Humberside	6.4	91
North West	7.1	93
East Midlands	4.7	98
East Anglia	4.9	99
West Midlands	6.2	94
London	8.1	146
South East	4.1	109
South West	4.5	100
Wales	7.3	84
Northern Ireland	8.8	82
Scotland	7.3	97

THE CHANGING MAP OF REGIONAL ASSISTANCE, 1979–1995

The three maps illustrate not only a change in where and how much aid is given but also how it is used. In the 1970s much aid was used to support British industries. Today assisted areas try to attract inward investment from foreign multinationals.

Policy: 1979
- large areas were given regional assistance (the North, Wales, Scotland, Northern Ireland)
- reflects an ideology that regional disparities will increase without financial aid (carrots) and disincentives (sticks)

Policy: 1984

1981 - policy of disincentives for location in the South abolished

1984 - reduction in the geographical scale of aid

1985 - reduction in the financial scale of aid (£584 million in 1985 to £384 million in 1992)

Policy: 1993
- further reduction of the regional extent of aid
- shift in regional policy focus towards the South
- more than half of the 32 new assisted areas are in the South, including areas of London (Lea Valley and Park Royal)

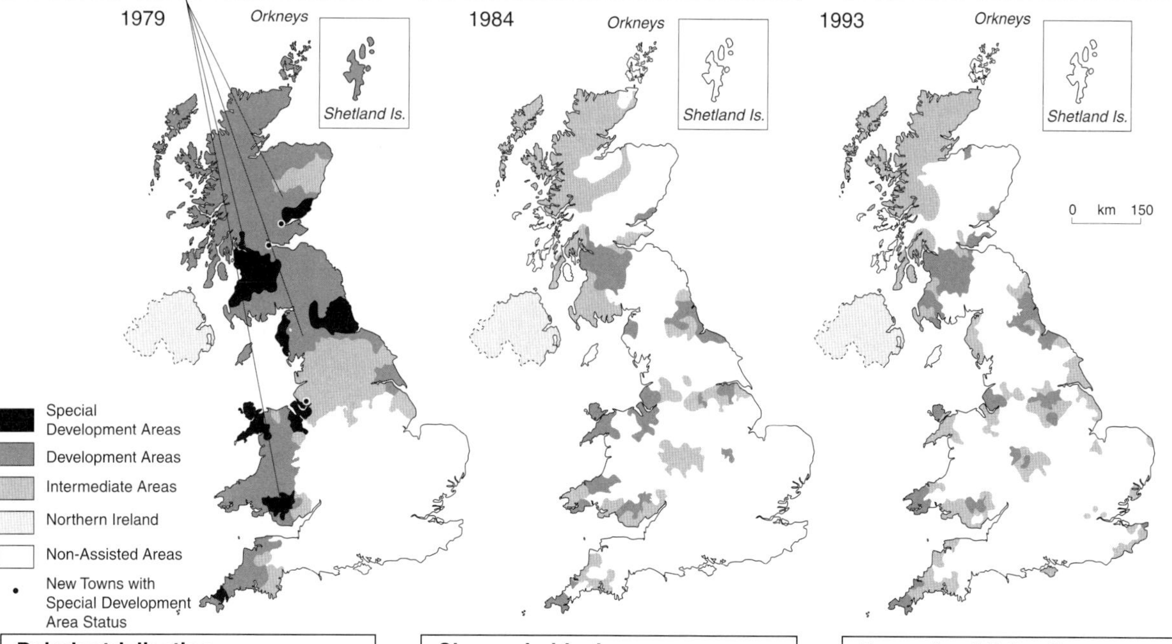

Deindustrialisation
- northern conurbations were tied to industrial regions based on coal and mature industries
- these were declining from the 1930s onwards but were particularly hit by the recession of the 1970s

Change in ideology
- reduction in scale and amount of aid was the result of a shift in political ideology
- regional aid still important, but now controlled by competing regional development agencies for Scotland, Wales, and Northern Ireland

Recession-hit South
- southern economy based on services was hit by the recession of the late 1980s
- policies which had encouraged industries to locate outside the South-East meant that the decline in services resulted in a growth in southern unemployment

Classification of countries

DEVELOPMENT

Development is a very general term. There are four main interrelated features:

- economic progress
- technological improvement
- social, cultural, and political freedom
- justice.

Sustainable development refers to long-term development that improves basic standards of living without compromising future living standards. As the term suggests, it involves management and conservation of resources.

A COMPARISON OF THE DEVELOPMENT IN THE UK AND INDIA

	UK	India
Human Development Index	75.8	59.7
Adult literacy (%)	99	49.8
GDP per head ($)	16340	1150
Life expectancy (years)	75.6	59.7
Daily calorie supply (% of needs)		105
Malnourished children ('000)	-	69 345
Malnourished children (%)	-	63
Infant mortality rate (per '000)	8	93
Cars per 100 people	41	0.7
% employment in agriculture	2	62
% employment in manufacturing	28	11
% employment in services	70	27
Income inequality: % of total Income that the poorest 40% have	17.3	21.3

WORLD BANK CLASSIFICATION

Countries are classified in a number of ways. According to the World Bank there are:

- **industrial market economies** such as the UK, USA, and Japan
- **centrally planned economies** (CPEs) such as the CIS
- **high income oil exporters** such as Saudi Arabia
- **more economically developing countries** (MEDCs) such as Mexico and Brazil
- **less economically** (LEDCs) developing countries such as Malawi and Rwanda.

Many characteristics are used to measure levels of development. These include levels of urbanisation, literacy, manufacturing workforce, life expectancy, infant mortality and income. No single indicator can give all the information; the indicators are usually used in combination. One of the most widely used indicators is gross national product (GNP) per head.

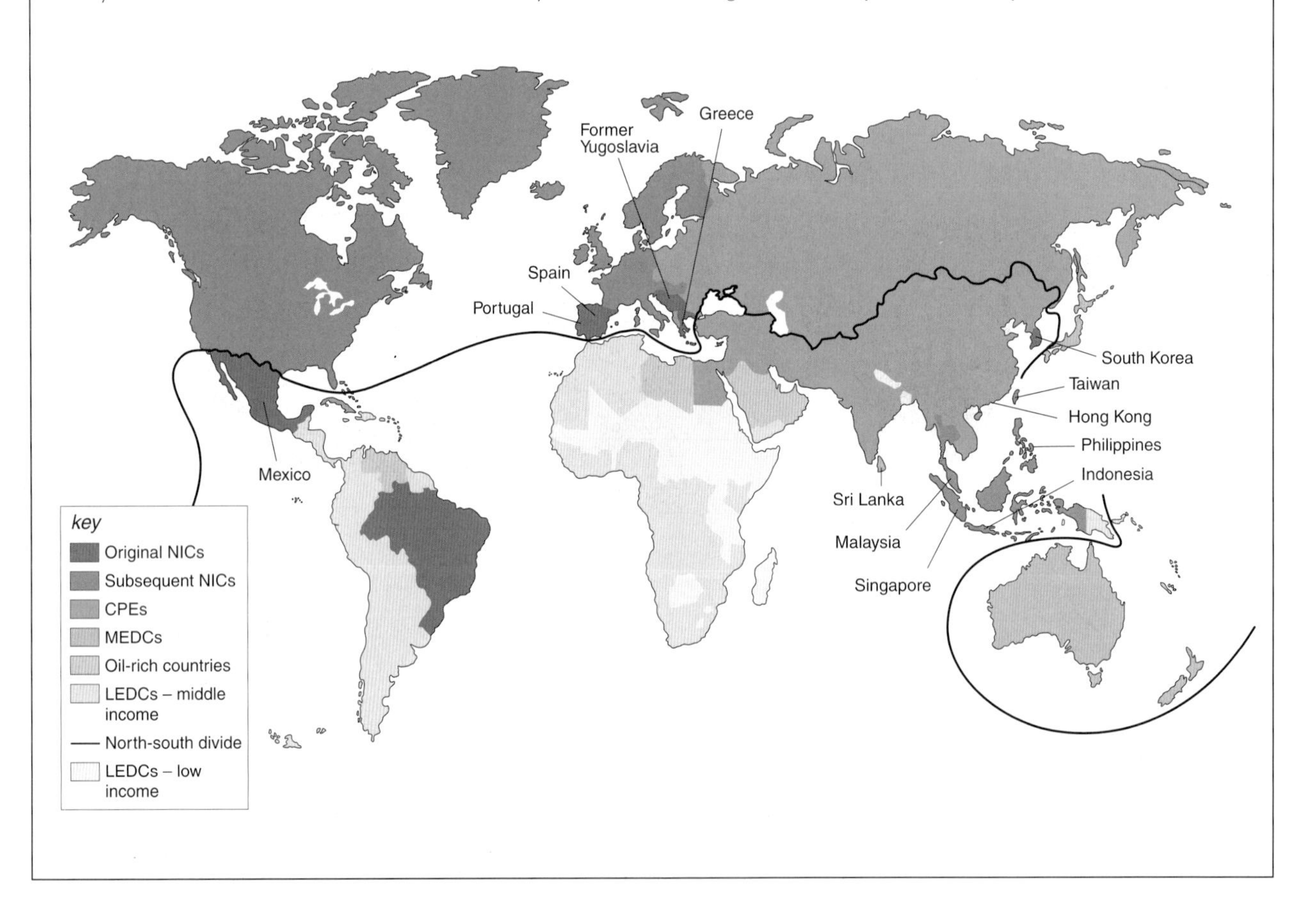

GNP and PPP

GROSS NATIONAL PRODUCT (GNP) PER CAPITA

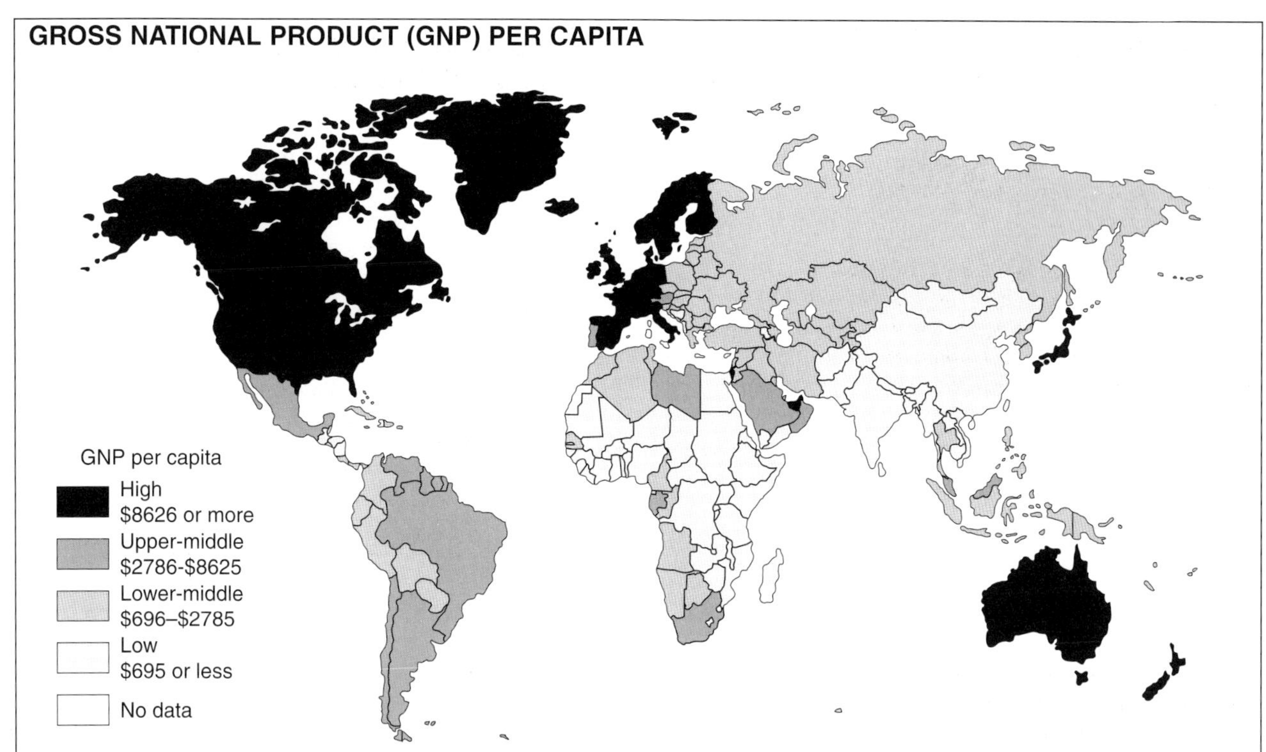

The map of GNP per capita shows a clear bias towards the **More Economically Developed Countries** (MEDCs). Western Europe, North America, Japan, and Australia come out on top according to GNP per capita: the highest values are Switzerland ($36 410), Luxembourg ($35 850), and Japan ($31 450). Approximately 15% of the world's population live in areas with a **high GNP per capita**. By contrast, 56% of the world's population live in areas classified as having a **low GNP per capita**. A number of countries have a GNP per capita of less than $200 per year: Rwanda, Burundi, Ethiopia, Tanzania, Uganda, Mozambique, Sierra Leone, and Vietnam.

PURCHASING POWER PARITY PER CAPITA

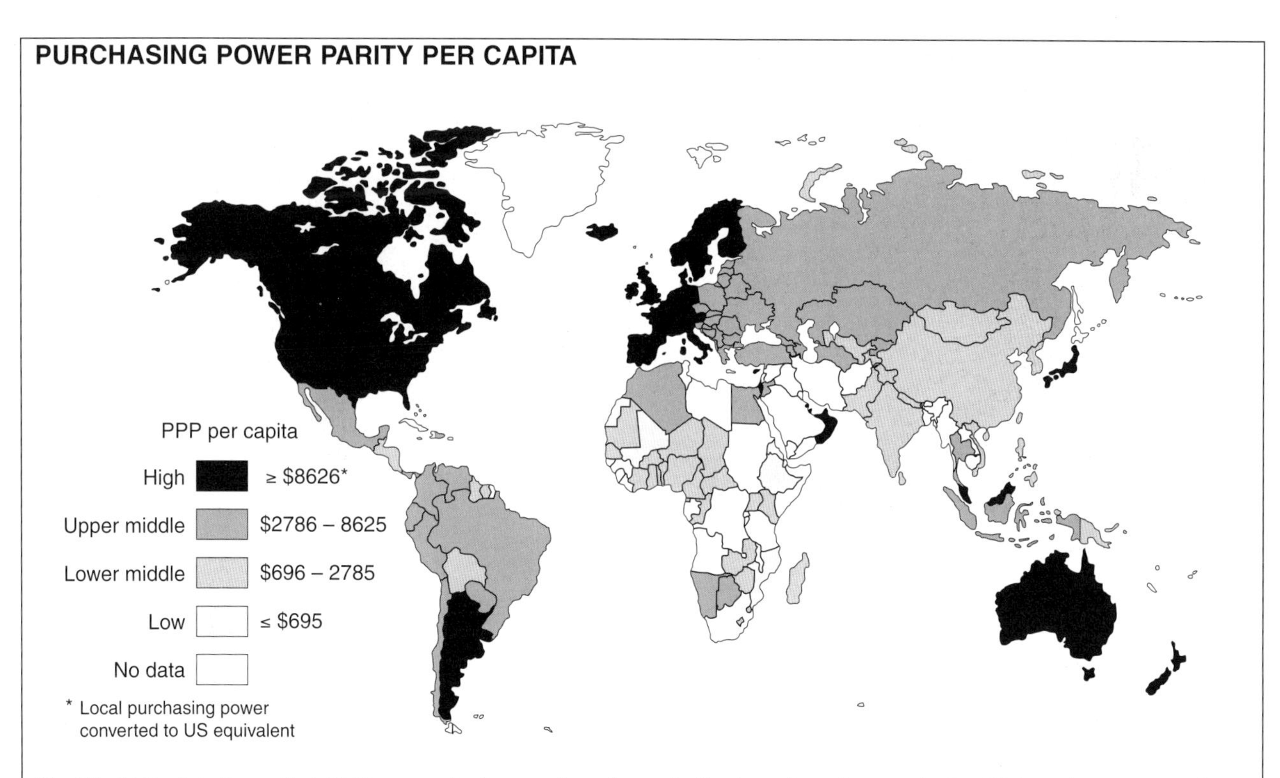

The World Bank believes that GNP per capita figures give a better indication of relative standards of living when converted into **purchasing power parity** (PPP). PPP relates average earnings to the ability to buy goods, i.e. how much you can buy for your money. For example, although wages in India are low compared to British wages, they can buy similar amounts of goods and services because local prices are also lower. In Switzerland, although GNP per head is very high, the high cost of goods and services lowers PPP. PPP lifts GNP per capita for most developing countries and former communist states but lowers it for developed countries.

HDI - regional and racial

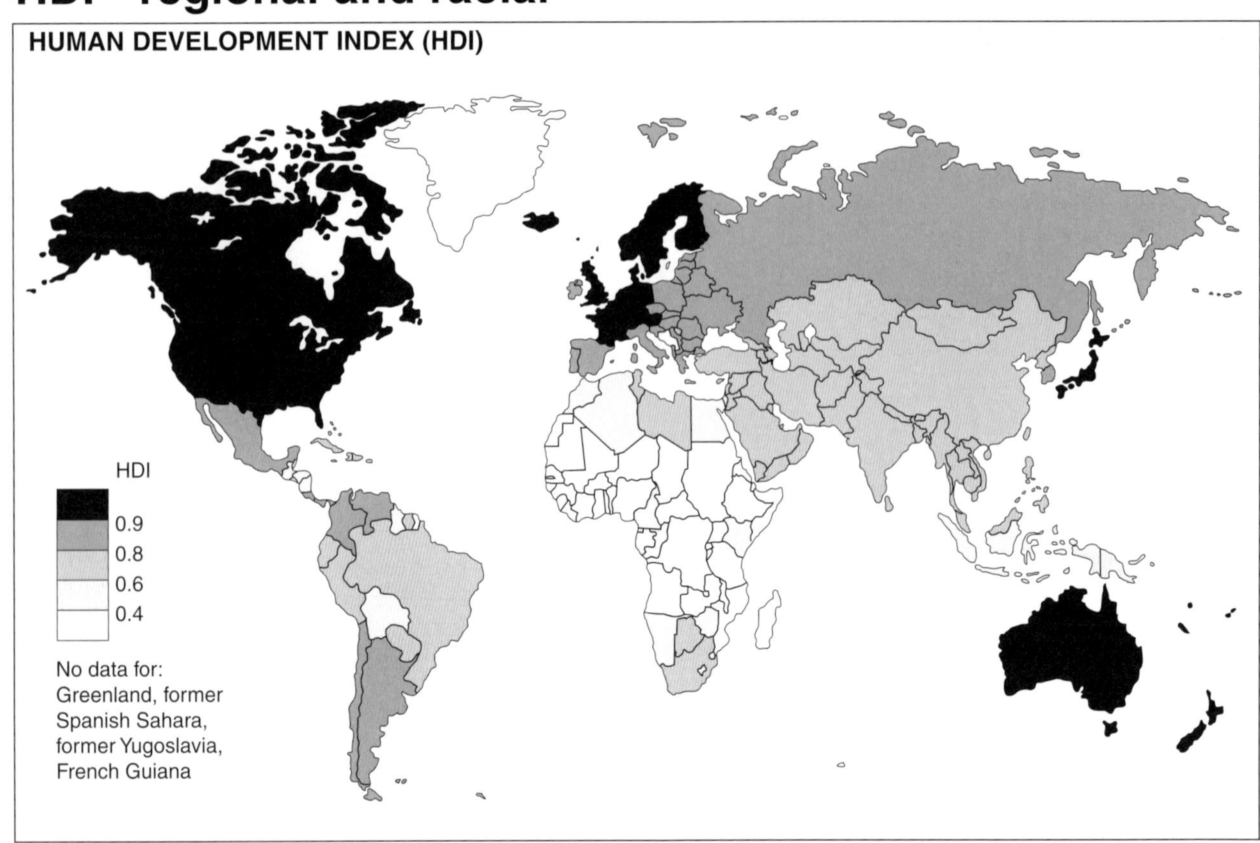

Since 1980 the United Nations (UN) has urged the use of the HDI as a measure of development. It is a more reliable and comprehensive measure of human development and well-being than GNP/head. It includes three basic components of human development:

- longevity (life expectancy)
- knowledge (adult literacy and average number of years schooling)
- standard of living (purchasing power adjusted to local cost of living).

The 1994 HDIs show Canada as the top HDI country, closely followed by Switzerland, Japan, and Sweden. The UK was placed 10th, ahead of Germany (11th), Ireland (21st), and Italy (22nd), but behind France (6th). At the other end, Guinea, Burkina Faso, and Afghanistan had the lowest HDI scores. Some countries, notably Saudi Arabia, Namibia, and the United Arab Emirates, had a higher GNP rank than HDI rank, suggesting scope to transfer oil or mineral revenue into human welfare projects.

National averages can conceal a great deal of information. HDIs can be created to show regional and racial variations, as shown in these diagrams.

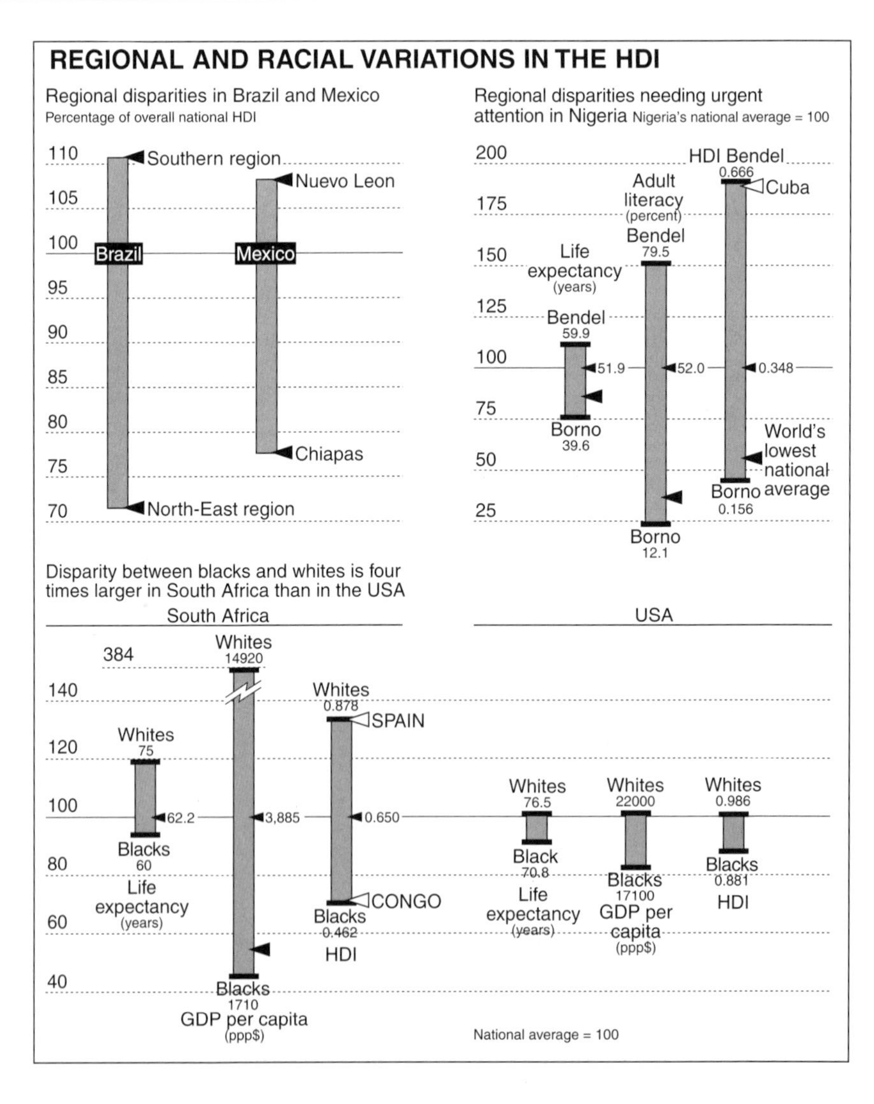

Clark and Rostow

CLARK'S SECTOR MODEL

All developed countries have progressed from agricultural societies to industrial and service economies. For some, such as the UK, the transition was early, mostly in the nineteenth century, whereas for others, such as Ireland, it occurred during the twentieth century. Clark's model clearly shows the **transition** from an economy dominated by the primary sector to one dominated in turn by the secondary and tertiary sector. Change occurs because success in one sector produces a surplus revenue which is invested into new industries and technologies, thereby increasing the range of industries in an area. For example, in the UK, the cotton industry encouraged textile machinery, other metallurgical industries, and service industries. The sector model is descriptive and offers only a crude level of analysis. It does not say how or why a country developed, nor does it show internal variations within a country.

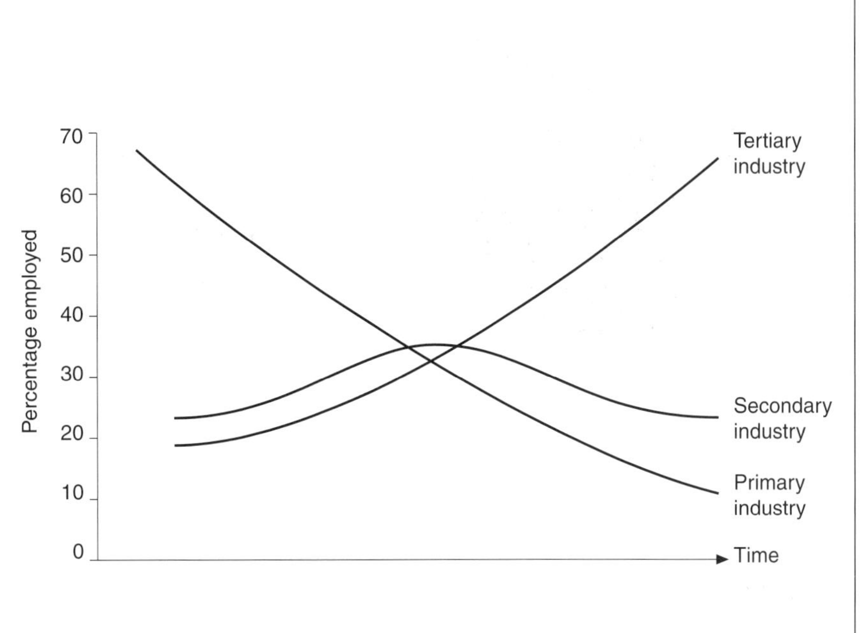

ROSTOW'S MODEL OF DEVELOPMENT

W. W. Rostow, a US economist, envisaged five stages in the development of an economy. His model is a useful starting point in describing and understanding levels of development. These levels can be described as:

1 **Traditional subsistence economy:** agricultural basis, little manufacturing, few external links, and low levels of population growth (stage 1 of the demographic transition model (DTM)). This stage is no longer present in the developed world.

2 **Preconditions for take-off:** external links are developed; resources are increasingly exploited, often by colonial countries or by multinational companies (MNCs); the country begins to develop an urban system (often with primate cities), a transport infrastructure, and inequality between the growing core and the underdeveloped periphery. The population continues to increase (stage 2 of the DTM). Again, this level has disappeared from developed countries.

3 **Take-off to maturity (sustained growth):** the economy expands rapidly, especially manufacturing exports. Regional inequalities intensify because of multiplier effects. This growth can be 'natural' (as in the case of most countries of the developed world), 'forced' (as in the former socialist countries of Eastern Europe), or planned (as in the Newly Industrialising Countries (NICs)).

4 **The drive to maturity:** diversification of the economy, and the development of the service industry (health, education, welfare, and so on). Growth spreads to other sectors and to other regions in the country. Population growth begins to slow down and stabilise (late stage 3 or early stage 4 of the DTM). Ireland, Greece, Spain, and Portugal are at this level.

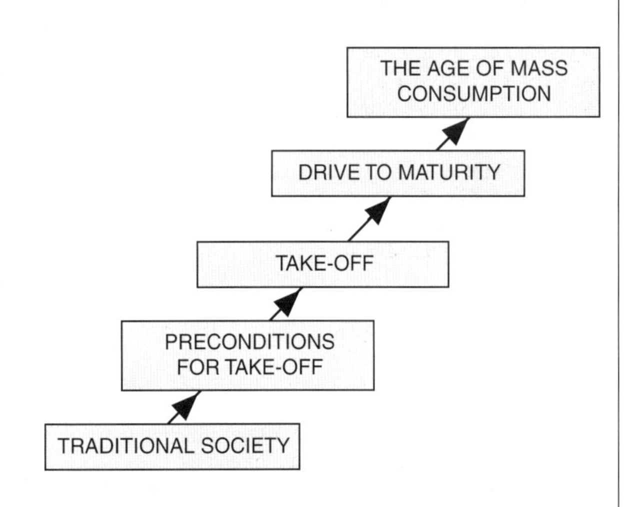

5 **The age of high-mass consumption:** advanced urban-industrial systems, with high production and consumption of consumer goods, such as televisions, compact disc players, dishwashers, and so on. Population growth slows considerably (stage 4 of the DTM). The UK and Germany characterise this level.

The main weaknesses of Rostow's model are:

- it is **anglo-centric**, based on the experience of North America and Western Europe
- it is **aspatial** and does not look at variations within a country. For example, within the UK, there are great disparities in the levels of development between the North and South - Rostow's model fails to pick this out.

Myrdal

MYRDAL'S MODEL OF CUMULATIVE CAUSATION

The core-periphery model, based on the work of **Gunnar Myrdal**, adopts a **spatial** outcome. It is seen as more useful than Rostow's model. In *Rich Lands and Poor Lands* (1957), Myrdal argued that, over time, economic forces increase regional inequalities rather than reduce them. He believed that development was caused by:

- Initial **comparative advantages**, e.g. resources such as location, minerals, or labour. These create the initial stimulus for an industry to develop in a particular location. In turn, a process of **cumulative causation** (multiplier effect) occurs as **acquired advantages**, such as improvements in infrastructure, skilled workforce, and increased tax revenues, are developed and reinforce the area's reputation, thereby attracting further investment, ensuring that it grows and stays ahead of other regions.
- Increased **spatial interaction**, i.e. skilled workers, investment, new technologies, and new developments gravitate to the growing area, the **core**, while the peripheral areas are inundated by manufactured goods from the core (the **backwash effect**), preventing the development of a local manufacturing base. As the core expands it may stimulate surrounding areas to develop due to increased consumer demand (the **spread effect**).

Three main stages can be identified in Myrdal's model:

- a traditional, pre-industrial stage, with few regional disparities (Rostow's stage 1)
- a stage of increased disparities caused by multiplier and backwash effects as the country industrialises (Rostow's stages 2 and 3)
- a stage of reduced regional inequalities as spread effects occur (Rostow's stages 4 and 5).

Myrdal's ideas have been used extensively in regional planning. In particular, they have been used in growth pole policies: places or districts favoured by their location, resources, labour, or market access are economically more attractive and are therefore developed by planners to form natural **growth poles**, expanding faster than other districts. Generally these are urban-industrial complexes which have good transport and accessibility, e.g. Dunkirk and Marseilles-Fos in France and Taranto in the Mezzogiorno, southern Italy.

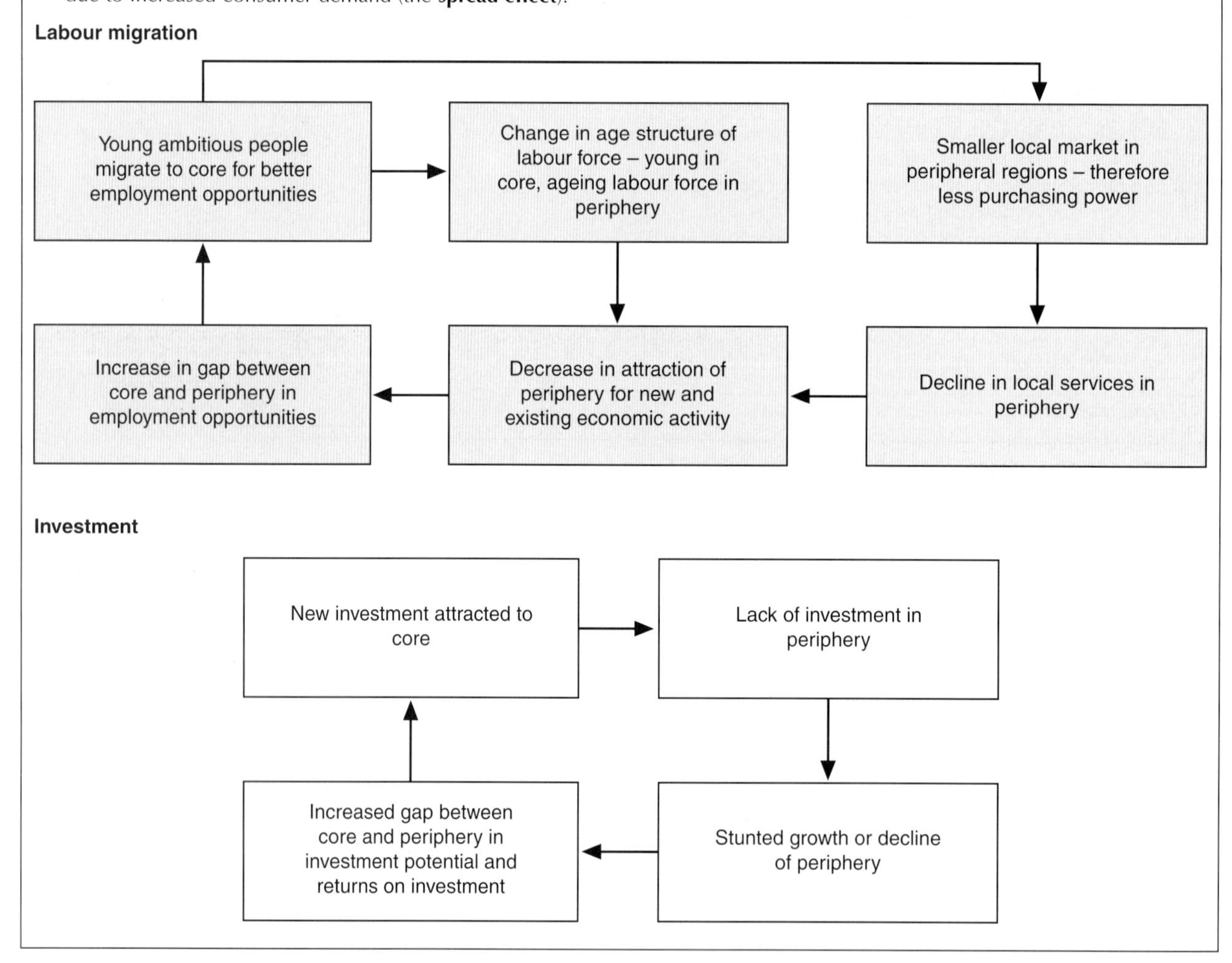

GROWTH POLE THEORY

A growth pole is a dynamic and highly integrated set of industries organised around a leading industry or industrial sector. It is capable of rapid growth and generating multiplier effects or spillover effects into the local economy. The idea was originated by **Perroux** (1955) and developed by **Boudeville** (1966). It has been widely used in regional and national planning as a means of regenerating an area. Growth poles can, however, increase regional inequalities by concentrating resources in favoured locations.

Friedmann and Frank

FRIEDMANN'S STAGES OF GROWTH

1 **Pre-industrial economy:** independent local centres, no hierarchy. Similar to Rostow's stage 1.

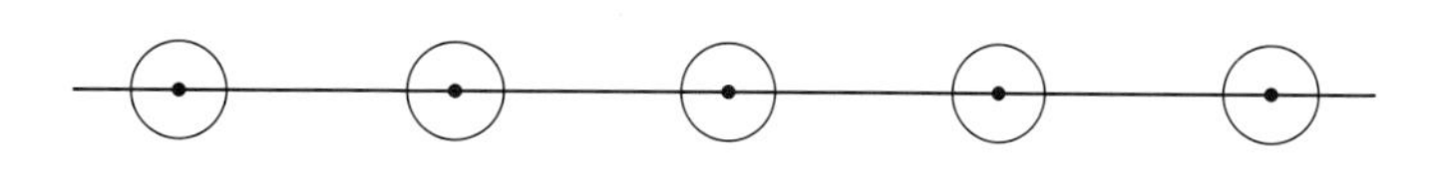

2 **Transitional economy:** a single strong centre emerges. This dominates the colonial society as the stage of pre-conditions begins. A growing manufacturing sector encourages concentration of investment in only a few centres - hence a core emerges with a primate city.

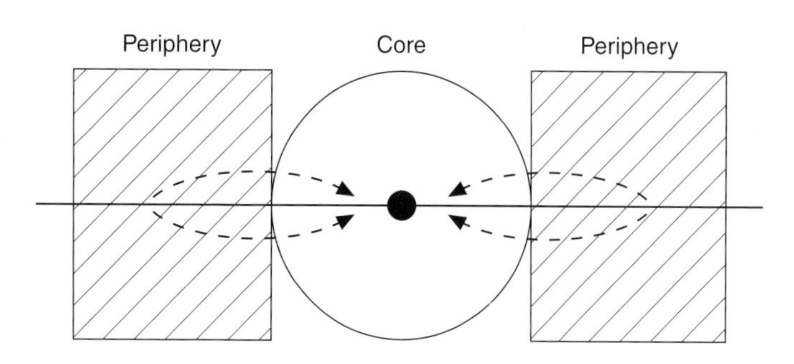

3 **Industrial economy:** a single national centre, strong peripheral sub-centres, increased regional inequalities between core and periphery; upward spiral in the core, downward spiral in the periphery (Myrdal's cumulative causation). In time, as the economy expands, more balanced national development occurs - sub-centres develop, forming a more integrated national urban hierarchy.

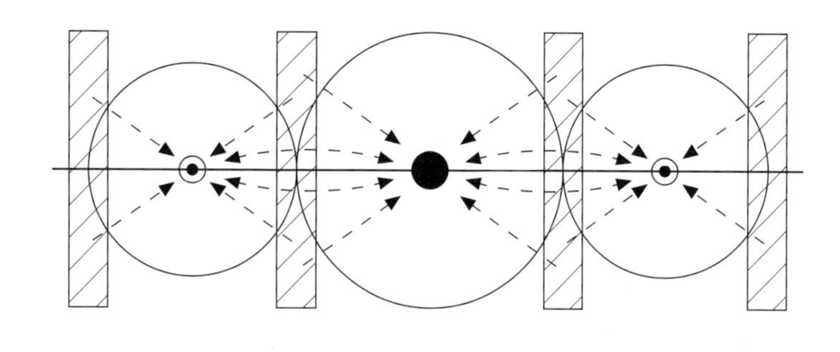

4 **Post-industrial economy:** a functionally interdependent urban system; the periphery is eliminated.

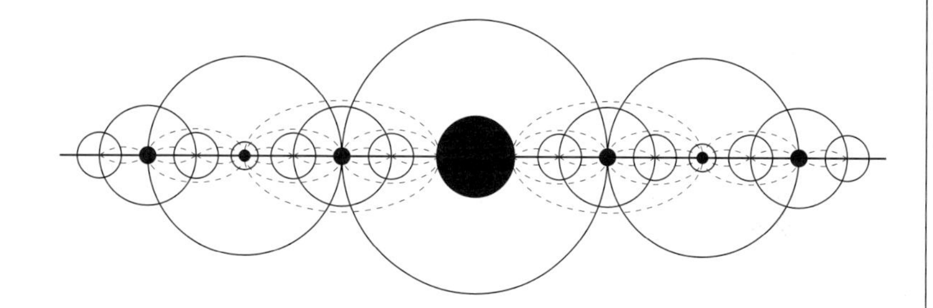

Friedmann believed that stage 4 has been reached in the USA, although there are still peripheral areas such as the Ozarks, Appalachians, and Alaska.

DEPENDENCY THEORY

According to **dependency theory**, countries become more dependent upon more powerful countries, frequently colonial powers, as a result of interaction and 'development'. As the more powerful country exploits the resources of its weaker colony, the colony becomes dependent upon the stronger power. Goods flow from the colony to support consumers in the overseas country.

Andre Frank (1971) described the effect of capitalist development on many countries as '**the development of underdevelopment**'. The problem of poor countries is not that they lack the resources, technical know-how, modern institutions, or cultural developments that lead to development, but that they are being exploited by capitalist countries.

Dependency theory is a very different approach to most models of development:

- it incorporates politics and economics in its explanation
- it takes into account the historical processes of how underdevelopment came about, i.e. how capitalist development began in one part of the world and then expanded into other areas (imperial expansion)
- it sees development as a revolutionary break, a clash of interests between ruling classes (the bourgeoisie) and the working classes (the proletariat)
- it stresses that to be developed is to be self-reliant and in control of national resources
- it believes that modernisation does not necessarily mean westernisation, and that underdeveloped countries must set goals of their own, appropriate to their own resources, needs, and values.

Newly industrialising countries (NICs)

CHARACTERISTICS

An NIC is characterised by:

- significant average annual growth in manufacturing production (4.5% in Portugal, over 15% in South Korea)
- an increasing share of world manufacturing output
- a significant growth in (manufactured) export production
- an increasing proportion of the workforce in manufacturing industries
- a significant increase in GDP provided by manufacturing.

Three main groups of NICs have been identified:

- 'Asian tigers' such as Hong Kong, Singapore, South Korea, and Taiwan
- Latin American NICs such as Brazil and Mexico
- European NICs - Spain, Portugal, Greece, and the former Yugoslavia.

Some of these NICs have been redefined, e.g. Spain was classified as an MEDC in 1983.

Multinational companies (MNCs) play a part in NIC development. They facilitate economic development through their economic and political power and through their access to capital, skills, and knowledge. They also influence the type, scale, and location of manufacturing.

STAGES IN THE EMERGENCE OF AN NIC

1 **Traditional society**
Labour intensive industries, low levels of technology. Local raw materials - food processing and textiles common.

2 **Import substitution industries (ISIs)**
Reduction of expensive imports by development of home industries. Protectionist policies, e.g. high trade tariffs on manufactured goods and in the car industry.

3 **Export-orientated industries (EOIs)**
High-technology, capital intensive industries. R&D functions. Rapid growth and development.

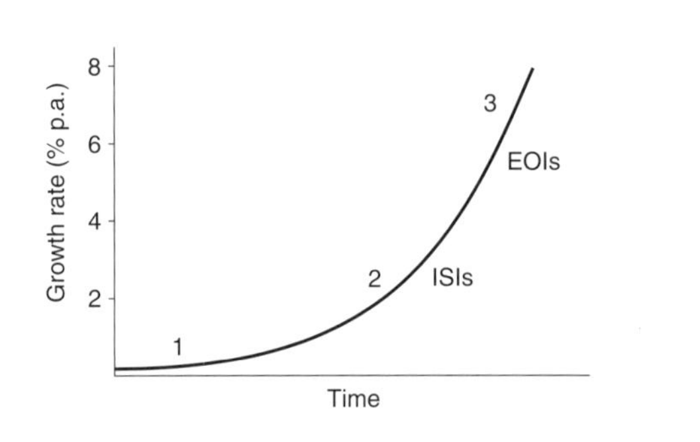

THE EFFECTS OF RAPID INDUSTRIALISATION

Economic effects

- Raises standards of living.
- Benefits the government rather than the people.
- Change in labour force:
  - (i) shift to manufacturing (secondary) industries, especially of young working population
  - (ii) decline in agricultural employment and productivity; growth of rural unemployment
  - (iii) increase in service industries, e.g. transport, retailing, and administration
- Growth of urban infrastructure.
- Increased competition for land, raising land prices.

Social effects

Rural areas

- Social imbalance between affluent industrial minority/commercial workers and the rest of the population.
- Population change, e.g. age-selective migration to urban areas.
- Decline of traditional values and lifestyles.
- Increased dependence on remittances from urban workers.
- Poor welfare systems.

Urban areas

- In-migration leads to rapid development of shanty towns, increased birth rates, and population growth.
- Concentration of unemployed and poor in shanty towns.
- High rates of crime, illiteracy, disease, and so on in areas of impoverished housing.
- Low levels of service provision.
- Unsatisfactory working conditions (sweatshops).
- Political and social unrest.

Environmental effects

- Resource exploitation can damage the natural environment and can lead to the destruction of whole habitats.
- Rivers polluted by industrial waste.
- Air pollution, e.g. in Taipei, Taiwan.
- Unsafe working practises may lead to environmental disasters, e.g. at Bhopal, India, in 1985, toxic gas from the Union Carbide factory lead to widespread blindness in the area.
- Urban blight, e.g. derelict buildings, contaminated land.
- Limited environmental legislation.

Trade

VISIBLE AND INVISIBLE TRADE

Visible trade involves the exchange of goods such as foodstuffs, fuel, manufactured, or raw materials. By contrast, **invisible trade** refers to the movement of finance such as the spending of finance and tourism.

Invisible trade is becoming important for a number of reasons:
- deindustrialisation in MEDCs has led to a reduction in visible trade
- the growth of service industries led to increased earnings from banking, insurance, tourism, and consultancy, for example
- new international divisions of labour (NIDL) is favouring the growth of visible trade from newly industrialising countries (NICs)
- multinational corporations (MNCs) are often based in MEDCs and their LEDC earnings contribute significant earnings
- increased self-sufficiency in foodstuffs has reduced foodstuffs being traded.

The **balance of trade** refers to the trade balance in visible goods (imports - exports). By contrast, the **balance of payments** refers to the balance of imports and exports in visibles and invisibles.

TRADE IN A DEVELOPED COUNTRY AND A DEVELOPING COUNTRY

UK exports
Total exports: $190 099 million

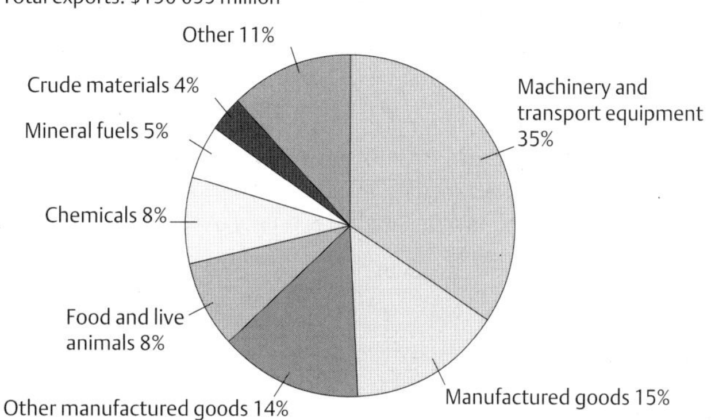

India exports
Total exports: $17 900 million

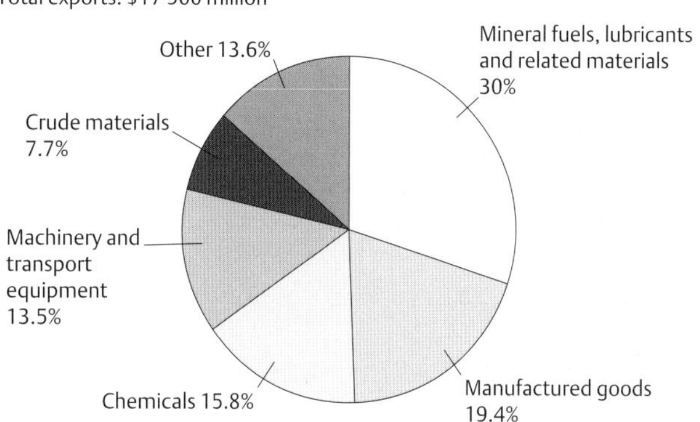

MEDCs and LEDCs have different export and import patterns. MEDCs mostly export machinery, transport equipment, chemicals, agricultural products, and services. Their range of imports is similar. By contrast, LEDCs have a much smaller range of exports. These are mostly agricultural products and raw materials. The range of imports is similar to MEDCs but is likely to be cheaper and less sophisticated.

FREE TRADE

Free trade allows a country to trade competitively with another country. There are no restrictions regarding what can be exported or imported.

What are the advantages of free trade?
- it allows countries to specialise and concentrate on their comparative advantages, that is, the things they do better than other countries.
- it allows them to obtain goods and services more cheaply than if they had to produce them themselves
- it allows them to obtain goods year-round
- it provides consumers with a better choice
- specialisation allows large-scale production
- it increases competition, promotes efficiency, and reduces waste
- it encourages economic, political, and cultural links between countries.

TRADE CONTROL AND PROTECTIONISM

There are a number of factors which limit trade. These include:
- lack of mobility of labour, land and capital
- different national currencies
- customs duties or tariffs are levied on imports at the point of entry
- financial subsidies for home producers
- exchange controls limit the amount of foreign currency into a country
- physical controls include bans and trade embargoes
- unofficial controls such as tradition, 'small print' and bureaucracy.

What are the reasons for protection?	Are there any reasons against protection?
• to improve trade balance • to protect new home industries against foreign ones • to reduce imports • to enable industries to restructure • to increase self-sufficiency • to improve political ties with members of trading blocs • to maintain employment levels.	• it prevents countries from specialising • it maintains inefficiency • it reduces the total volume and value of trade • it raises inflation.

COMPARATIVE ADVANTAGE

According to the economist Adam Smith, if a foreign country can supply a product cheaper than another country, the second country should buy it off them with part of the profit made from their own products. The theory is that if each country specialises in what it does best and exports some of the results, it can buy a greater and wider variety of goods from the rest of the world with the income it has made, compared with what it could produce. This idea was refined by the economist David Ricardo who developed the idea of **comparative advantage**. It makes sense for countries to specialise in what they are better at than other countries.

Aid

Aid can be in the form of money, equipment, staff, goods, or services. Most aid is from MEDCs to LEDCs but sometimes the recipient is a poorer region within an MEDC; for example the EU has given regional aid to Northern Ireland and Scotland.

There is a widespread disparity between who receives aid and who actually needs it. Frequently the poorest countries receive the least amount of aid per person.

Aid to LEDCs is generally divided into **short-term aid, long-term aid**, or **development aid**. Short-term aid normally follows a natural disaster and is provided to keep people alive. Long-term aid or development aid is given to try to improve the long-term standards of living in an area. Examples include improvements to water and sanitation, and programmes to increase agricultural productivity in rural areas.

ALLOCATION OF AID

The amount of aid given varies between countries. The United Nations originally suggested that most countries donate up to 2% of GNP. However, only Norway and Sweden reached this target. A revised target of 0.7% has since been set but many countries are still failing to match this amount.

Official aid as a percentage of GNP

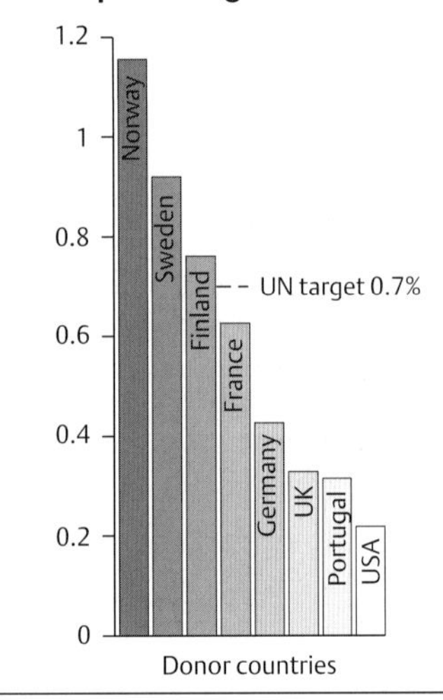

COUNTRIES RECEIVING AID FROM THE UK 1993-4

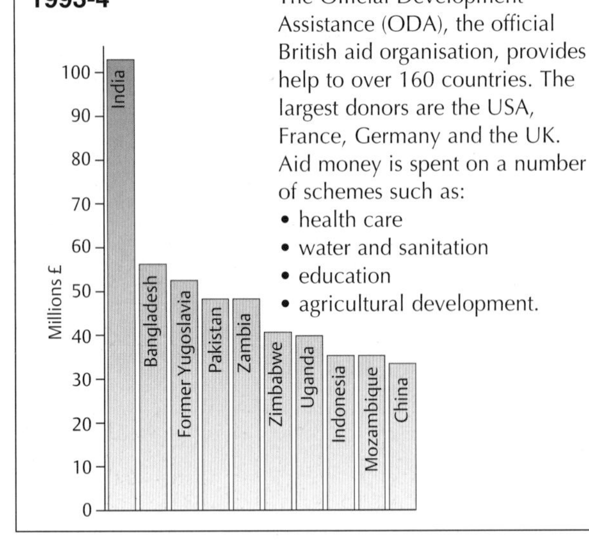

The Official Development Assistance (ODA), the official British aid organisation, provides help to over 160 countries. The largest donors are the USA, France, Germany and the UK. Aid money is spent on a number of schemes such as:

- health care
- water and sanitation
- education
- agricultural development.

TYPES OF AID

Bilateral aid: Aid is given from one country to another. The transfer is generally between countries with political ties. The MEDC often uses it to its own advantage; dictating the conditions of the exchange.

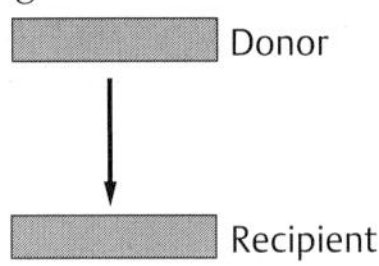

Multilateral aid: A number of countries receive aid from more than one country. The amount received by each country may be low and the interest rates are often very high.

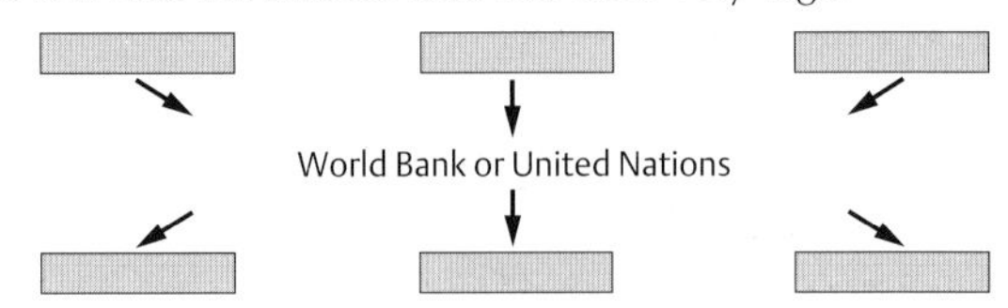

Charities: The amounts involved are considerably less than those received through other types of aid, but there are no political ties (charities are **non-government organisations**). Charities are more flexible and have specialist knowledge and skills. Examples include Oxfam and Save the Children.

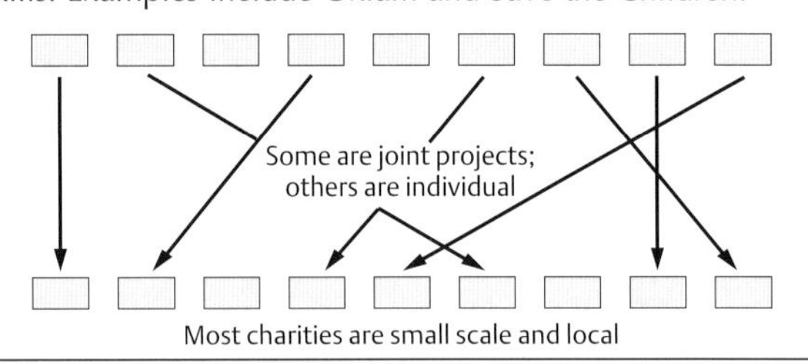

THE IMPACT OF AID

When aid is effective

- it provides humanitarian relief
- it provides external resources for investment and finances projects that could not be undertaken with commercial capital
- project assistance helps expand much needed infrastructure
- aid contributes to personnel training and builds technical expertise
- aid can support better economic and social policies

When aid is ineffective

- aid might allow countries to postpone improving economic management and mobilisation of domestic resources
- aid can replace domestic saving, direct foreign investment and commercial capital as the main sources of investment and technology development
- the provision of aid might promote dependency rather than self-reliance
- some countries have allowed food aid to depress agricultural prices, resulting in greater poverty in rural areas and a dependency on food imports; it has also increased the risk of famine in the future
- aid is sometimes turned on and off in response to the political and strategic agenda of the donor country, making funds unpredictable; this can result in interruptions in development programmes
- the provision of aid might result in the transfer of inappropriate technologies or the funding of environmentally unsound projects
- emergency aid does not solve the long-term economic development problems of a country
- too much aid is tied to the purchase of goods and services from the donor country, which might not be the best nor the most economical
- a lot of aid does not reach those who need it, that is, the poorest people in the poorest countries.

Tourism

The tourist industry is now one of the world's largest industries. The number of foreign holidays rose from approximately 25 million in 1950 to over 350 million in the mid-1990s. The number of holidays at home is even greater. The rise in tourism is related to a number of economic and social trends:

- increased leisure time
- cheaper, faster forms of transport, especially air travel
- an increase in disposable income and a broadening of lifestyle expectations
- the growth of the package holiday
- greater media exposure, travel programmes on television, etc.
- a rise in the number of second holidays, short breaks, weekend trips, etc.

GROWTH OF WORLD TOURISM

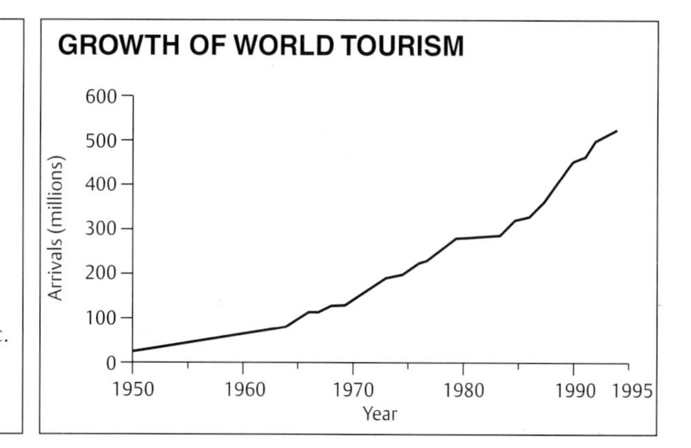

TOURISM IN EUROPE

The main destinations in Europe can be divided into six broad, overlapping types of destination:

1 **Coastal**, e.g. the Costa del Sol, Blackpool, the Algarve
2 **Scenic** landscapes, e.g. South-West Ireland, Scotland
3 **Mountain and ski** centres, e.g. the Alpine resorts
4 **Capital cities and heritage** centres, e.g. London, Paris, Bruges, Oxford
5 **Leisure** centres, e.g. EuroDisney, Tivoli Gardens, Center Parcs
6 **Business-conference** centres, e.g. Korpilampi near Helsinki

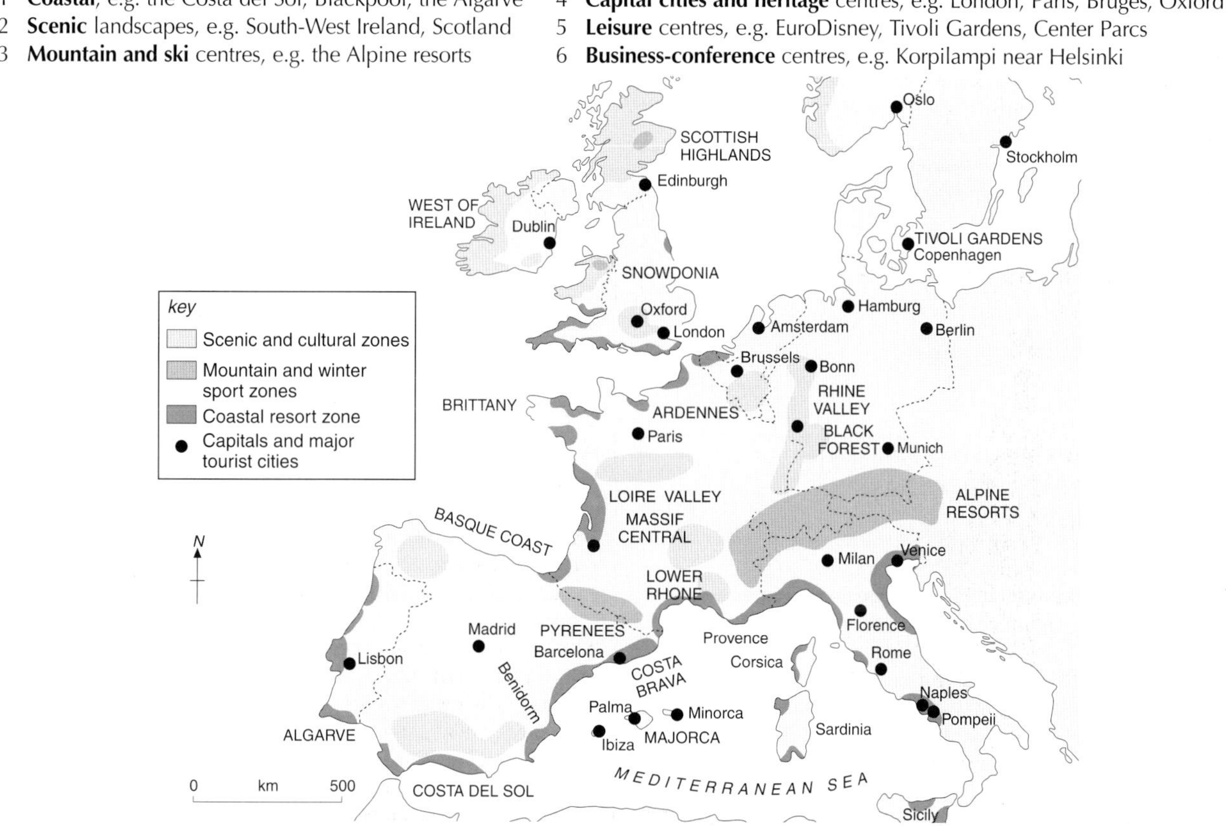

THE IMPACT OF TOURISM

Area of effect	Negative impacts	Positive impacts
Biodiversity	Disruption of breeding patterns; killing of animals for leisure (hunting) or to supply souvenir trade; loss of habitats and change in species composition; destruction of vegetation	Encouragement to conserve animals as attractions
Erosion and physical damage	Soil erosion; damage to sites through tramping; overloading of key infrastructure, e.g. water supply networks	Tourism revenue to finance ground repair and site restoration; improvement to infrastructure prompted by tourist demand
Pollution	Water pollution through sewage or fuel spillage from pleasure boats; air pollution, e.g. vehicle emissions; noise pollution, e.g. from vehicles or tourist attractions, such as discos, etc.; littering	Cleaning programmes to protect the attractiveness of location to tourists
Resource base	Depletion of ground and surface water; diversion of water supply to meet tourist needs, e.g. golf courses or pools; depletion of local fuel sources; depletion of local building-material sources	Development of new and improved sources of supply
Visual and structural change	Land transfers to tourism, e.g. from farming; detrimental visual impact on natural and non-natural landscapes through tourism development; introduction of new architectural styles	New uses for marginal or unproductive lands; landscape improvement, e.g. to clear urban dereliction; regeneration and/or modernisation of built environment; re-use of disused buildings

Tourism in LEDCs

TOURISM IN SOUTH AFRICA

Tourists are attracted to developing countries, such as South Africa, Kenya, and Sierra Leone, for a number of reasons. South Africa, in particular, for example, is rapidly becoming a popular destination for foreign visitors. The reasons for this include:

- its rich and varied **wildlife** and world-famous **game reserves**, e.g. the Kruger National Park
- a **warm and sunny** climate, especially in December and January
- glorious **beaches** in Natal
- the **cultural heritage** and tradition of the Zulu, Xhosa, and Sotho peoples
- it is relatively **cheap** compared to developed countries
- **English** is widely spoken and there are many links with the UK
- it is perceived as a **safe destination** since the collapse of apartheid and the election of the new ANC government.

There are a number of **benefits** that tourism can bring to a developing economy such as that of South Africa:

- **Foreign currency**: the number of foreign tourists has increased by 15% per annum during the 1990s and contributes 3.2% of South Africa's GDP.
- **Employment**: thousands are employed in formal (registered) and informal (unregistered) occupations ranging from hotels and tour operators to cleaners, gardeners, and souvenir hawkers.
- It is a more **profitable** way to use semi-arid grassland: estimates of the annual returns per hectare range from R60-80 for pastoralism to R250 for dry-land farming and R1000 for game parks and tourism.
- **Investment**: over R5 billion was invested in South Africa's tourist infrastructure in 1995, upgrading hotels, airlines, car rental fleets, roads, and so on.

However, there are a number of **problems** that have arisen as a result of the tourist industry:

- There is undue **pressure** on natural ecosystems, leading to soil erosion, litter pollution, and declining animal numbers.
- Much tourist-related employment is **unskilled**, **seasonal**, **part-time**, **poorly paid**, and lacking any rights for the workers.
- **Resources** are spent on providing for tourists while local people may have to go without.
- A large proportion of **profits** goes to overseas companies, tour operators, and hotel chains.
- **Crime** is increasingly directed at tourists; much is petty crime but there have been some very serious incidents, such as rape.
- Tourism is very **unpredictable**, varying with the strength of the economy, cost, safety, alternative opportunities, stage in the family life-cycle, and so on.

ECOTOURISM

Ecotourism is a specialised form of tourism where people want to see and experience relatively untouched natural environments. Typical destinations include game reserves, coral reefs, mountains, and forests. A classic example would be gorilla-watching in Central Africa. It has recently been widened to include 'primitive' indigenous people. It is widely perceived as an 'acceptable' type of tourism and a form of **sustainable development**, i.e. a type of tourism that can be developed without any ill-effects on the natural environment. However, much that passes for ecotourism is merely an expensive package holiday cleverly marketed with the 'eco' label.

Ecotourism developed as a form of specialised, flexible, low-density tourism. It emerged because mass tourism was seen as having a negative impact upon natural and social environments, and because there was a growing number of wealthy tourists dissatisfied with package holidays. In the original sense, people who took an 'ecotourism' holiday were prepared to accept quite simple accommodation and facilities. This is sustainable and has little effect upon the environment. However, as a location becomes more popular and is marketed more, the number of tourists increases resulting in more accommodation and improved facilities, e.g. more hotels with showers and baths, air conditioning, bars, sewage facilities, and so on. This leads to an unsustainable form of ecotourism as it destroys part of the environment and/or culture that the visitors want to experience.

Tourism in a national park

National Parks were introduced into England and Wales in 1949
- to preserve and enhance the countryside and
- to promote enjoyment of the National Parks by the public.

Ten National Parks were designated in 1955 and they are now managed by the National Parks Authority (NPA).

National Parks are run by local boards, committees and county councils. One-third of the Board are appointed by the government. However, much of the land in National Parks is privately owned by farmers and others.

The duties of the NPA include to
- gain access for the public
- plant woodlands
- set up information centres
- provide car parks and picnic sites.

Interest groups in the National Parks include Farmers, Landlords, Water authorities, Forestry Commission, National Trust, Ministry of Defence, and County Councils.

PROBLEMS IN NATIONAL PARKS

- honeypot sites - increasing numbers of visitors destroy the sites they come to see
- landowners - do not like people roaming over their land
- visitors want car parks, visitor centres and so on
- crops get trampled on, gates are left open
- congestion leads to noise and air pollution
- second homes cause an increase in local house prices
- agricultural developments remove dry stone walls, heathlands, hedgerows, meadows and woodlands

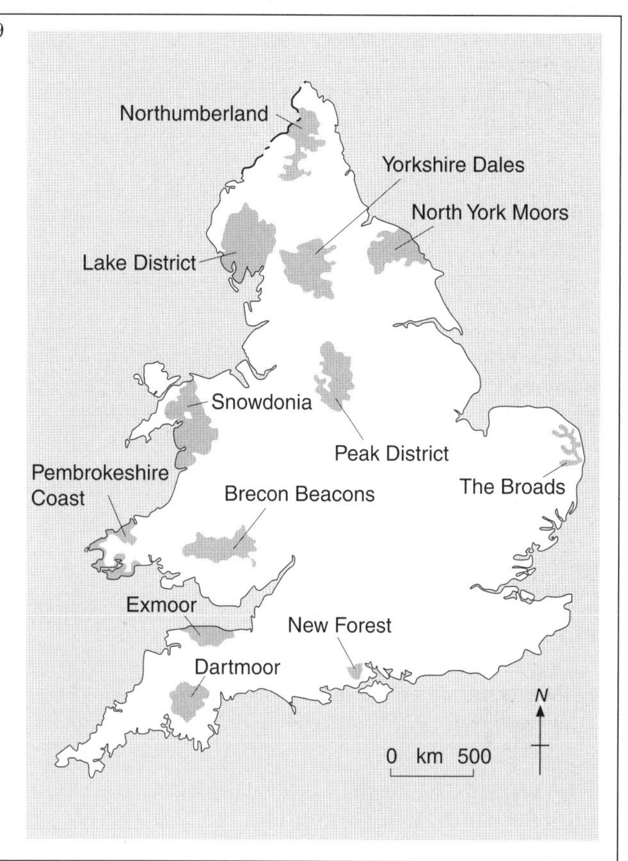

THE CONFLICTS ON DARTMOOR

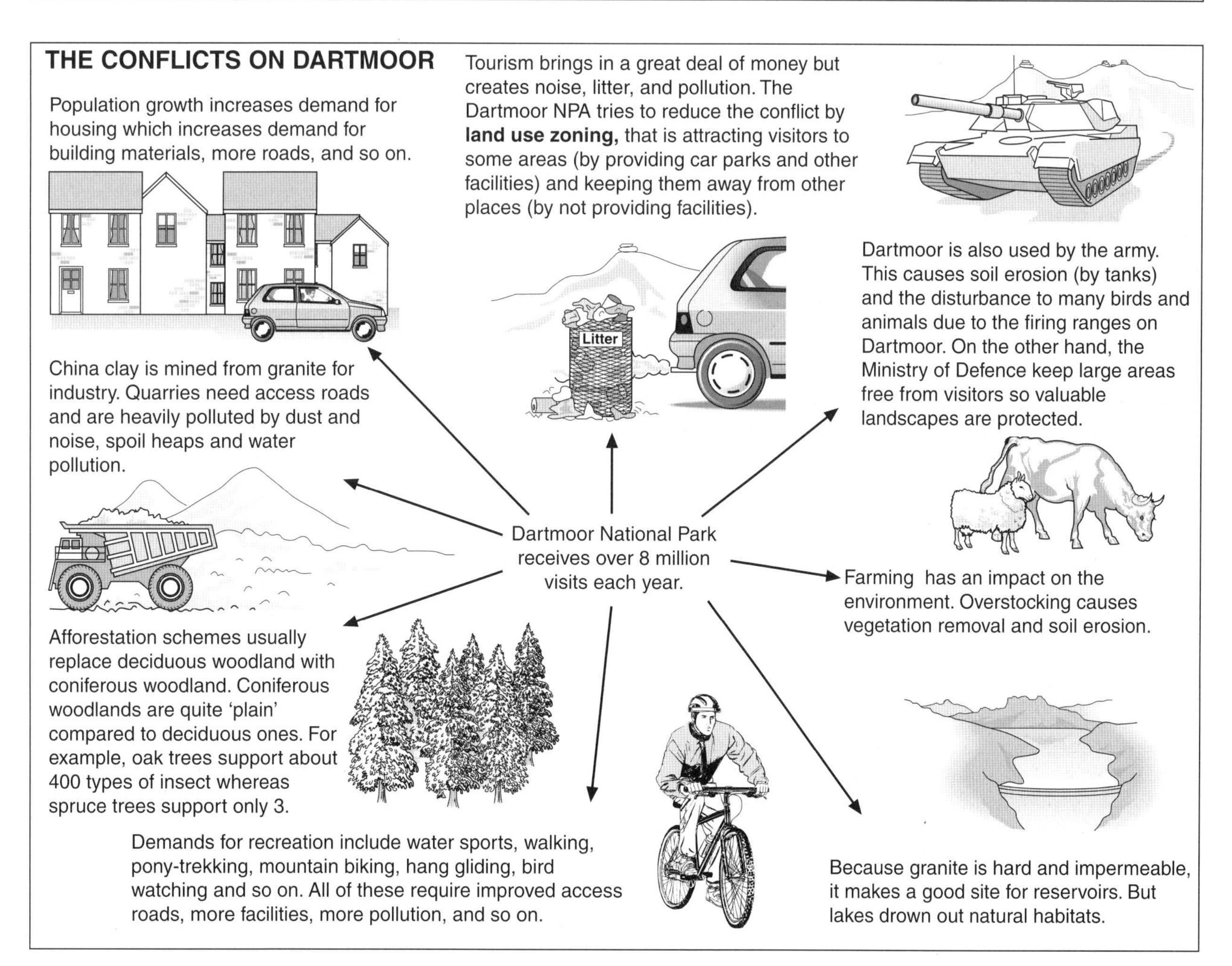

Revision tips

Revision involves pulling all the elements of the course together. To be successful revision needs to be planned, and it needs to be geared towards the needs of the exam.

PLANNING YOUR REVISION

1 You must know what the exam consists of. If you do not have past exam papers, you can order some from the exam boards. For this you need to know your Exam Board and the syllabus number. It is useful to have a selection of past papers, and if you write to the Board (you'll find the address below), ask for the most recent papers. There have been many changes to the syllabus, for example, 2001 A/S and 2002 A2 examinations; ask for a copy of the specimen paper.

2 Study the past papers and familiarise yourself with the layout of the paper.

3 Revise in short, manageable chunks; do not attempt to do all of the subject in one go, but take each topic in turn. It is important to revise topics more than once. Two or three revisions of a topic improve memory and recall considerably. Use the revision checklist on pages 3 to 7 to structure your revision.

4 When you revise, use whichever method or methods you feel most comfortable with. These could include:

- highlighter pens
- lists and rhymes
- note cards
- Mnemonics (the first letter of words, e.g. CASH standing for corrosion, abrasion, solution, and hydraulic impact, i.e. the types of erosion in a river or at the coast).

5 Take regular, varied breaks; it is difficult to concentrate for more than 40 minutes at a time. Have a 15 minute break first of all. After the next 40 minutes take a longer break. Then after the next 40 minutes take a shorter break, and so on.

6 Test yourself. This could be writing answers to past questions, drawing sketch maps, learning facts and figures, identifying symbols on an Ordnance Survey map, etc. Ask a teacher, parent, or friend to assess you. If you ask a friend, then two of you are revising and helping each other's work. This is very often the best method.

7 Reward yourself with a treat.

8 Do not work too late, get plenty of sleep, and try to stay fresh.

9 Work to a revision timetable - it is best if you have a timetable plotted and keep to it. This needs to take into account when your exams are in all other subjects.

10 Make sure you revise a topic more than once. It may take a number of goes before you remember it all (see the 'forgetting curve').

THE FORGETTING CURVE

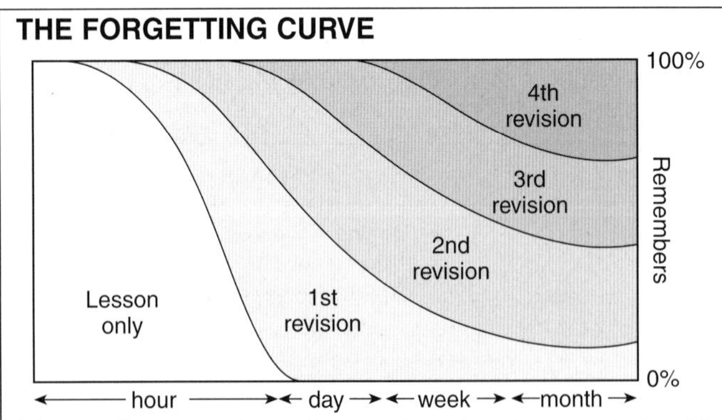

Most pupils memory is good over a short time, however, to be able to recall information a day later, or a week later generally requires one or two revisions of the material. To recall information a month later may require an extra revision period.

Essentially the more often you revise a topic the more information you can recall.

CONTACTING YOUR EXAM BOARD

Edexcel: 32 Russell Square, London WC1B 5DN
Tel 0870 240 9800

AQA: Assessment and Qualifications Alliance, Stag Hill House, Guilford, Surrey GU2 7XJ
Tel 01223 553 998

WJEC: Welsh Joint Education Committee, 245 Western Avenue, Cardiff CF5 2YX
Tel 02920 265 000

SQA: Scottish Qualifications Authority, Ironmills Road, Dalkeith, Midlothian EH22 1LE
Tel 0141 242 214

CCEA: Northern Ireland Council for the Curriculum Examinations and Assessment, Clarendon Dock, 29 Clarendon Road, Belfast BT1 3BG
Tel 02890 261 200

IB: International Baccalaureate Organization, Peterson House, Malthouse Avenue, Cardiff Gate, Cardiff CF23 8GL
Tel 02920 547 777

Essay plans

There are a number of ways of answering an essay question. In general, the essay title and the material to be included will suggest what type of structure should be used. However, for all essays you need to:

- examine the wording of the question closely
- plan your answer.

It is better to spend time thinking and planning, so that you do not waste time writing about irrelevant material. The 25 minutes of relevant material is better than 35 minutes of irrelevant material.

TYPES OF ESSAY

There are three main types of essay – **description**, **explanation**, and **evaluation**, although not all essays fit into this classification. Descriptive essays are the easiest and require factual recall. By contrast, explanation requires you to give reasons and account for why a particular object is the way it is. Evaluation expects an opinion based on the evidence presented throughout the essay. Alternatively, you may be given a ready-made evaluation and asked to say how far you agree.

COMMAND WORDS

These tell you what to do in the essay or how to use the material. There are a number of such command words:

- describe — show the details and characteristics of
- evaluate — show the relative importance
- examine — investigate in detail
- explain — show in detail how something works

WRITING AN ESSAY

It is vital to think about the essay and to plan it. Quality is more important than length. One way of planning an essay is using the points-group-order method. Write down a list of points that are relevant to the essay and then group them; finally put them into order of importance.

Mind maps or branching maps are a useful way of selecting and presenting information. There is more than one way of answering most essays. As long as the structure is logical and clear, and it answers the questions, they are all acceptable. To find a structure requires a general overview of the subject as well as the small details to support your views.

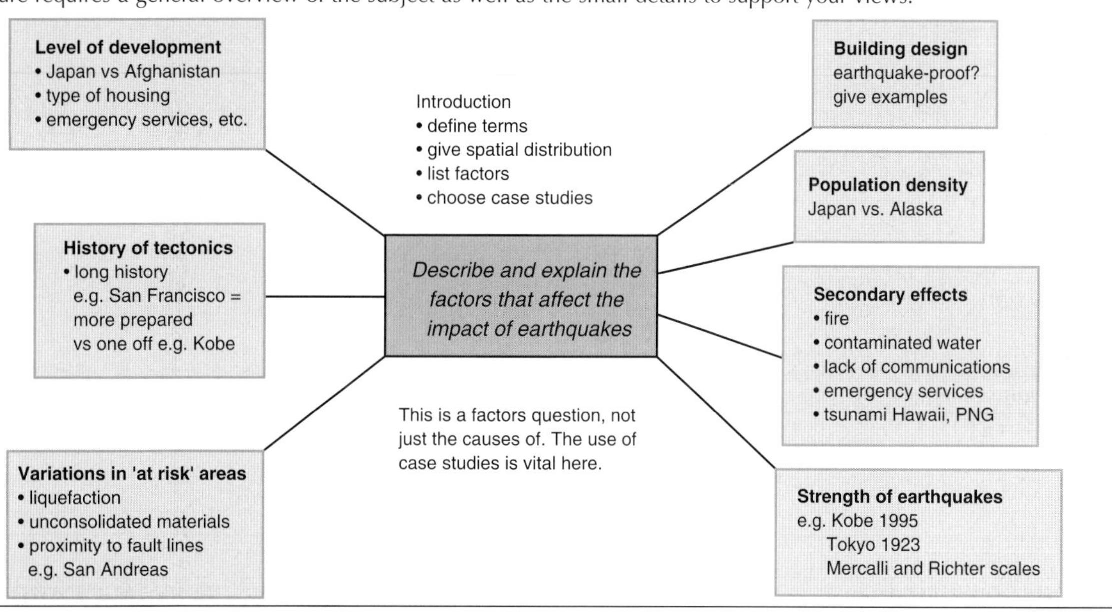

ESSAY STRUCTURE

Writing the introduction is a key skill. Most examiners have a good idea of the grade a candidate will achieve after they have read the introduction. The introduction needs to be clear and full of impact. It should:

- define the terms used in the title
- show the line of argument that will be taken
- list the factors that are important
- state which examples and case studies will be used.

The main body of the essay develops the arguments. Each paragraph should have a key sentence or key point. The rest of the paragraph explains and provides evidence. Paragraphs must be linked, this is done in a variety of ways:

- referring back to the point above (e.g. … the result of this is to cause)
- linking in a time sequence (e.g. next, after, etc.)
- comparisons (e.g. there are also environmental problems in shanty towns …)
- contrasts (e.g. by contrast, the UK's economy is based on services).

Throughout the essay the quality of language needs to be high. It is important to use key words and phrases that make the essay read well, while allowing the text to flow.

The conclusion is more than just a summary. It may:

- assess the changing nature of the topic
- examine the changing importance of factors involved
- draw out the uniqueness of the material used (every example is different)
- look at the contrasts between developed countries and developing countries
- look to the future (how will the subject change in the next 25 years).
- end with a question, e.g. 'Even if we can predict earthquakes and volcanoes, can we stop people from living in hazardous areas?'

Structured questions and map descriptions

STRUCTURED QUESTIONS

When answering a structured question, it is vital you are clear and concise. Look for the maximum, the minimum, the trend and any exceptions in the data supplied. Use your data in your answers; include names and give figures to support your statements.

The effect of common salts on sandstone under laboratory conditions

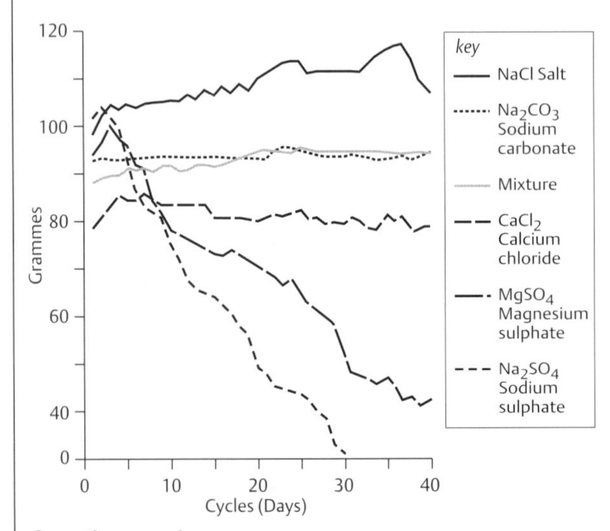

The effects of salt weathering on rock types

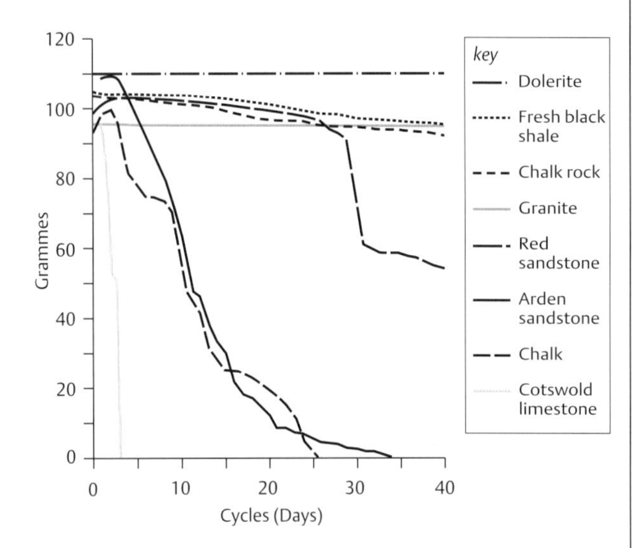

Questions and answers.

Study the two graphs above and attempt to answer the following questions:

1 Which salts have the most effect when added to sandstone?

sample answer: The most effective salts are, in descending order, sodium sulphate (Na_2SO_4), magnesium sulphate ($MgSO_4$) and calcium chloride ($CaCl_3$). Sodium sulphate caused a 100-g block of stone to break down to about 30 g, a loss of 70%. Similarly, magnesium sulphate reduced a 95-g block to just over 40 g, a loss of over 50%. The least effective salts were common salt (NaCl) and sodium carbonate (Na_2CO_3).

2 Which rock type is most susceptible to breakdown when treated with sodium sulphate? How long do the changes take?

sample answer: Chalk is the most susceptible rock. Within four days a 100-g block had completely disintegrated. Cotswold limestone took just 25 days to disintegrate and Arden sandstone about 34 days. By contrast, dolerites and granites showed the least change.

3 Describe the changes that take place to a block of red sandstone.

sample answer: Red sandstone was only slightly changed during the first four weeks – from 105 g to 95 g. However, it broke down very rapidly soon after – from 95 g to 60 g in a matter of days. From then on, its rate of breakdown slowed down.

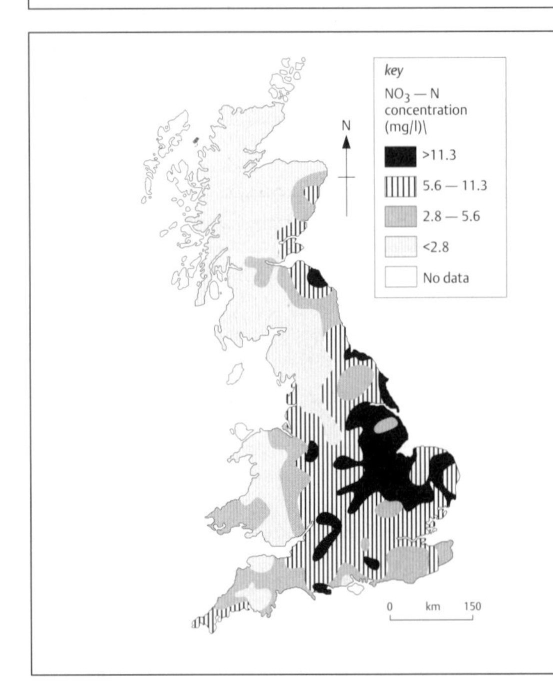

MAP DESCRIPTIONS

There are certain key elements to look for on a map. These include:

- maximum values
- minimum values
- trends or patterns
- exceptions.

It is important to name places and use figures from the map. For example, the map of the UK shows the concentration of nitrates in drinking water in Britain. A good description could be as follows:

There is a very distinct geographic pattern in the level of nitrates in Britain's water. Highest levels, over 11.3 mg/l are mostly found in eastern parts of England whereas the lowest levels, less than 2.8 mg/l are found in Scotland, north-west England and central Wales. In general, there is an east-west trend with higher values in the east and lower values in the west. There are, however, certain anomalies. Parts of south-east England have very low values (<2.8 mg/l) while there are quite high rates (5.6 – 11.3 mg/l-1) in the south-west and in parts of north-west England.

Building a case study

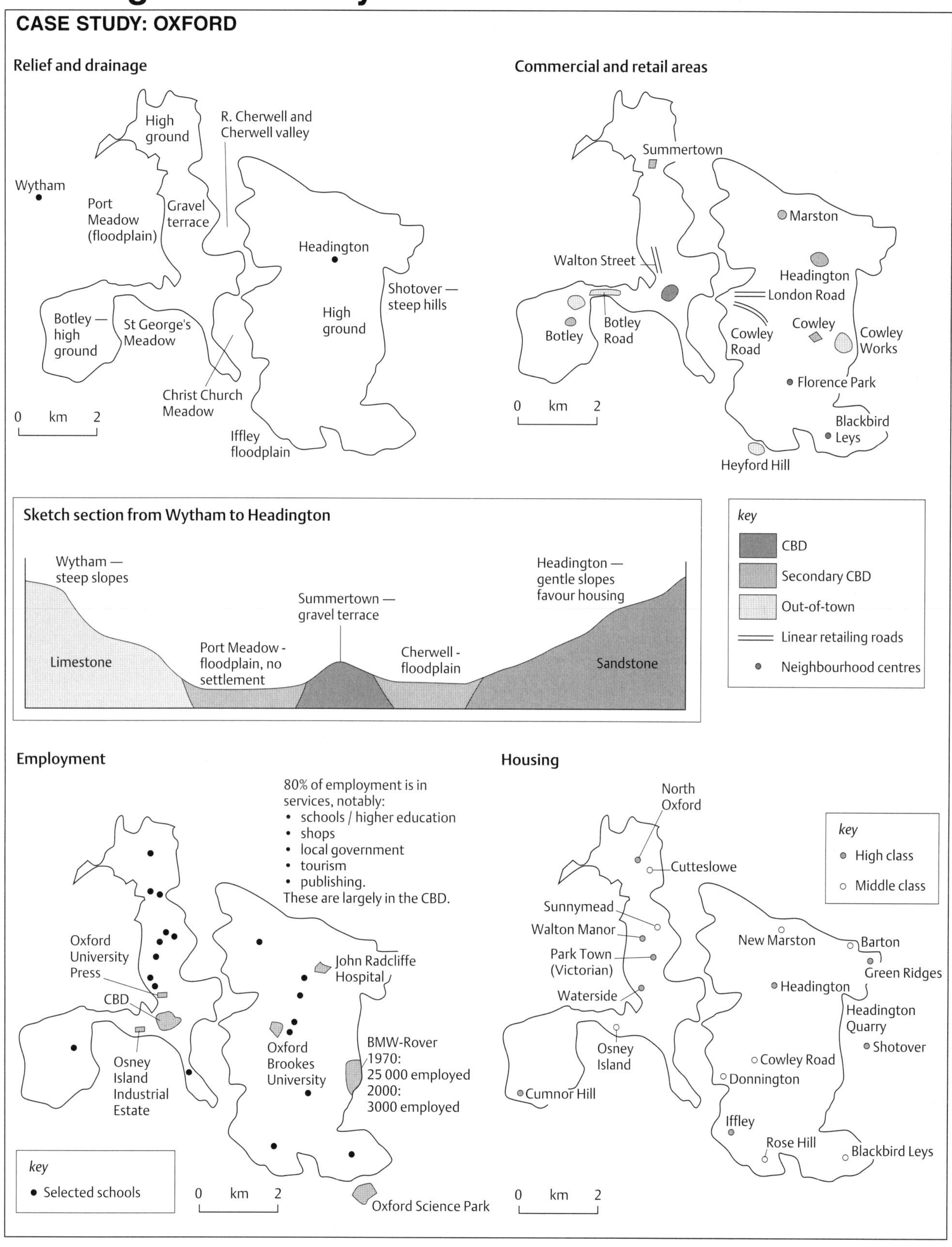

This page shows you how to build up your own case study.
On outline maps of the city, mark on:
(a) the physical features
(b) the commercial and retail areas
(c) the main places and types of employment
(d) variations in residential areas.
A sketch section can be used to show how physical factors influence human geography.

Using data

An answer that is supported by evidence is more likely to earn a good grade. Geography examiners are looking for evidence in your answers.

EXPECTATIONS AT A-LEVEL

At GCSE level you will have come across the reasons why birth rates are higher in developing countries than developed countries. They may include:

- a high infant mortality rate
- the low status of women
- the need for children to work on the farm
- the need for children to look after their parents when they are old
- the need for a male heir
- the lack of contraceptives.

At A-level, however, you will expected to show evidence.

- What is the infant mortality rate and the death rate?
- What proportion of people are employed in farming?
- What percentage of the population has access to contraceptives?
- What is the age-structure of the population? and so on.

In addition, there is an implicit suggestion that circumstances are different in developed countries, so you should be aware of the statistics for a developed country. Many of these key data are shown in the table below. From this can be produced an answer that is far more informative and that begins to evaluate (compare and contrast).

SELECTED STATISTICS

Factor	Brazil	India	Japan	South Africa	UK	USA
Population (millions)	159.1	943.0	125.1	44.0	58.3	263.6
Area (thousand km^2)	8512	3287	378	1221	244	9809
Population density (people /km^2)	19	317	332	36	241	28
Population growth (% p.a.)	1.8	2.1	0.4	2.2	0.2	0.5
Birth rate (per 1000)	26	31	12	31	14	14
Death rate (per 1000)	8	10	8	9	12	9
Infant mortality rate (per 1000 live births)	57	88	5	62	8	8
Life expectancy	66	60	78.6	62	75.8	75.6
Urban population (%)	76	26	77	50	89	76
Fertility rate (no. of children per woman)	3	4	2	4	2	2
Use of contraceptives (%)	66	43	64	50	81	74
Age structure						
0—14	35	37	18	37	19	21
15—59	58	56	64	57	61	62
≥ 60	7	7	17	6	21	17
Adult literacy (%)	81	50	99		99	99
Employment						
Agriculture	25	62	7	13	2	3
Industry	25	11	34	25	28	25
Services	50	27	59	62	70	72
Average income ($US)	2920	290	31450	2900	17970	24750
Energy use (tonnes/person/year)	0.44	0.35	4.74	2.49	5.4	10.74
Literacy (%)	81	50	99	81	99	99
Spending on education (as % of GNP)	3.7	3.5	5.0	3.8	5.3	7.0
Spending on military (as % of GNP)	1.2	2.2	1.0	3.0	4.0	5.3

SAMPLE A-LEVEL ANSWERS

India is a developing country which has a very rapid population growth of 2.1 per annum. It is characterised by a high birth rate, 31 per thousand, and a low death rate, 10 per thousand. This gives it a high natural increase. There are many reasons for such a high birth rate.

- India has a young population structure. Moreover, 37% of the population are under the age of 15 years so population growth is likely to be high for many years to come.
- India is an agricultural society. Nearly two-thirds of the population are employed in farming so there is a need for children to work on the farm.
- India has a high infant mortality rate, 88 per thousand, so there is a higher birth rate to compensate for those that die early. The use of contraceptives is relatively low, 43%, and with low levels of literacy and high levels of poverty (average annual income is only $290) people continue to have large families – the average fertility rate is four children for each woman.

By contrast, the UK is a developed country and is characterised by a low population growth, 0.2% per annum, low birth rates, 14 per thousand, and low infant mortality rates, 8 per thousand. Most of the population are employed in services. Less than 3% are employed in farming so there is not the need to have children to help with farm work. Moreover, the UK has an ageing population with over 20% of the population over the age of 60 years. Thus there is a smaller reproductive group and greater use of family planning (over 80% of families use some form of family planning).

The Chi-squared test

The Chi-squared test is used to test whether there is a significant difference between data. For example, it can be used to test whether there is any difference between altitude and the number of cirques, or orientation and the number of cirques. It is also widely used in human geography. A common test is to see whether there are significant differences in levels of well being between areas.

DATA CHARACTERISTICS

The Chi-squared test can only be used on data which has the following characteristics:

- The data must be in the form of frequencies counted in a number of groups.
- Data must be on the interval or ratio scale (it has a precise numerical value) and can be grouped into categories.
- The total number of observations must be greater than 20.
- The expected frequency in any one category must be greater than five.

METHOD

1 State the hypothesis being tested e.g. there is a significant difference between two or more sample groups. A null hypothesis (a negative test) could be given, saying that there is no significant difference between the samples.
2 Tabulate the data. The data being tested for significance is known as the observed frequency, and the column is headed **O**.
3 Calculate the expected number of frequencies that you would expect to find. These should be written in column **E**.
4 Calculate the Chi-squared statistic using the formula

$X^2 = \Sigma(O-E)^2/E$

where X^2 is the Chi-squared statistic
Σ is the sum of
O refers to the observed frequencies, and
E are the expected frequencies.

5 Calculate the degrees of freedom. This is a value one less than the total number of observations (N), that is, N – 1.
6 Compare the calculated figure with the critical values in the significance tables using the appropriate degrees of freedom. Read off the probability that the data frequencies you are testing could have occurred by chance.

EXAMPLE

The following figures provide data on the number of cirques and their orientation.

Orientation	Number of cirques
North-east	40
South-east	15
South-west	5
North-west	12

What is the probability that the number of cirques is related to orientation?

1 The null hypothesis (Ho) states there is no significant variation in the frequency of cirques with orientation.
The alternative hypothesis (H1) states that there is a significant difference in the frequency of cirques and orientation.
2/3 If there is no difference in the frequency of cirques they should all have roughly the same frequency.
That means they will all have about the average.
The expected frequency is thus the same as the average frequency which is (40 + 15 + 5 + 12) / 4 = 72/4 = 18

4

Orientation	(Number	(Average)	of cirques)		
	O	E	(O-E)	$(O-E)^2$	$(O-E)^2/E$
North-east	40	18	22	484	26.89
South-east	15	18	3	9	0.5
South-west	5	18	13	169	9.39
North-west	12	18	6	36	2
					Σ 38.28

5 Degrees of freedom (df) = (N -1) = (4-1) = 3
6 The critical values for 3 df are

0.10	0.05	0.01	0.001
6.25	7.82	11.34	16.27

EXPLAINING THE RESULTS

The computed value of 38.28 is greater than the critical values even at the 0.001 level of significance. This means that there is less than one in a thousand (0.001) chance that, given the figures above, there is no variation in the frequency of cirques and orientation. Therefore we would reject the null hypothesis and accept the alternative hypothesis. This means that there is a significant difference in the frequency of cirques and their orientation.

Sometimes we know a lot of local detail e.g. the amount of land above a certain level. We may know that in a given area:

- 10% of the land is above 1100 m
- 20% of the land is between 900 m and 1100 m
- 30% of the land is between 700 m and 899 m
- 40% of the land is between 500 m and 699 m

In this case, if there were no variation in the frequency of cirques and altitude, we would expect 10% of cirques above 1100 m, 20% between 900 m and 1100 m, 30% between 700 m and 900 m, and 40% between 500 m and 700 m. In a survey the following results were found:

Altitude	Number of cirques
> 1100 m	36
900 – 1100 m	18
700– 899 m	11
500– 699 m	5
Total	70

The total number observed at each level is the Observed (O). The expected will be 10% of the observed for the > 1000 m group, 20% of the total for the 900 – 1100 m group, 30% of the total for the 700 – 899 m group, and 40% for the 500 – 699 m group.

Spearman's rank correlation coefficient

This test is called a 'rank' correlation because only the ranks are correlated, not the actual values. The use of ranks (Rs) allows us to decide whether there is a significant relationship between two sets of data. In some cases it is obvious whether a correlation exists or not. However, in most cases, it is not as clear and to avoid subjective comments, Rs can be used to bring in a degree of statistical accuracy.

Spearman's Rank Correlation Coefficient is one of the most widely used tests in geography. It is relatively quick and easy to do. It requires that data are available on the ordinal (ranked) scale, although other data can be transformed into ranks very simply.

Ranked data

Sample	Organic content (OC)	Moisture content (MC)
1	3.8	15
2	4.7	22
3	6.2	30
4	3.9	18
5	5.4	24
6	7.1	29
7	6.2	26
8	4.6	20
9	4.6	25
10	5.1	20

PROCEDURE

1 State the null hypothesis (Ho) that is there is no significant relationship between organic content (OC) and moisture content (MC). The alternative hypothesis (H1) is that there is a significant relationship between the two variables.

2 Rank both sets of data from high to low i.e. highest value gets rank 1, second highest 2, and so on. In the case of joint ranks, find the average rank e.g. if two values occupy positions two and three they both take on rank 2.5; if three values occupy positions four, five, and six, they all take rank 5.

3 Using the formula

$$\mathbf{Rs} = 1 - \frac{6\sum d^2}{n^3 - n}$$

work out the correlation, where **d** refers to the difference between ranks and **n** to the number of observations.

4 Compare the computed Rs with the critical values in the statistical tables.

Levels of significance

N	Significance level 95%	99%
4	1.00	–
5	0.90	1.00
6	0.83	0.94
7	0.71	0.89
8	0.64	0.83
9	0.60	0.78
10	0.56	0.75
12	0.51	0.71
14	0.46	0.65
16	0.43	0.60
18	0.40	0.56
20	0.38	0.53
22	0.36	0.51
24	0.34	0.49
26	0.33	0.47
28	0.32	0.45
30	0.31	0.42

Working out Spearman's rank correlation coefficient

Sample	OC	MC	Rank OC	Rank MC	Difference in ranks (d)	d^2
1	3.8	15	10	10	0	0
2	4.7	22	6	6	0	0
3	6.2	30	2.5	1	1.5	2.25
4	3.9	18	9	9	0	0
5	5.4	24	4	5	–1	1
6	7.1	29	1	2	–1	1
7	6.2	26	2.5	3	–0.5	0.25
8	4.6	20	7.5	7.5	0	0
9	4.6	25	7.5	4	3.5	10.25
10	5.1	20	5	7.5	-2.5	6.25
						$\sum d^2 = 21$

$$\mathbf{Rs} = 1 - \frac{6\sum d^2}{n^3 - n} = 1 - \frac{6x21}{0^3-10} = 1 - \frac{126}{990} = 1-0.13 = 0.87$$

WORKED EXAMPLE

Once we have the computed value, we compare it to the critical values. For a sample of 10, these values are 0.564 for 95% significance and 0.746 for 99% significance. In this example it is clear that the relationship is very strong i.e. there is more than 99% chance that there is a relationship between the data. The next stage would be to offer explanations for the relationship.

It is important to realise that Spearman's rank has its weaknesses.

- It requires a sample of not less than seven observations.
- It tests for linear relationships and would give an answer of 0 for data such as river discharge and frequency. Such data follows a curvilinear pattern, with few very low or very high flows and a large number of medium flows.
- It is easy to make meaningless correlations, as between the success of English cricket teams and IMR in India.
- The question of scale is always important. For example, a survey of river sediment rates and discharge for the whole of a drainage system may give a strong correlation, whereas analysis of just the upper catchments gives a much lower result.

As always, statistics are tools to be used. They are only part of the analysis, and we must be aware of their limits.

The Nearest Neighbour Index (NNI)

Much of geography is concerned with distributions in space. Some of the more important distributions include rural settlements and the distribution of functions in an urban area. The **spatial distribution** of settlements in an area can be described by looking at a map. We could describe the settlements as scattered, dispersed, or concentrated. However, the main weakness with the visual method is that it is subjective and individuals differ in their interpretation of the pattern. An objective measure is required and this is provided by the **NNI**.

NEWEST NEIGHBOUR DISTRIBUTIONS

There are three main types of pattern that can be distinguished: **uniform or regular**; **clustered or aggregated**, and **random**. The points may represent settlements or any feature which can be regarded as being located at a specific point.

Clustered NNI = 0.0

Regular NNI = 2.15

Random NNI = 1.0

The technique most commonly used to analyse these patterns is the Nearest Neighbour Index (NNI). It is a measure of the spatial distribution of points, and is derived from the average distance between each point and its nearest neighbour. This figure is then compared to computed values that state whether the pattern is regular (NNI = 2.15), clustered (NNI = 0) or random (NNI = 1.0). Thus a value below 1.0 shows a tendency towards clustering; a value above 1.0 shows a tendency towards uniformity.

PATTERNS AND THEIR NNI

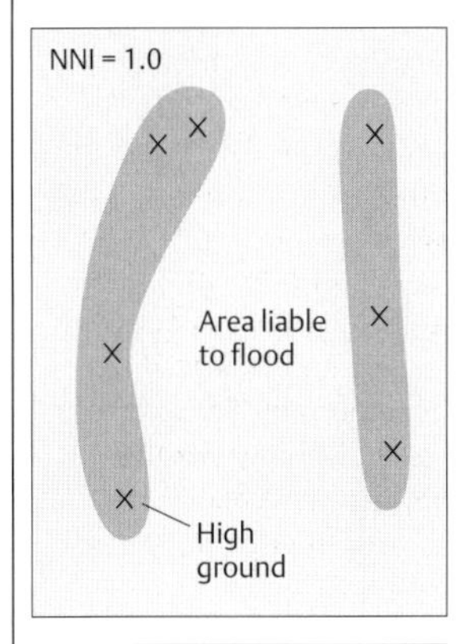

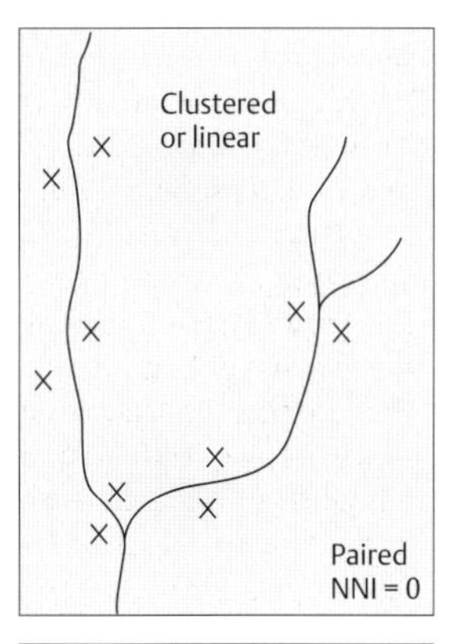

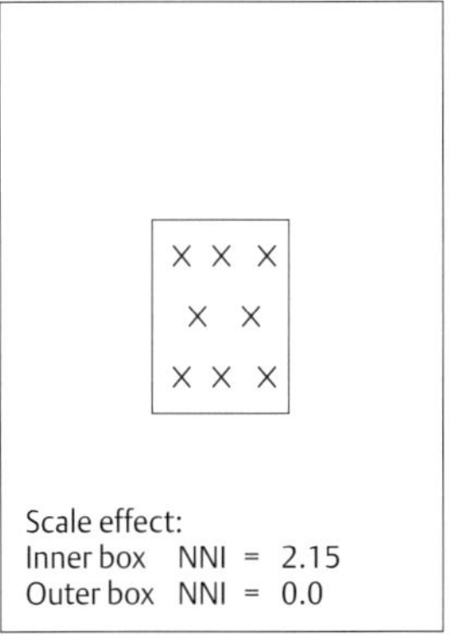

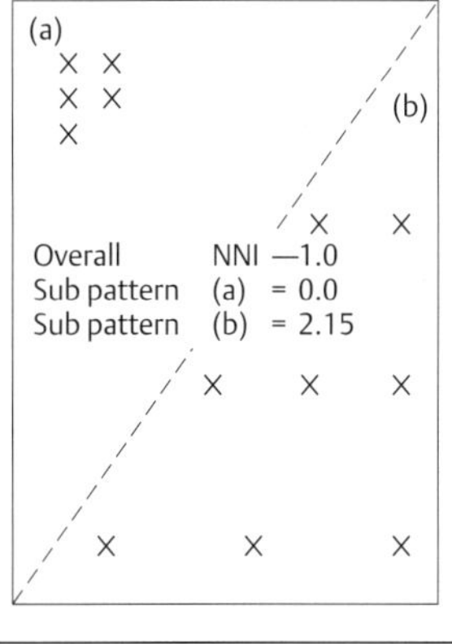

THE FORMULA FOR THE NNI

NNI or Rn = $2\bar{D}\sqrt{(N/A)}$

where $\bar{D}$ is the average distance between each point and its nearest neighbour, and is calculated by finding Σ**d/N**
(**d** refers to each individual distance).
N is the number of points under study.
A is the size of the area under study. It is important that you use the same units for distance and area e.g. metres or km but not a mixture.
For example, a survey of the nine villages of Otmoor produced the following results:

Village	Nearest neighbour	Distance (km)	Formula
Merton	Fencott	1.5	
Islip	Noke	2.0	NNI or Rn = 2D $\sqrt{(N/A)}$
Noke	Islip	2.0	D = Σd/N = 20.5/11 = 1.9
Beckley	Horton	3.0	Rn = 2 x 1.9 x $\sqrt{(11/64)}$
Horton	Beckley	3.0	Rn = 1.57
Fencott	Murcott	1.0	
Murcott	Fencott	1.0	
Charlton	Oddington	1.5	
Oddington	Charlton	1.5	
Woodeaton	Elsfield	2.0	
Elsfield	Woodeaton	2.0	
		Σd 20.5	

The results vary anywhere between 0 and 2.15. There is a continuum of values and any distribution lies somewhere between the two extremes. The answer above suggests a significant degree of clustering.

USING NNI

There are important points to bear in mind when using the NNI:

- Where do you measure from – the centre or the edge of a settlement? Some discretion is needed.
- Do you measure with a straight line or by road (or water)?
- What is the definition of the feature being studied, e.g. a settlement? Do you include all settlements – and individual houses – or only those above a certain size?
- Why do we take the nearest neighbour? Why not the third or fourth nearest?
- What is the effect of paired distributions?
- One overall index may obliterate important sub-patterns.
- The choice of the area, and the size of the area studied, can completely alter the result and make a clustered pattern appear regular, and vice-versa.
- Although the NNI may suggest a random pattern it may be that the controlling factor, e.g. soil type or altitude, is itself randomly distributed, and that the settlements are not located in a random fashion.

Sample questions

These are specimen questions. You are advised to contact your exam board, and get hold of past papers and marking schemes. The mark allocation per question varies from board to board. This is true for both A2 and AS questions.

PLATE TECTONICS (PAGES 10-15)

AS

1. **Figure 1** shows the global pattern of tectonic plates.
   (a) (i) Name plate A. (1 mark)
   (ii) What type of plate boundary is located at Y? (1 mark)
   (b) Draw arrows on the diagram to show the direction of plate movement at location X. (1 mark)
   (c) Draw a fully labelled sketch section to show the processes that are operating at X and Z. (8 marks)
   (d) What is the evidence for plate tectonics? (5 marks)
   (e) Why do people live in tectonically active areas? (4 marks)
2. **Figure 2** shows the Pacific Ring of Fire.
   (a) (i) Describe the distribution of volcanoes. (4 marks)
   (ii) Explain the distribution of volcanoes. (6 marks)
   (b) Draw a fully labelled diagram to show the main features of a volcano. (4 marks)
   (c) Explain with references to examples that you have studied
   (i) the advantages of living in tectonically active areas
   (ii) the disadvantages of living in tectonically active areas. (6 marks)

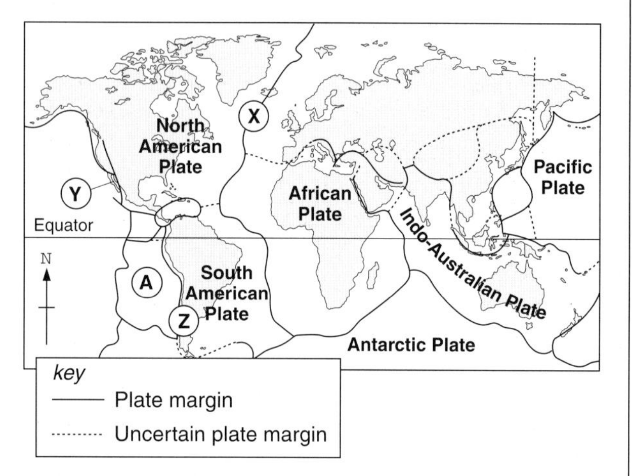

Figure 1

A2

3. 'It is not earthquakes that kill people but buildings.' Discuss, with reference to specific examples. (20 marks)
4. Describe and explain the global distribution of either volcanoes, or earthquakes. (20 marks)
5. Using examples, explain why either volcanic activity or earthquake activity occurs in areas away from plate tectonic boundaries. (20 marks)

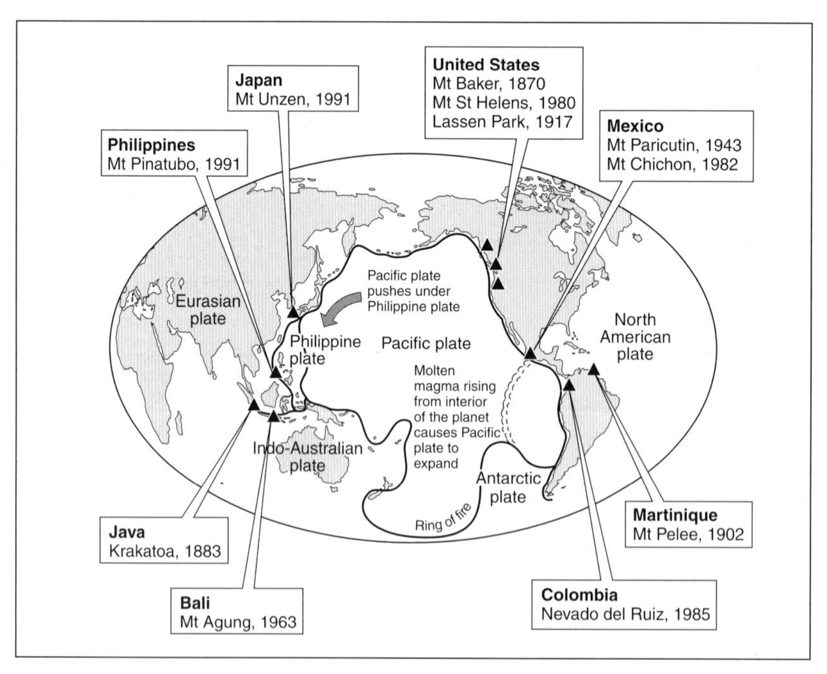

Figure 2

WEATHERING, ROCKS, AND RELIEF (PAGES 16-26)

AS

6. **Figure 3** shows how weathering is related to mean annual temperature and precipitation.
   (a) Define the term weathering. (2 marks)
   (b) Name and explain one possible type of physical weathering process found at location A and one at location C. (6 marks)
   (c) Suggest one reason why frost action is likely to be more intense at location B than at A. (2 marks)
   (d) Why is there limited weathering at D? (2 marks)
7. **Figure 4** shows how mass movement is related to mean annual temperature and precipitation.
   (a) Define the term mass movement. (4 marks)
   (b) Name and explain one type of mass movement found at location E. (3 marks)
   (c) How might the climate influence mass movements at F? (3 marks)

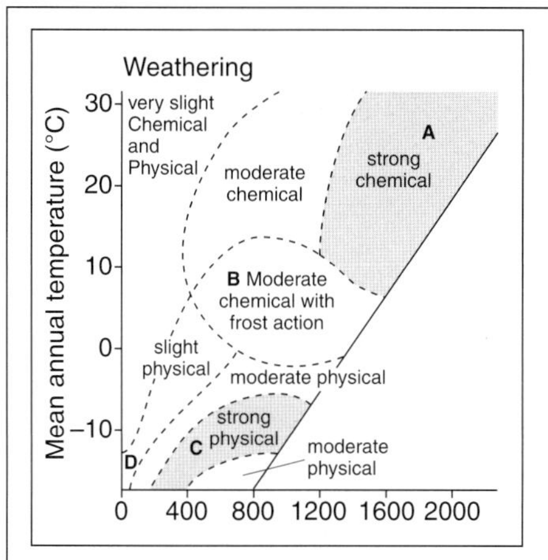

Figure 3

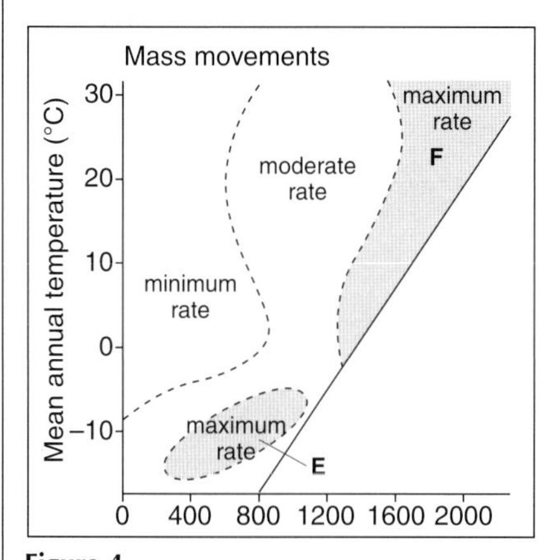

Figure 4

A2

8. Explain the effect on slope development of
   (i) climate
   (ii) rock type. (20 marks)
9. Compare and contrast the major landforms associated with carboniferous limestone in tropical and temperate areas. (20 marks)
10. Evaluate the role of rock type and climate in the development of granite tors. (20 marks)
11. With the use of examples, explain why mass movements occur. (20 marks)

RIVERS AND HYDROLOGY (PAGES 27-38)

AS

12. **Figures 5a** and **5b** show the hydrographs for two different streams and their responses to the same rainfall event.
    (a) Calculate the time lags and peak flows for both streams. (4 marks)
    (b) On hydrograph 5a, shade in the area that represents baseflow. (2 marks)
    (c) Compare the two hydrographs in terms of rising limbs, peak flows, and recessional limbs. (6 marks)
    (d) Suggest reasons for the differences you have described in (c). (6 marks)
    (e) If a new housing development were built in the drainage basin of the stream in 5a, describe and explain the likely effects on
        (i) the time lag
        (ii) the amount of surface run-off
        (iii) the peak flow. (12 marks)
13. (a) Define the following terms: infiltration; evapotranspiration; interception. (6 marks)
    (b) Briefly describe the factors that effect the infiltration rate. (6 marks)
    (c) With the use of examples, show how human activity can modify the global hydrological cycle. (8 marks)

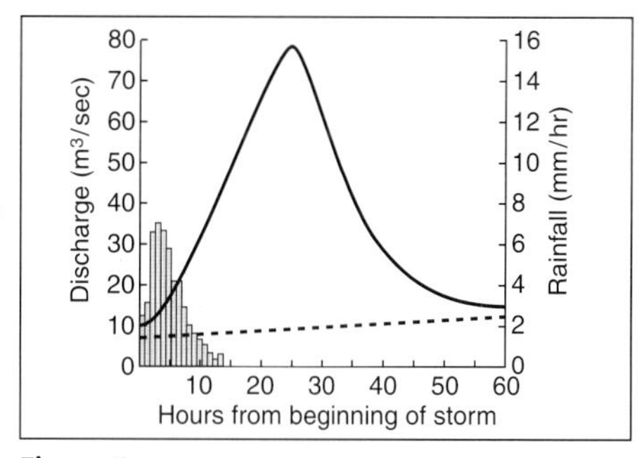

Figure 5a

A2

14. Explain, using diagrams, why river channels meander. (20 marks)
15. Examine the factors that influence storm hydrographs. (20 marks)
16. With the use of examples, explain how deltas and levées are formed. (20 marks)
17. Explain how the work of a river varies with its velocity and the size of the load it carries. (20 marks)
18. In what ways can human activity affect river environments? (20 marks)
19. How do rivers affect human activity? (20 marks)

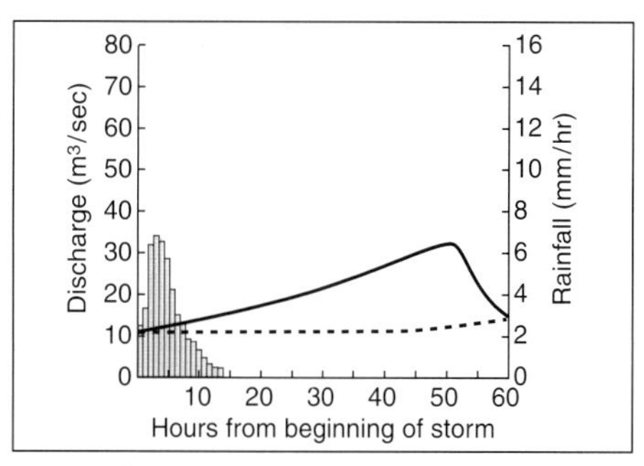

Figure 5b

COASTS (PAGES 39-46)

AS

20. **Figure 6** shows a coastal spit and associated features.
    (a) (i) Define the term longshore drift. (2 marks)
    (ii) Draw a labelled diagram to show how longshore drift operates. (3 marks)
    (iii) State the direction of longshore drift in the diagram. (1 mark)
    (iv) Explain how you can identify the direction of longshore drift from the diagram. (2 marks)
    (b) (i) Describe the main features of a spit. (4 marks)
    (ii) Outline the main processes that are likely to be taking place in area B. (4 marks)
    (iii) Suggest how plants are able to survive in area B. (4 marks)

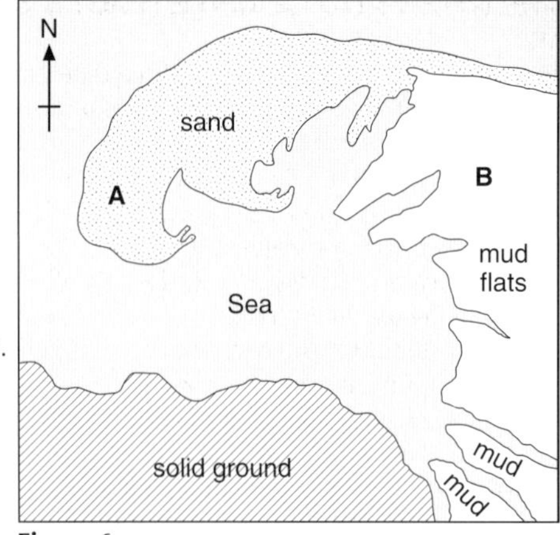

Figure 6

A2

21. **Figure 7** is a geological map of part of the Dorset coastline. Using only information in **Figure 7** suggest how the shape of coastline has been influenced by the geology of the area. (20 marks)
22. With reference to one or more stretches of coastline
    (a) explain the reasons for the rapid erosion of some coastlines.
    (b) discuss the strategies of coastal management in such an area or areas. (20 marks)

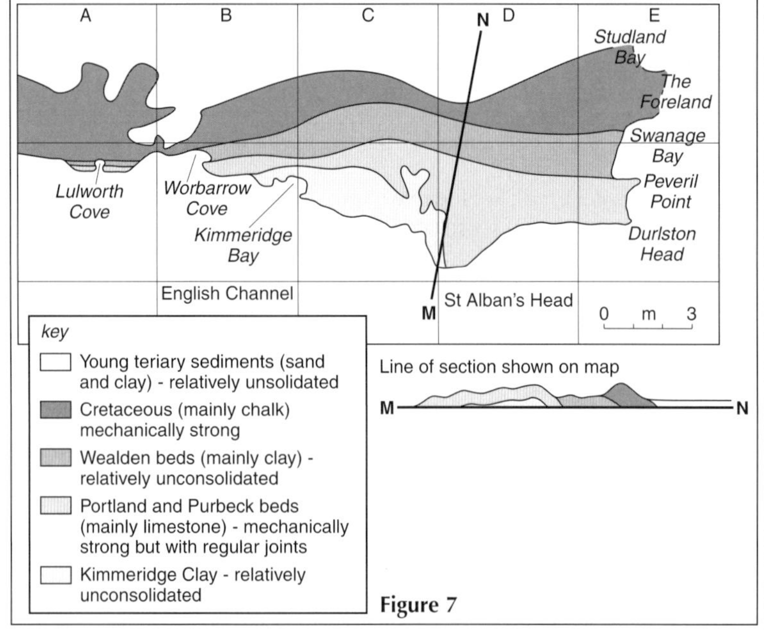

Figure 7

GLACIATION AND PERIGLACIATION (PAGES 47-57)

AS

23. **Figure 8** shows the main processes operating in a cirque.
    (a) Describe the main characteristics of a cirque. (6 marks)
    (b) Suggest how cirques are formed. (6 marks)
    (c) Briefly explain why most cirques in Britain face an easterly or northerly direction. (4 marks)
    (d) What factors, other than aspect, are likely to have an impact on cirque formation? (4 marks)
24. (a) How would you distinguish between those depositional landforms in a glacial environment produced by ice and those produced by meltwater? (6 marks)
    (b) Describe and explain the formation of any three landforms produced by glacial deposition. (8 marks)
    (c) Examine the role of permafrost in the development of landforms in periglacial environments you have studied. (6 marks)

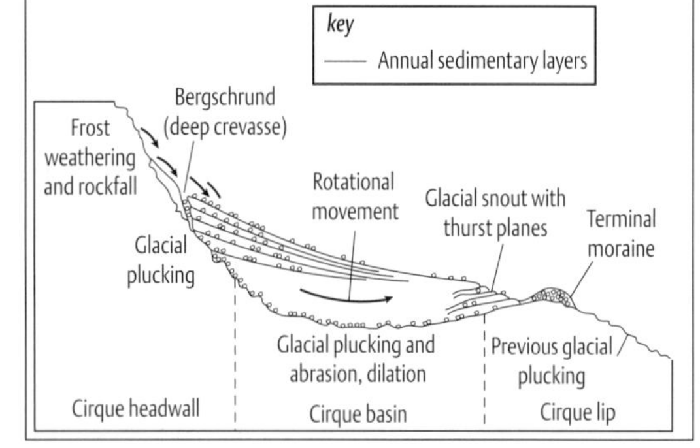

Figure 8

25. (a) Describe the main features of
    (i) drumlins
    (ii) terminal moraine. (4 marks)
    (b) How are drumlins and terminal moraine formed? (6 marks)

A2

26. What is permafrost? What are its thermal and hydrological characteristics? (20 marks)
27. Describe the main features of a periglacial climate. (20 marks)
28. Explain how glacial landforms have affected human activity. (20 marks)
29. Using diagrams, explain
    (i) how cirques are formed
    (ii) the characteristics of a hanging valley. (20 marks)

WEATHER AND CLIMATE (PAGES 58-78)

AS

30. **Figures 9a**, **9b**, and **9c** show three different rainfall events over the British Isles. The general pressure patterns and relief are shown but not frontal systems even if they were present. Rainfall intensities are indicated in each case but not rainfall amounts.
    (a) What type of rainfall is occurring at **Figure 9a**? Explain why rainfall would take place. (4 marks)
    (b) Explain the likely reasons for rainfall in **Figure 9b**. (4 marks)
    (c) (i) Describe the distribution of rainfall in **Figure 9c**. (4 marks)
    (ii) Suggest reasons for the rainfall in **Figure 9c**. (4 marks)
    (d) On a copy of **Figure 9b**, mark on the positions of any fronts. Justify your answer. (4 marks)

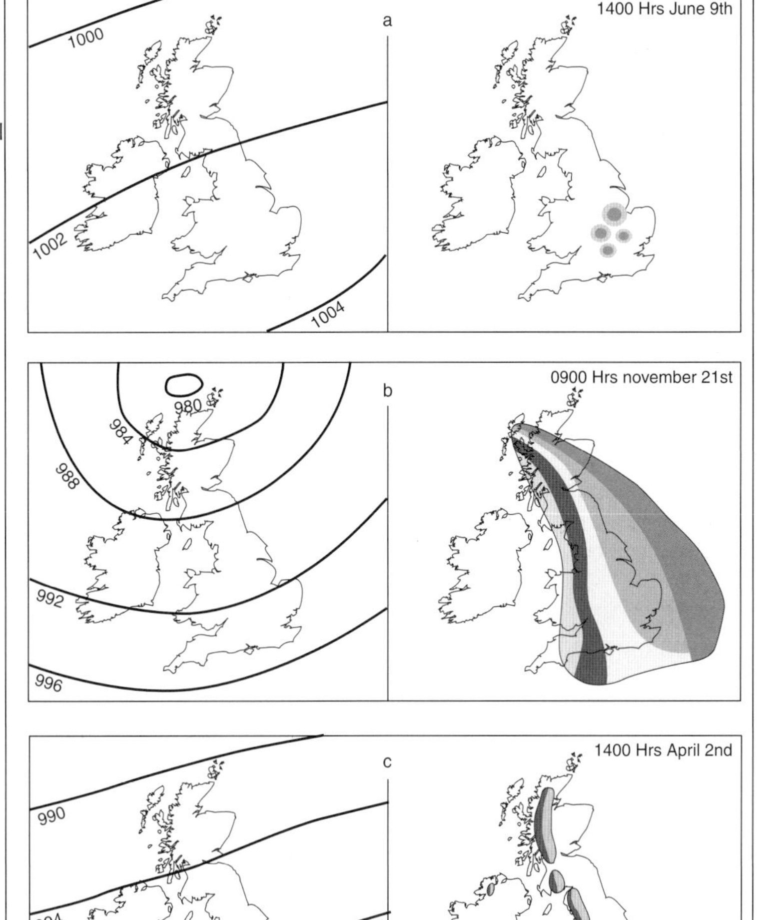

Figure 9

A2

31. Describe how the energy budget of an urban area differs from that of its rural surroundings. How do urban microclimates differ from nearby rural ones? (20 marks)
32. With the use of examples illustrate how human activity modifies climate processes. (20 marks)
33. Describe and explain the impact of air masses on the weather of the British Isles. (20 marks)

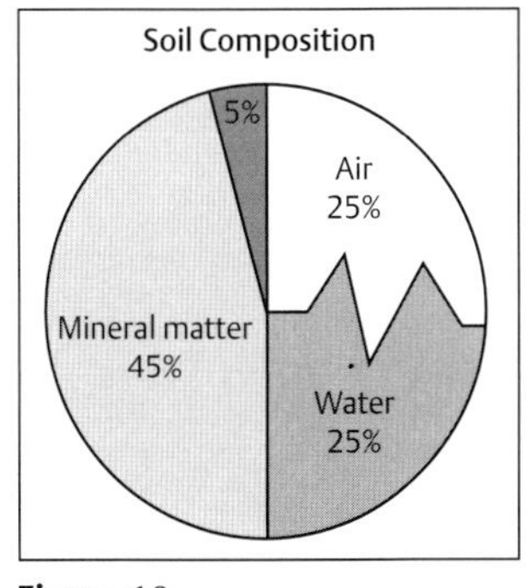

Figure 10

ECOSYSTEMS AND SOILS (PAGES 79-94)

AS

34. **Figure 10** shows the composition of a soil.
    (a) (i) Define the term soil. (2 marks)
    (ii) Describe the composition of the soil shown in **Figure 10**. (4 marks)
    (b) (i) What is a soil horizon? (2 marks)
    (ii) Briefly explain why podzols have such clear horizons. (4 marks)
    (c) (i) Under what natural conditions would you expect podzols to be formed? (3 marks)
    (ii) What can be done to make podzols more suitable for farming? (5 marks)
35. (a) What is meant by the term biomass? (4 marks)
    (b) How is biomass measured? (4 marks)
    (c) (i) How are nutrients incorporated into the soil? (3 marks)
    (ii) Suggest ways in which nutrients in the soil are lost from the ecosystem. (3 marks)
    (d) With reference to an ecosystem you have studied, describe how human activity can affect nutrient cycling. (6 marks)

(ECOSYSTEMS AND SOILS CONTINUED)

36. Define the term succession. (5 marks)
37. **Figure 11** shows the global pattern of ecosystems.
    (a) (i) Define the term biome. (2 marks)
        (ii) Describe the distribution of tropical rainforests as shown in **Figure 11**. (4 marks)
        (iii) Compare the distribution of the boreal forest with that of temperate deciduous forest. (4 marks)
    (b) (i) Define the term climatic climax vegetation. (2 marks)
        (ii) What is the climatic climax vegetation for Britain? (2 marks)
        (iii) Using examples, explain how human activities have affected vegetation. (6 marks)

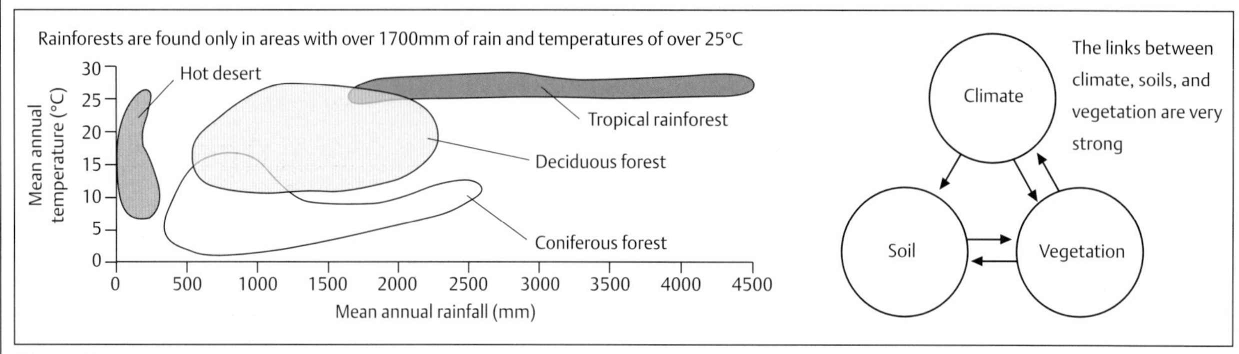

Figure 11

A2

38. With reference to specific examples explain how and why plant communities change over time. (20 marks)
39. Compare and contrast temperate grasslands with tropical grasslands in terms of their climate, soils, and vegetation. (20 marks)

POPULATION (PAGES 95-104)

AS

40. **Figure 12** shows the demographic transition model (DTM).
    (a) (i) Label the birth rate and the death rate. (1 mark)
        (ii) Draw on the first, second, and fifth stages of the DTM. (3 marks)
        (iii) Complete the line on the diagram to show total population change over the whole period. (2 marks)
    (b) Describe the changes in the rate of natural increase over time. (3 marks)
    (c) Suggest, using real examples, why
        (i) crude birth rate varies over time
        (ii) crude death rate varies from place to place. (6 marks)
    (d) What are the advantages and disadvantages in using the demographic transition model to explain population growth in developing countries? (5 marks)

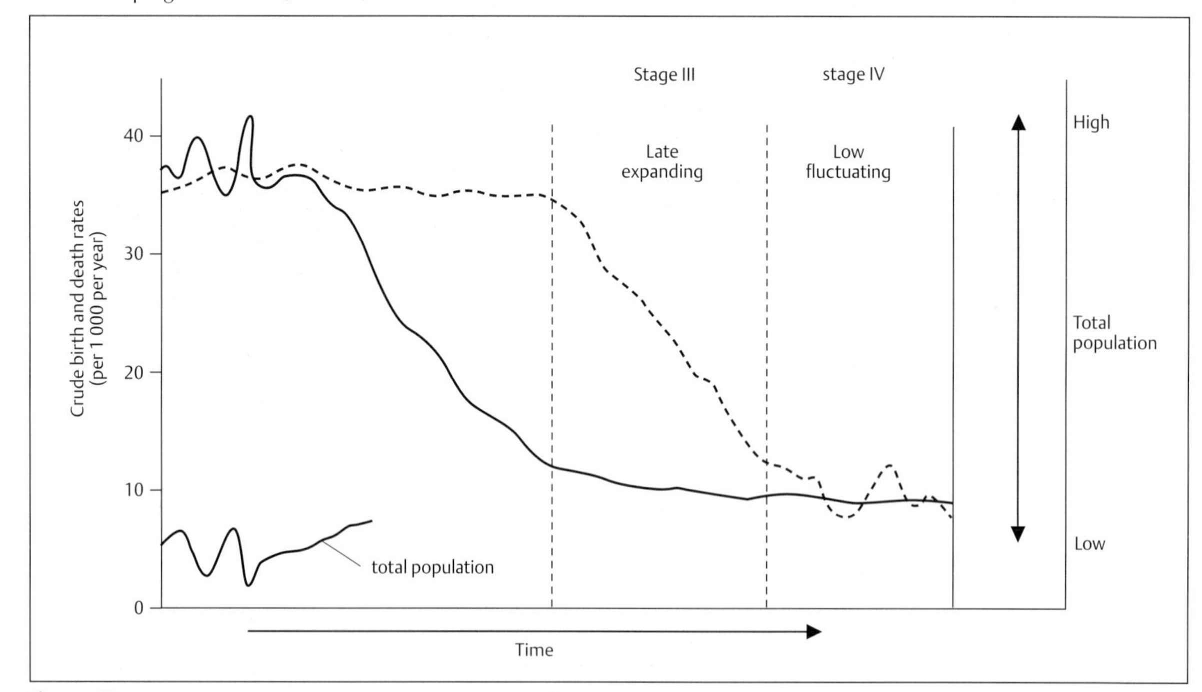

Figure 12

(POPULATION CONTINUED)

41. **Figure 13** shows the origins and destinations of migrants to and from the United Kingdom in 1996.
    (a) (i) Define the term 'permanent migration'. (1 mark)
        (ii) Describe the pattern of immigration into the UK. (3 marks)
        (iii) Describe the pattern of outmigration from the UK. (3 marks)
        (iv) Suggest possible reasons why an MEDC such as the UK might experience out-migration as well as immigration. (4 marks)
    (b) Explain why international migrations may be selective in terms of
        (i) the age of the migrants
        (ii) the sex of the migrants
        (iii) the educational qualifications of the migrant. (9 marks)

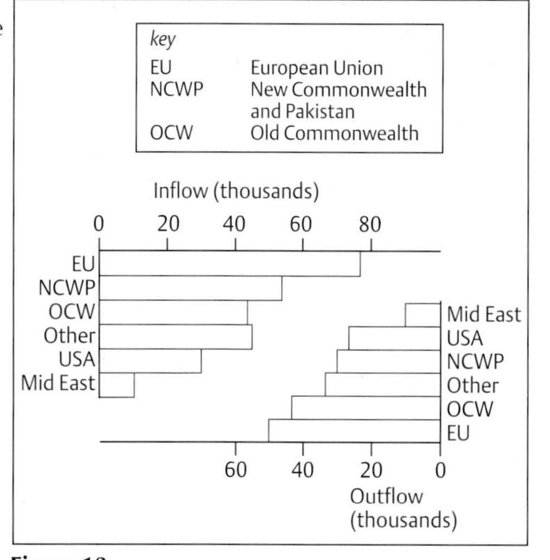

Figure 13

A2

42. Using examples, explain the impact of migration on
    (i) source areas
    (ii) destinations. (20 marks)
43. What are the world's main population problems? What can be done to manage these problems? Support your answers with examples. (20 marks)

SETTLEMENT (PAGES 105-120)

AS

44. **Figure 14** shows the relationship between settlement size and the number of services in a number of villages.
    (a) What is the relationship between settlement size and number of services? (2 marks)
    (b) (i) Define the terms range, threshold, low order goods, and high order goods. (8 marks)
        (ii) Explain why the number of services in a settlement normally increases with population size. (2 marks)
        (iii) Why might some settlements have fewer services than some settlements with a smaller population? (3 marks)
    (c) (i) What is meant by the term 'sphere of influence'? (2 marks)
        (ii) How can spheres of influence be measured? (3 marks)

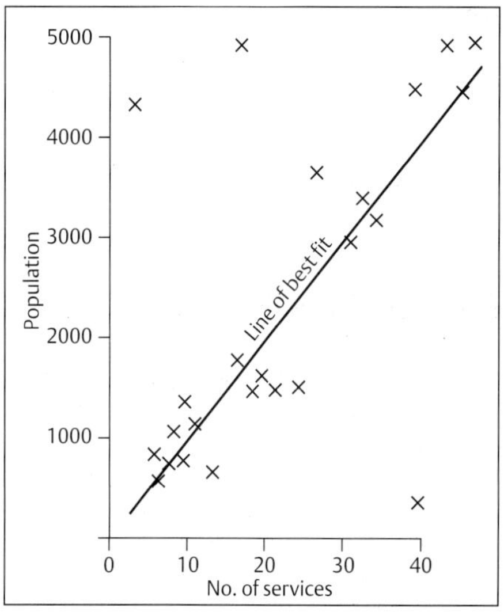

Figure 14

A2

45. For an area that you have studied show how rural settlements have changed over time. (20 marks)
46. **Figure 15** shows population growth in the world's fastest growing cities, 1985 - 2000.
    (a) Define the term urbanisation. (2 marks)
    (b) Describe the global pattern of population as shown in **Figure 15**. (4 marks)
    (c) (i) How have urban populations changed between 1985 and 2000? (3 marks)
        (ii) Suggest reasons to explain these changes. (4 marks)
    (d) With reference to a named urban settlement in a LEDC
        (i) Outline the reasons for its growth. (4 marks)
        (ii) Describe and explain the problems that have resulted from its growth. (8 marks)

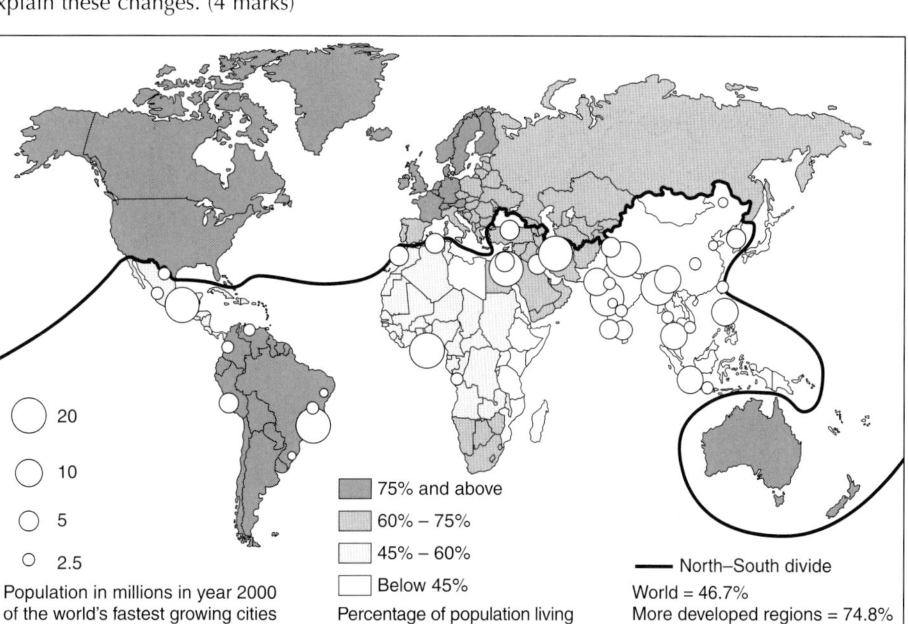

Figure 15

AGRICULTURE (PAGES 121-130)

AS

47. Study **Figures 16a**, **16b**, and **16c** which show world fertiliser use for selected countries, 1950-94
    (a) Describe the trend in fertiliser use. (4 marks)
    (b) Which country increased its use of fertilisers the most from 1950? (2 marks)
    (c) How did the pattern for fertiliser use in the USA differ from that in the CIS? (4 marks)
    (d) Briefly explain two reasons why fertiliser use in China was greater than in the CIS? (4 marks)
    (e) What are the most economic and environmental consequences of the increasing use of fertilisers? (6 marks)

A2

48. 'An agricultural system is a simplified and controlled ecosystem where the farmer is trying to direct more of the energy and nutrient flows of the system into food products'. Discuss the ways in which the farmer attempts to do this, and the impacts of such manipulation. (20 marks)
49. Evaluate the impact of the Green Revolution. (20 marks)
50. With the use of specific examples, describe and explain the changing role of government policy on agricultural activities. What are the economic, social, and environmental effects of government policies on agriculture? (20 marks)

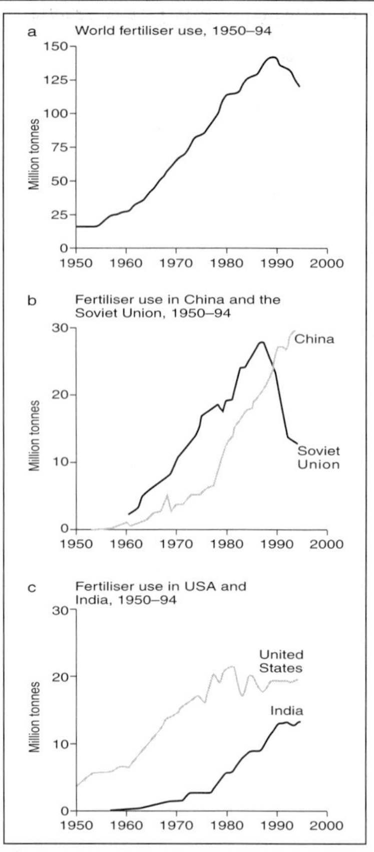

Figure 16

Figure 17

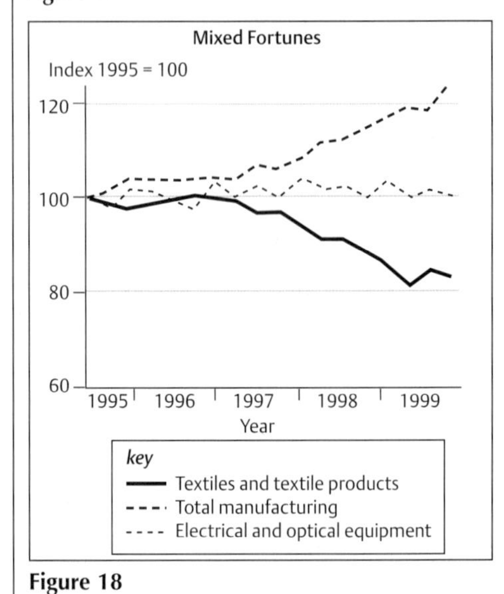

Figure 18

INDUSTRY AND SERVICES (PAGES 131-143)

AS

51. **Figure 17** shows changes in the number of employees in manufacturing industry in the United Kingdom between 1978 and 1996. **Figure 18** gives some additional information about manufacturing industry in the UK
    (a) Describe the changes in the number of people employed in manufacturing as shown in **Figure 17**. (3 marks)
    (b) Compare the fortunes of different sectors of manufacturing as shown in **Figure 18**. Which ones have
        (i) fared best
        (ii) declined the most? (2 marks)
    (c) Suggest reasons for these changes. (5 marks)
    (d) What are the economic and social consequences for areas that have experienced a loss of jobs in manufacturing? (6 marks)
    (e) Outline the environmental impacts of the decline of manufacturing industry. (4 marks)
52. (a) Why do governments attempt to influence the location of industry? (10 marks)
    (b) With reference to one named country explain how the government, or government agencies, have attempted to influence the location of manufacturing industry. (10 marks)

A2

53. What are the causes and consequences of deindustrialisation? (20 marks)
54. For an industry you have studied, describe and explain its spatial division of labour. (20 marks)

REGIONAL INEQUALITIES (PAGES 144-147)

AS

55. Explain the terms
    (i) core and periphery
    (ii) regional disparities (4 marks)
56. Regional inequalities are the result of structure and location. Discuss. (20 marks)
57. **Figure 19** shows regional unemployment in the UK between 1983-93
    (a) Which region experienced the peak of unemployment? When did it do so? (5 marks)
    (b) What is the trend in regional unemployment in the UK? (5 marks)
    (c) Suggest reasons to explain the pattern described in (b). (5 marks)
    (d) Why is the unemployment rate often used as a measure of regional inequality? (5 marks)

A2

58. 'The prosperity of any region is largely determined by its geographical location and its physical environment.' Discuss this statement with reference to the regions of the European Union. (20 marks)

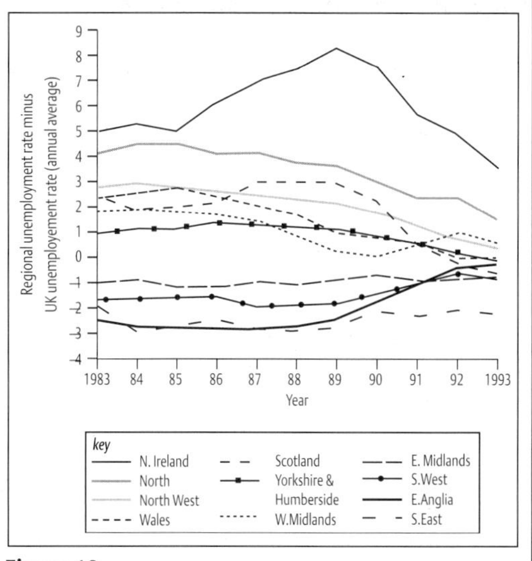

Figure 19

	1	2	3	4	5	6	7
Botswana	1.5	3260	56	70	4.7	51	0.670
Brazil	164	4720	36	83	2.4	67	0.756
India	961	390	65	48	3.1	62	0.382
Malaysia	21	4680	11	84	3.4	72	0.768
Mozambique	16.2	80	113	33	6.5	47	0.252
Singapore	3	32 940	4	92	1.7	76	0.836
South Africa	41.5	3400	49	82	2.9	65	0.650
UK	59	20 710	6	99	1.7	77	0.919
USA	268	28 740	7	99	2.1	77	0.927
Zimbabwe	11.0	750	55	85	4.6	56	0.474

1 Population (million)
2 GNP per head, 1997
3 IMR per '000
4 Literacy (%)
5 Fertility rate per woman
6 Life expectancy (years)
7 HDI

Figure 20

DEVELOPMENT (PAGES 148-156)

A2

59. Study **Figure 20** that shows selected indicators of development.
    (a) Which country has
        (i) the highest GNP
        (ii) the lowest IMR? (2 marks)
    (b) What relationship is there between GNP and fertility as shown in **Figure 20**. How do you explain this? (7 marks)
    (c) What information does not tell us anything about development? (3 marks)
60. Briefly describe two ways in which educational levels affect development. (4 marks)
61. What is the HDI (human development index)? (3 marks)
62. Compare the levels of development in Mozambique with those in the UK. (6 marks)

TOURISM (PAGES 157-159)

A2

63. Tourist environments are very dynamic. Explain how and why coastal resorts and nature reserves have changed in popularity since the Second World War. How useful is the resort life-cycle model as a means of explaining these changes? (20 marks)
64. **Figure 21** lists the world's top 20 tourist destinations in 1997.
    (a) Describe the geographical pattern of tourist destinations? (10 marks)
    (b) How do you explain this pattern? (10 marks)

Rank 1985	Rank 1990	Rank 1997	Country	Arrivals (thousands) 1997	% change 1997/96	% of total 1997
1	1	1	France	66 864	7.1	10.9
4	2	2	USA	48 409	4.1	7.9
2	3	3	Spain	43 378	7.0	7.1
3	4	4	Italy	34 087	3.8	5.6
6	7	5	UK	25 960	2.6	4.2
13	12	6	China	23 770	4.4	3.9
22	27	7	Poland	19 514	0.5	3.2
9	8	8	Mexico	19 351	-9.6	3.2
7	10	9	Canada	17 610	1.6	2.9
16 (1)	16	10	Czech Republic	17 400	2.4	2.8
11	5	11	Hungary	17 248	-16.6	2.8
5	6	12	Austria	16 646	-2.6	2.7
8	9	13	Germany	15 837	4.2	2.6
18 (2)	17 (2)	14	Russian Federation	15 350	5.2	2.5
10	11	15	Switzerland	11 077	4.5	1.8
19	19	16	China, Hong Kong	10 406	-11.1	1.7
14	13	17	Greece	10 246	11.0	1.7
15	14	18	Portugal	10 100	3.8	1.7
28	24	19	Turkey	9 040	13.5	1.5
25	21	20	Thailand	7 263	1.0	1.2
20	20	21	Netherlands	6 674	1.4	1.1

Figure 21

INDEX